**Introduction to
Molecular Spectroscopy**
Theory and Experiment

Introduction to
Molecular Spectroscopy
Theory and Experiment

E. F. H. BRITTAIN
W. O. GEORGE
C. H. J. WELLS

Kingston Polytechnic
Kingston-upon-Thames, England

1970

ACADEMIC PRESS LONDON AND NEW YORK

CARL A. RUDISILL LIBRARY
LENOIR RHYNE COLLEGE

ACADEMIC PRESS INC. (LONDON) LTD
Berkeley Square House
Berkeley Square,
London, W1X 6BA

U.S. Edition published by
ACADEMIC PRESS INC.
111 Fifth Avenue,
New York, New York 10003

Library of Congress Catalog Card Number: 71–109035
SBN: 12–135050–9

PRINTED IN GREAT BRITAIN BY
SPOTTISWOODE, BALLANTYNE & CO. LTD
LONDON AND COLCHESTER

Preface

Molecular spectroscopy has been the subject of a great deal of interest and activity in the past two decades. The early practitioners were mainly physicists but the subject has occupied an increasing share of the literature of chemistry both in relation to detailed fundamental studies of small molecules and to the necessarily more empirical studies of larger molecules.

The object of the present book is to present a balanced treatment of the principal methods of molecular spectroscopy together with selected practical work. The book is intended primarily for undergraduate courses in chemistry and represents the authors' experience in teaching the principles and applications of molecular spectroscopy to L.R.I.C., G.R.I.C. and B.Sc. students at both Ordinary and Honours level. Some of the subject matter is, however, covered in more detail than would normally be presented to undergraduates. This additional material, taken in conjunction with more advanced texts, has been used by the authors in a number of specialist short courses and in an M.Sc. course in molecular spectroscopy.

The book is limited to consideration of electronic, vibrational, nuclear magnetic resonance and mass spectrometry because of the general applicability of these methods and their availability to students. The first chapter is concerned with material which is common to different branches of molecular spectroscopy and an attempt has been made to present a unified approach to the underlying principles of the subject. Subsequent chapters deal with the different chosen branches in turn, and each chapter is divided into three sections—principles, instrumentation and experiments. The experiments have been designed largely to illustrate topics discussed in the principles and instrumentation sections, and cross-referencing to the appropriate experiments is given throughout the text.

We would like to express gratitude to many students and colleagues who have shared in the development of much of the material of this book and to Mr V. G. Mansell, Mr N. Falla and Miss J. Grice for valuable technical assistance. We are grateful also to Mr B. Haynes for the encouragement he has provided for the development of molecular spectroscopy within various courses in Kingston Polytechnic. Our thanks are also due to the staff of Academic Press who have provided considerable assistance and shown much patience.

April, 1970

E. F. H. BRITTAIN
W. O. GEORGE
C. H. J. WELLS

Contents

Chapter 3

Vibrational Spectroscopy

Chapter 4

Nuclear Magnetic Resonance

Chapter 5

Mass Spectrometry

Appendix

Units Used in This Volume

The book has been written utilizing SI units with a few exceptions where it is felt that the non-SI usage is likely to continue for some time. The three major exceptions are the use of electron volts as an energy unit in the chapter on mass spectrometry, the use of g cm^{-3} for density throughout the book rather than kg m^{-3} and, since dm^3 is equivalent to the litre, the use of mol dm^3 instead of mol m^3 for concentration. Since almost all existing texts utilize the cgs system of units, we give (see Appendix, p. 367) a table showing the relationships between the two systems of units and a table quoting the values of some of the common physical constants expressed in SI units.

To Our Wives

I. THE ELECTROMAGNETIC SPECTRUM

A spectrum of light from the Sun is observed in nature in the well-known form of the rainbow. In 1672 Newton was able to show that the splitting of visible radiation from the Sun into component colours could be simulated using a glass prism instead of a moist atmosphere. A suitable arrangement for performing this experiment is shown in Fig. 1.1.

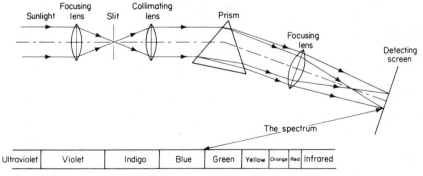

Ultraviolet	Violet	Indigo	Blue	Green	Yellow	Orange	Red	Infrared

Fig. 1.1. The formation of the visible spectrum

1

It may be shown that a continuous visible spectrum can be obtained from sources other than the Sun. For instance the passage of an electric current through a filament made of a material such as tungsten produces an incandescent source which emits visible radiation. The radiation emitted from a source may be detected by the human eye if the radiation lies in the visible region of the spectrum, but other detecting systems have to be used if the radiation lies outside this region. As can be seen from Fig. 1.2 the visible region of the spectrum extends over a very small range of the entire spectrum; different types of sources and detectors have to be used in different regions.

The regions of the electromagnetic spectrum are classified in an arbitrary way as γ-rays, X-rays, ultraviolet, visible, infrared, microwave and radiowave with some further subdivisions such as near or far, short or long.

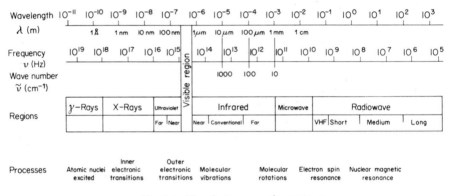

Fig. 1.2. The electromagnetic spectrum

A. WAVE THEORY OF ELECTROMAGNETIC RADIATION

The nature of the various radiations shown in Fig. 1.2 has been explained by Maxwell's classical theories of electro- and magneto-dynamics. Radiation is considered as constituted of electric and magnetic fields which are mutually perpendicular to each other and to the direction of propagation and which oscillate sinusoidally. The sum effect of these fields is termed *electromagnetic* radiation and is represented diagrammatically in Fig. 1.3 as a plane polarized wave.

In Fig. 1.3 E and H are vectors representing the electric and magnetic fields respectively. The number of cycles per second is the frequency, ν, the distance between adjacent peaks (or troughs) is the wavelength, λ, and the height of the wave is the amplitude, A. The velocity of the wave is given by the product of the frequency and the wavelength. It has been found that the velocity is constant for all electromagnetic radiations in vacuum; the value is $c = 2 \cdot 998 \times$

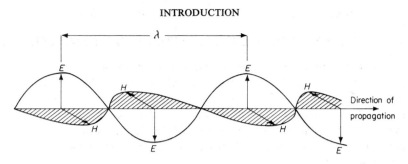

Fig. 1.3. The electromagnetic wave

10^8 m s^{-1}. In any homogeneous medium of refractive index, n, the velocity is c/n. Therefore the frequency and the wavelength are interrelated by the expression

$$c = n\nu\lambda \qquad (1.1)$$

If the wavelength is known, the frequency may be calculated from equation 1.1 (the recommended S.I. unit for frequency is the hertz, 1 Hz = 1 cycle per second), where $n = 1$ for measurements made in air. A unit which is particularly useful in optical regions of the spectrum (ultraviolet, visible and infrared) is the wave number which is the number of waves contained in a length of one centimetre. The wave number value, $\tilde{\nu}$, for an electromagnetic wave is given by the relationship

$$\tilde{\nu} = \frac{1}{\lambda} = \frac{\nu}{c} \qquad (1.2)$$

B. QUANTUM THEORY OF ELECTROMAGNETIC RADIATION

The work of Einstein, Planck and Bohr led to the idea that electromagnetic radiation could be regarded for many purposes as a stream of particles (quanta) for which the energy, ϵ, is given by the Bohr equation

$$\epsilon = h\nu \qquad (1.3)$$

where h is Planck's constant and ν is a parameter equivalent to classical frequency. Equation 1.3 relates the energy of electromagnetic radiation to the classical concept of frequency of an electromagnetic wave and shows that high-frequency radiation (γ-rays, X-rays) has considerably more energy than low-frequency radiation (microwaves, radiowaves). This result is also apparent from the effect of these radiations on various forms of matter; thus all radiation with higher frequency than that of visible radiation has to be regarded as potentially hazardous to organic tissues whereas lower-frequency radiation is physiologically harmless.

II. INTERACTION OF RADIATION WITH MATTER

Molecular spectroscopy is concerned with the interaction of electromagnetic radiation with molecular systems. A spectrometer produces a record of variation in intensity of electromagnetic radiation with frequency (or wavelength) after interaction with the molecular system. There are two principal ways in which the interactions are observed, firstly that in which the sample itself *emits* radiation and secondly where the sample *absorbs* radiation from a continuous source.

A. ATOMIC AND MOLECULAR SPECTRA

Matter in its atomic form is readily stimulated to emit radiation by the application of electrical or thermal energy. The resultant atomic emission spectrum may be of the ionized or non-ionized form of the sample according

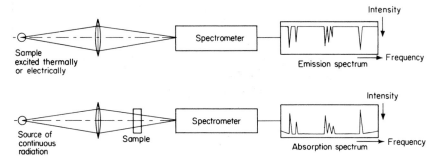

Fig. 1.4. Arrangement for obtaining emission and absorption spectra

to its mode of excitation. Atomic absorption spectra can also be observed if the atoms are sufficiently far apart to allow transmission of radiation. Both emission and absorption spectra are used in the analysis of atoms and atomic ions.

Matter in its molecular form is frequently decomposed by the conditions used to excite emission spectra and consequently molecular spectroscopy is nearly always concerned with absorption spectra. When both absorption and emission spectra are obtainable from a sample without changing its nature, it is observed that the frequencies at which absorption and emission occur coincide except when phenomena such as phosphorescence also occur (p. 65). This correspondence signifies that the sample undergoes the same energy change during the interaction. Energy, corresponding to a particular molecular process, may be emitted or absorbed by the sample according to the experimental arrangement. The nature of the atomic and molecular processes associated with the various regions of the spectrum are summarized in Fig. 1.2.

B. CLASSICAL THEORY OF INTERACTION OF RADIATION WITH MATTER

Atomic and molecular processes may involve magnetic or electrical dipolar changes within the atom or molecule. These may be partially understood in terms of classical ideas which have evolved from observations of macro systems of matter.

Interaction between the magnetic part of electromagnetic radiation in the radio-frequency region of the spectrum and the magnetic properties of certain nuclei (particularly protons) is the basis of nuclear magnetic resonance spectroscopy. This magnetic interaction is described in classical terms in Chapter 4.

Interaction between the electrical component of electromagnetic radiation and electrical dipolar motions within molecules is the basis of rotational, electronic and vibrational spectroscopy. Because bands in these spectra are observed at specific frequencies it follows that the rotational, vibrational and electronic motions leading to electrical dipole changes must occur at specific *natural* frequencies.

On the basis of a classical model it is not possible to explain the existence of natural frequencies of rotation of molecules. The classical model gives some explanation for the existence of natural frequencies at which electronic motions occur (Maccoll, 1947) but there is a considerable body of experimental data which cannot be explained adequately on this basis. By contrast a classical model provides considerable information on the existence of natural frequencies of vibrations of atoms within molecules and it is informative to examine how far spectroscopic observations in the infrared spectrum of diatomic molecules can be explained in terms of a classical model.

In the case of diatomic molecules it is observed that heteropolar molecules (A—B) absorb radiation in the infrared region at a particular frequency characteristic of the molecule, but that homopolar diatomics (A—A) do not absorb infrared radiation. This can be understood if a diatomic molecule is considered as two masses m_a and m_b joined by a bond which maintains the atoms at some equilibrium separation and has spring-like properties. The "stiffness" of the bond can be characterized by a force constant, f, which is the force required to produce unit displacement in accordance with Hooke's Law. This law states that for small displacements of a spring there is an opposing restoring force, F, which is proportional to the magnitude of the displacement.

$$F = -fx \tag{1.4}$$

For a diatomic molecule, x represents the displacement of the atoms from the equilibrium internuclear separation, r_e, to some value r, and is given by $r - r_e$.

If the masses m_a and m_b are not identical they must attract the electrons constituting the bond to different extents. If A has a greater electron attracting power than B the molecule will have a permanent dipole moment. The dipole moment is represented in Fig. 1.5 by an electric vector p. It can readily be shown by combining Hooke's Law with Newton's second Law of Motion

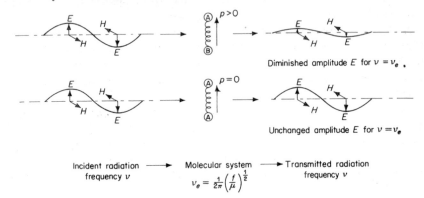

Incident radiation   ⟶   Molecular system   ⟶ Transmitted radiation
frequency ν $\nu_e = \frac{1}{2\pi}\left(\frac{f}{\mu}\right)^{\frac{1}{2}}$ frequency ν

Fig. 1.5. Classical model for the interaction of e.m. radiation with a diatomic oscillator

that a small displacement of one of the masses relative to the other will cause the system to vibrate in simple harmonic motion. The frequency at which the system vibrates is termed the equilibrium frequency, ν_e, and is given by

$$\nu_e = \frac{1}{2\pi}\left(\frac{f}{\mu}\right)^{1/2} \tag{1.5}$$

where μ is the reduced mass of the system and is given by

$$\frac{1}{\mu} = \frac{1}{m_A} + \frac{1}{m_B} \tag{1.6}$$

A mechanism for displacing one of the masses relative to the other is by coupling between the electric vector of the incident radiation and the electrical dipole moment vectors. If the frequency, ν, of radiation incident on the molecule (Fig. 1.5) is "tuned" to ν_e then the electric vector, E, of the incident radiation will couple with the dipole moment p and induce the molecule A—B to vibrate. Energy is taken up from the incident radiation and a "hole" or absorption band will appear in the spectrum of the source, at the particular frequency corresponding to ν_e. If ν is not equal to ν_e no interaction can take place. If the diatomic molecule has no permanent dipole moment (A—A in Fig. 1.5) no electrical coupling can take place and it is apparent why molecules of this category do not absorb in the infrared region.

On the basis of simple classical dynamics, a heteropolar diatomic molecule should possess an infrared spectrum which consists of a single absorption band. As an example the vibrational spectrum of carbon monoxide under low resolution is shown in Fig. 1.6(a).

The strong band at 2143 cm^{-1} in this spectrum may be assigned to the fundamental stretching vibration of carbon monoxide. Two features of this spectrum are in conflict with the classical explanation; firstly there is a second

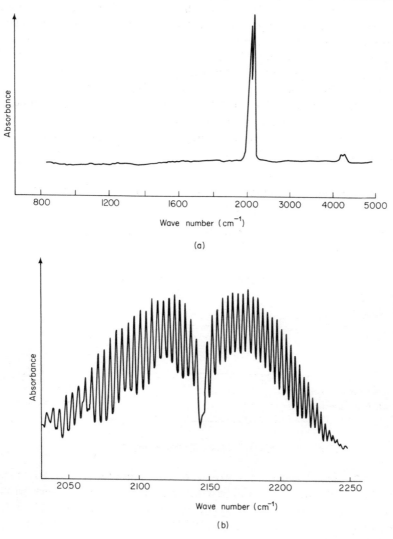

(a)

(b)

Fig. 1.6. Infrared spectrum of CO (a) at low resolution (b) at high resolution

weaker band at 4260 cm^{-1} which is significantly less than double the wave number of the fundamental vibration, and secondly if the spectrum is determined at higher resolution [Fig. 1.6(b)] the fundamental absorption band is seen to have a detailed fine structure.

Since classical theory is incorrect in its detailed predictions on vibrational spectra, and cannot be applied adequately to electronic spectra, and also cannot explain rotational spectra, explanation of these phenomena are sought in terms of the Quantum Theory. The limitations of the classical theory are inherent in the tacit assumption on which the theory rests; that the laws which govern the motion of particles which are large enough to be studied by direct observation are also applicable to matter at the molecular level. There is a logical fallacy in applying classical concepts to molecular systems. Classical mechanics is based on the idea of continuity of space, time and matter, and hence energy. However, the atomic hypothesis and the theory of molecular structure demand recognition of discontinuous, discrete particles of matter and a logical extension is to regard energy also as being discontinuous, discrete or *quantized*.

C. QUANTUM THEORY OF INTERACTION OF RADIATION WITH MATTER

Quantum theory expresses the energy of a molecule in terms of a series of discrete energy levels E_0, E_1, E_2, etc. (Fig. 1.7).

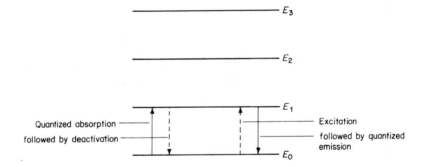

Fig. 1.7. Molecular energy levels

Each molecule in a system must exist in one or other of these levels. In a large assembly of molecules there will be a distribution of all the molecules between these various energy levels. The relative population of molecules, (N_i/N_j), in any two levels E_i and E_j is given by the Maxwell–Boltzmann equation

$$N_i/N_j = (g_i/g_j) \cdot e^{-\Delta E/kT} \tag{1.7}$$

where g_i and g_j are the number of permitted states with energy E_i and E_j (the degeneracy of the state), ΔE is the difference in energy between the states $(E_j - E_i)$, k is the Boltzman constant and T is the absolute temperature.

The energy levels are functions of an integer n (the quantum number) and the function is related to the particular molecular process undergone by the molecule, e.g. a change in electronic configuration, vibrational energy, rotational energy or spin orientations. In each case the energy of the quantum of radiation associated with any molecular process must exactly equal the energy gap $E_1 - E_0$ or $E_2 - E_1$, etc. The magnitude of the quantum of energy is related to the frequency of the radiation by equation 1.3. Hence the frequency of emission or absorption of radiation for a transition between the energy states E_0 and E_1 is given by

$$\nu = \frac{(E_1 - E_0)}{h} \qquad (1.8)$$

The concept of quantized energy levels thus explains the observation that the frequencies of bands or lines in absorption spectra are the same as the frequencies of bands or lines in emission spectra, in the absence of decomposition or processes involving phosphorescence.

D. POTENTIAL ENERGY FUNCTIONS

The vibrational potential energy, E_x, of a diatomic molecule (considered as a harmonic oscillator) is given by

$$E_x = -\int F\,dx = \int fx\,dx = \tfrac{1}{2}fx^2 + \text{const} \qquad (1.9)$$

If the values of f and r_e are known, then since $x = r - r_e$, the vibrational potential energy of the systems may be plotted if suitable values for r are chosen (Fig. 1.8a).

It may be shown from physical considerations that the expression for the potential energy based on the harmonic oscillator model is not valid for large positive or negative values of x. At large positive values of x $(r > r_e)$ the energy rises to some limiting value which represents complete dissociation of the atoms with zero interaction between them. The value of the potential energy will, therefore, always be *less* than the value predicted solely on the basis of the forces of extension of the bond (equation 1.4). At large negative values of x $(r < r_e)$ the energy increases to a very high value because the forces of repulsion between the atoms become large. The value of the potential energy will, therefore, always be *greater* than the value predicted solely on the basis of the forces of compression of the bond. These considerations suggest the model should be modified to take account of departure from harmonic

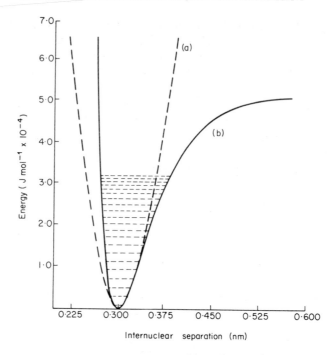

Fig. 1.8. Potential energy of an excited state iodine molecule (a) harmonic oscillator (b) anharmonic oscillators. Data computed and drawn by Dr. A. J. Bowles

behaviour. A number of mathematical expressions for the potential energy of an *anharmonic* oscillator have been suggested.

The simplest expression for the potential energy of an anharmonic oscillator is a power series

$$E_x = \tfrac{1}{2}f_1 x^2 - \tfrac{1}{2}f_2 x^3 + \tfrac{1}{4}f_3 x^4 - \ldots \tag{1.10}$$

where f_1, f_2, f_3, etc. are constants. The negative sign before the second term leads to an increase in the value of E_x for negative values of x and a decrease in the value of E_x for positive values of x compared with the expression based on a harmonic model. Inclusion of the third and higher terms in equation 1.10 gives a potential energy function which accords more closely with the experimental behaviour of a molecular system.

An alternative function for the potential energy of a diatomic molecule was proposed by Morse in 1929. This function has the form

$$E_x = D_e(1 - e^{-\alpha x})^2 \tag{1.11}$$

where D_e is the dissociation energy of the molecule and α is a constant characteristic of the internuclear bond. The dissociation energy is defined by the difference in energy between the system when the distance between the atoms is equal to the equilibrium internuclear separation (minimum of potential energy curve) and when the distance between the atoms is large (asymptotic region of potential energy curve).

A third function for the potential energy for a diatomic molecule was proposed by Lippincott in 1955. Whereas the Morse function was based on empirical considerations, the function proposed by Lippincott was based on quantum mechanical principles. This latter function has the form

$$E_x = D_e(1 - e^{-nx^2/2r}) \qquad (1.12)$$

where n is a constant characteristic of the internuclear bond.

The three potential energy functions given by equations 1.10, 1.11 and 1.12 may be plotted and give similar curves if suitable values for the constants are chosen. Figure 1.8b shows the curve for an excited electronic state of the I_2 molecule based on the Morse potential function. Potential energy functions based on an anharmonic oscillator model approximate to that based on the harmonic oscillator model for small values of x^2.

The potential energy functions for a diatomic molecule which have been considered above, are based on classical theory and represent energy variation to be continuous. In terms of quantum theory, vibrational energy can be described by a series of discrete vibrational energy levels which are defined by a vibrational quantum number V. These levels are shown as broken horizontal lines in Fig. 1.8 and they represent the values to which vibrational energy is limited. It may be noted that the lowest vibrational quantum level does not correspond to zero classical vibrational energy. Furthermore, the significance of the coordinate r in quantum terms has not been considered. These matters are considered in the wave mechanics section of this chapter.

The absorption of ultraviolet or visible radiation by a molecular system can cause the promotion of an electron from a low energy state to one of higher energy with a resultant change in the electronic configuration of the system. The potential energy of each electronic configuration of a molecule will vary with internuclear separation. It is possible therefore, to represent each electronic configuration by means of a potential-energy function. At room temperature nearly all the molecules will be in the lowest vibrational level and electronic transitions will mainly occur from this level.

An electronic transition is very rapid compared with the time required for the nuclei to move appreciably and so the internuclear separation can be regarded as fixed during the transition (the Franck–Condon principle). Electronic transitions can thus be represented on a potential energy diagram by vertical lines connecting the initial and final states.

III. WAVE MECHANICS

A rationale of the existence of energy levels in molecular systems and of other aspects of the properties of matter on a sub-molecular scale has been developed by the methods of wave mechanics. These methods are based on the work of Planck, de Broglie, Einstein and Schrödinger. This section will outline some of the main principles and results of wave mechanics which are relevant to the more detailed discussions in later chapters.

A. THE SCHRÖDINGER EQUATION

One of the basic principles which leads to the use of wave mechanical methods is that of the dual wave-particle nature of radiation and atomic particles. De Broglie postulated that all moving particles have an associated wave motion with a wavelength which is given by

$$\lambda = \frac{h}{mv} = \frac{h}{p} \tag{1.13}$$

where h is Planck's constant, m is the mass of the particle, v is the velocity of the particle and p the momentum of the particle.

The wave character of particles has been demonstrated for electrons, protons and α-particles. It is helpful in formulating the Schrödinger equation to consider one particular example of wave motion: the diffraction of light at a slit. When monochromatic light is diffracted at a slit the amplitude of the diffracted wave at any point depends on the distance, x, of the point from the slit image and the time, t, after diffraction. The total amplitude function, $F(x,t)$, can thus be written as a product of a distance-dependent function and a time-dependent function

$$F(x, t) = f(x) \cdot g(t) \tag{1.14}$$

The intensity I of the wave at any point is proportional to the square of the amplitude function

$$I \propto f(x)^2 \cdot g(t)^2 \tag{1.15}$$

If the wave motion is considered to be that of a standing wave then the amplitude will have maximum and minimum values (antinodes and nodes) at the same points independent of time. It then follows that for a standing wave the intensity at any point is independent of time. As a consequence the intensity of the wave will be proportional to the square of the time-independent part, $f(x)^2$, of the total amplitude function.

When light is considered as corpuscular motion, the intensity I on a small element of area is a measure of the energy of the radiation falling on that small element per second. This is represented by

$$I = n_x \epsilon = n_x h\nu \tag{1.16}$$

where n_x is the number of photons incident upon the element of area per second at a distance x from the centre of the diffraction pattern, ϵ is the energy of each photon, h is Planck's constant and ν is the frequency of the radiation.

Since I is proportional to $f(x)^2$, it follows from equation 1.16 that n_x is also proportional to $f(x)^2$. Now the probability of a photon reaching the position x is proportional to n_x and therefore this probability is also proportional to $f(x)^2$.

As the relationship postulated by de Broglie (equation 1.13) is valid for photons, it would suggest that expressions analogous to equations 1.14, 1.15 and 1.16 may be applied to any other particle. On this basis, the amplitude function $\Psi(x,t)$ for any particle moving in one dimension, x, can be written as

$$\Psi(x,t) = \psi(x) . g(t) \tag{1.17}$$

where $\psi(x)$ and $g(t)$ are the distance-dependent and time-dependent functions respectively.

If the particle is considered to be in a stationary state, i.e. the energy of the particle is held constant, then the system is equivalent to a standing wave on the wave concept. Thus for a stationary state, the probability of finding a particle at a position x is proportional to $\psi(x)^2$. Also, the expression for the time-dependent function $g(t)$ of a stationary state is given by the equation representing a wave in simple harmonic motion

$$g(t) = A e^{-2\pi i \nu t} \tag{1.18}$$

where A is a constant which is independent of time or position and ν is the frequency of the wave motion.

Thus the amplitude function $\Psi(x,t)$ for a stationary state system is given by

$$\Psi(x,t) = \psi(x) . A e^{-2\pi i \nu t} \tag{1.19}$$

Now the amplitude function $\Psi(x,t)$ for a wave is related to the velocity of propagation v of the wave in the x-direction by Maxwell's equation

$$\frac{\partial^2 \Psi(x,t)}{\partial x^2} = \frac{1}{v^2} . \frac{\partial^2 \Psi(x,t)}{\partial t^2} \tag{1.20}$$

Double differentiation of equation 1.19 will enable expressions for $\partial^2 \Psi(x,t)/\partial x^2$ and $\partial^2 \Psi(x,t)/\partial t^2$ to be obtained which may be substituted into the Maxwell equation (equation 1.20) to obtain the following expression:

$$\frac{\partial^2 \psi(x)}{\partial x^2} = \frac{-4\pi^2 v^2 \psi(x)}{v^2} \tag{1.21}$$

Now since the wavelength, λ, of wave motion is equal to v/ν, equation 1.21 can be written as

$$\frac{\partial^2 \psi(x)}{\partial x^2} = \frac{-4\pi^2}{\lambda^2} \cdot \psi(x) \tag{1.22}$$

If equation 1.13 is substituted into equation 1.22, the latter equation becomes

$$\frac{\partial^2 \psi(x)}{\partial x^2} = \frac{-4\pi^2 p^2}{h^2} \cdot \psi(x) \tag{1.23}$$

For a three-dimensional wave form, the equation corresponding to equation 1.23 is

$$\frac{\partial^2 \psi(x,y,z)}{\partial x^2} + \frac{\partial^2 \psi(x,y,z)}{\partial y^2} + \frac{\partial^2 \psi(x,y,z)}{\partial z^2} = \frac{-4\pi^2 p^2}{h^2} \cdot \psi(x,y,z) \tag{1.24}$$

Equation 1.24 may be written in abbreviated form as

$$\nabla^2 \psi(x,y,z) = \frac{-4\pi^2 p^2}{h^2} \cdot \psi(x,y,z) \tag{1.25}$$

where ∇^2 is the Laplacian operator and represents $\partial^2/\partial x^2 + \partial^2/\partial y^2 + \partial^2/\partial z^2$.

The total energy E of a particle is the sum of its kinetic energy ($\frac{1}{2}mv^2$) and potential energy (V)

$$E = \tfrac{1}{2}mv^2 + V \tag{1.26}$$

Since the momentum, p, of the particle is related to the velocity v of the particle by the expression $p = mv$, equation 1.26 can be written as

$$E = \frac{p^2}{2m} + V \tag{1.27}$$

Thus

$$p^2 = 2m(E - V) \tag{1.28}$$

Substitution of equation 1.28 into equation 1.25 gives

$$\nabla^2 \psi(x,y,z) = \frac{-4\pi^2 \, 2m(E - V)}{h^2} \cdot \psi(x,y,z) \tag{1.29}$$

Rearrangement of equation 1.29 gives the Schrödinger equation in a form in which it is often expressed

$$\nabla^2 \psi(x,y,z) + \frac{8\pi^2 m}{h^2}(E - V)\psi(x,y,z) = 0 \tag{1.30}$$

The Schrödinger equation can be written more succinctly in the form

$$H\psi(x, y, z) = E\psi(x, y, z) \tag{1.31}$$

where H is the Hamiltonian operator and is represented by

$$H = \left(-\frac{h^2}{8\pi^2 m} \cdot \nabla^2 + V \right) \tag{1.32}$$

The Schrödinger equation is of importance in spectroscopy since it can be used to determine the wave functions and energy levels of molecular systems. In order to solve the Schrödinger equation an expression for the potential energy function, V, and values of m are first substituted in the equation. A function ψ is then found which satisfies the equation. It is generally the case that the functions ψ which satisfy the equation (termed "eigenfunctions") only exist for certain values of E and these values are the allowed energy values for the system (termed "eigenvalues"). The use of the Schrödinger equation becomes apparent when applied to a specific problem (Expt. E.5).

The functions ψ can only be proper solutions of the Schrödinger equation if the following conditions are obeyed. Firstly ψ must be single valued for any one set of coordinates since there can be only one probability of occurrence, ψ^2, of a particle in a small volume element. Secondly ψ must be finite since if ψ were to equal infinity there would be an infinite probability of occurrence in a small volume element. This would contradict the Heisenberg Uncertainty Principle. Thirdly ψ must be continuous in order that the probability of occurrence has no discontinuity.

The probability of finding a particle in a small volume element dx, dy, dz is proportional to $\psi^2 d\tau$ where $d\tau = dx, dy, dz$ represents a small volume element in space. The probability of locating a particle in total space is equal to unity. This can be expressed in the form

$$\int_{-\infty}^{+\infty} \psi^2 \, d\tau = 1 \tag{1.33}$$

It is often easier to handle the mathematics involved in the solution of the Schrödinger equation if the Cartesian coordinates, x, y, z, are transformed into other coordinate systems, e.g. spherical polar coordinates, r, θ, ϕ. Such coordinates are shown, superimposed on Cartesian coordinates, in Fig. 1.9. The position of the point P in Fig. 1.9 can then be expressed in terms of r, θ, ϕ instead of x, y, z.

Solutions of the Schrödinger equation of particular importance in spectroscopy are those for a one-dimensional translator, rigid rotators, non-rigid rotators, simple harmonic oscillators, anharmonic oscillators and solutions

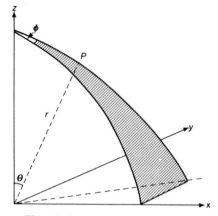

Fig. 1.9. Spherical polar coordinates

leading to various types of atomic orbitals. The potential energy functions relevant to each of these systems together with the quantized energy levels are shown in Table 1.1.

Each wave function in Table 1.1 is a function of some appropriate quantum number which is indicated as a subscript. For example, the wave function of the hydrogen atom is a function of n, l and m, which are the principal, orbital angular momentum and magnetic quantum numbers respectively. The principal quantum number, n ($n = 0$, 1, 2, 3, etc.), defines the energy levels within the atom. The orbital angular momentum quantum number, l ($l = 0$, 1, 2... $n - 1$), defines the shape of the orbital and the magnitude of the orbital angular momentum. The magnetic quantum number, m ($m = -1...0...+1$), specifies the orientation of the orbital due to the precessional motion of the orbital angular momentum in a magnetic field.

B. SIGNIFICANCE OF WAVE FUNCTIONS

The wave functions relevant to molecular spectroscopy concepts developed in subsequent chapters are the vibrational wave functions ψ_V and the electronic wave functions $\psi_{n, l, m}$.

If vibrational wave functions are considered first, then the simplest examples are those for a harmonic oscillator. The equations representing the wave functions ψ_V for $V = 0$, 1, 2 of a simple harmonic oscillator are given on page 118, and graphical representations of the functions ψ_V for $V = 0$, 1, 2 and ψ_V^2 for $V = 0$, 1, 2 are shown in Fig. 1.10. The classical values of the displacement, x, of the atoms from the equilibrium internuclear separation are also included in this figure. The probability of the molecule having a particular displacement in a particular state is given by the square of the wave function, ψ^2, representing that state. It can be seen from Fig. 1.10 that the function ψ^2 has several maxima and that the number of maxima is equal to $V + 1$.

It is of value to consider the infrared absorption of carbon monoxide [Fig. 1.6(a)] in terms of wave-mechanical and classical concepts and in relation to Fig. 1.10. In wave-mechanical terms, the band at 2143 cm^{-1} represents a transition from a state $V = 0$ in which there is maximum probability of having

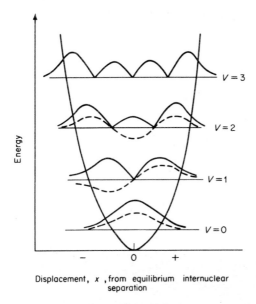

Displacement, x ,from equilibrium internuclear separation

Fig. 1.10. ψ_V (broken line) and ψ_V^2 (solid line) for a harmonic oscillator

x equal to zero, to a state $V = 1$ in which there is zero probability of having x equal to zero, but equal probabilities of x having a particular selected value. In classical terms, the band at 2143 cm^{-1} represents the fundamental stretching vibration which is induced in the molecule by electrical perturbation. The amplitude, x, of the vibration depends on the magnitude of the perturbation. The potential energy of the molecule varies between zero and the value E_x which is determined for particular values of x by equation 1.9. Inspection of Fig. 1.10 shows that, for a given vibrational level, the maximum classical value of x is always smaller than the maximum wave-mechanical value.

If carbon monoxide -is considered as an anharmonic oscillator, then the vibrations induced in the molecule may result in the appearance of overtone bands in the infrared spectrum. These bands are generally weak as instanced by the band at 4260 cm^{-1} in Fig. 1.6(a). In wave-mechanical terms, overtone

Table 1.1. Parameters associated with various energy systems

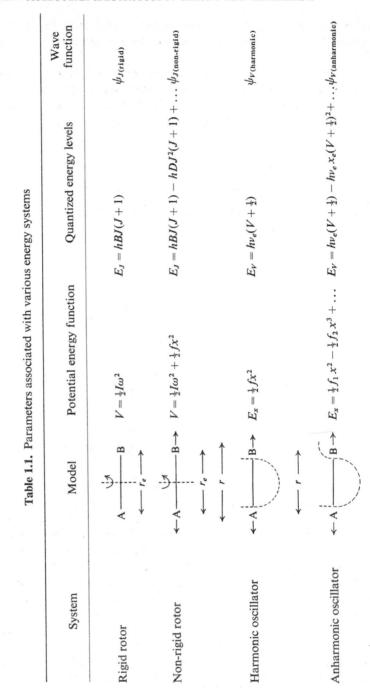

System	Model	Potential energy function	Quantized energy levels	Wave function
Rigid rotor		$V = \frac{1}{2}I\omega^2$	$E_J = hBJ(J+1)$	$\psi_{J(\text{rigid})}$
Non-rigid rotor		$V = \frac{1}{2}I\omega^2 + \frac{1}{2}fx^2$	$E_J = hBJ(J+1) - hDJ^2(J+1) + \ldots \psi_{J(\text{non-rigid})}$	
Harmonic oscillator		$E_x = \frac{1}{2}fx^2$	$E_V = h\nu_e(V + \frac{1}{2})$	$\psi_{V(\text{harmonic})}$
Anharmonic oscillator		$E_x = \frac{1}{2}f_1x^2 - \frac{1}{4}f_2x^3 + \ldots$	$E_V = h\nu_e(V + \frac{1}{2}) - h\nu_e x_e(V + \frac{1}{2})^2 + \ldots \psi_{V(\text{anharmonic})}$	

One-dimensional translator

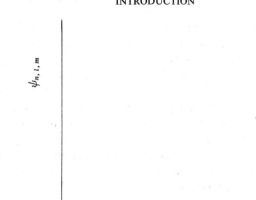

$V = 0$ or ∞

$E = \dfrac{nh^2}{8ml^2}$

$\psi_{(1\text{-dimensional})}$

Hydrogen atom

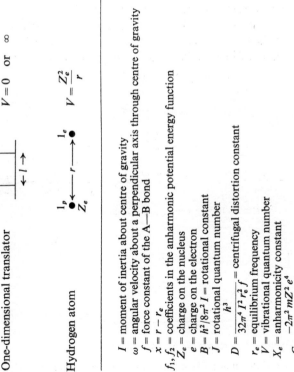

$V = \dfrac{Z_e^2}{r}$

$E_n = \dfrac{C}{n^2}$

$\psi_{n,\,l,\,m}$

I = moment of inertia about centre of gravity
ω = angular velocity about a perpendicular axis through centre of gravity
f = force constant of the A—B bond
$x = r - r_e$
f_1, f_2 = coefficients in the anharmonic potential energy function
Z_e = charge on the nucleus
e = charge on the electron
$B = h^2/8\pi^2 I$ = rotational constant
J = rotational quantum number
$D = \dfrac{h^3}{32\pi^4 I^2 r_e^2 f}$ = centrifugal distortion constant
r_e = equilibrium frequency
V = vibrational quantum number
X_e = anharmonicity constant
$C = \dfrac{-2\pi^2 mZ^2 e^4}{h^2}$

where m = mass of the electron
n = principal quantum number
l = orbital angular momentum quantum number
m = magnetic quantum number
h = Planck's constant

bands would represent transitions from the state $V = 0$ to the states $V = 2, 3$, etc. There is no obvious physical analogy between the classical concept of overtones of a fundamental stretching vibration and the wave-mechanical concept of a transition from a state with $V = 0$ to states with V greater than unity.

The classical state corresponding to zero potential energy would occur when the amplitude x of the vibration is zero and is a state in which the position and momentum of the nuclei would be completely defined. This would be contrary to the Heisenberg Uncertainty Principle and hence the state $V = 0$ is the lowest accessible state. This is the zero-point energy state (p. 116). According to wave mechanics, this state has a maximum probability of occurrence when x is equal to zero, and so this state cannot be thought of in terms of a classical vibrating system of small amplitude.

Turning now to electronic wave functions the electronic wave function for an atomic orbital can be expressed as a function of three variables R, Θ and Φ which depend upon the quantum numbers n, l and m

$$\psi(r, \theta, \phi) = R_{n,l}(r) . \Theta_{l,m}(\theta) . \Phi_m(\phi) \qquad (1.34)$$

The first variable R depends only on the distance r for particular values of n and l, the second and third variables Θ and Φ depend on the directions θ and ϕ for particular values of l and m. The above expression can be stated more succinctly by combining the last two variables so that the total wave function can be expressed as a radial function $R(r)$ and an angular function $A(\theta, \phi)$.

$$\psi(r, \theta, \phi) = R_{n,l}(r) . A_{l,m}(\theta, \phi) \qquad (1.35)$$

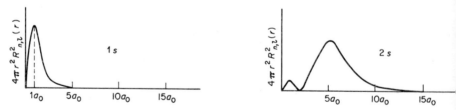

Distance from nucleus $(a_0 = 0.0529 \text{ nm})$

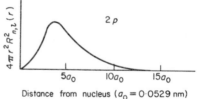

Distance from nucleus $(a_0 = 0.0529 \text{ nm})$

Fig. 1.11. Radial distribution functions for the hydrogen atom

The value of the square of the radial function is of interest since it is a measure of the probability of an electron being in an element of unit length at a certain distance r from the nucleus. A more useful expression is the radial distribution function, $4\pi r^2 . R^2_{n,\,l}(r)\,dr$, which gives the probability of an electron being contained in a spherical shell of thickness dr at a distance r from the nucleus. Some plots of radial distribution functions are shown in Fig. 1.11 (note: s, p, d and f correspond to the l-quantum numbers 0, 1, 2 and 3 respectively). It can be seen from the figure that the probability of an electron being at the nucleus is zero, and that the radius at which the electron probability is highest increases as n, the principal quantum number, increases.

In contrast to the radial function, the angular functions depend on direction and are independent of the distance from the nucleus. They are normally plotted as polar diagrams as in Fig. 1.12.

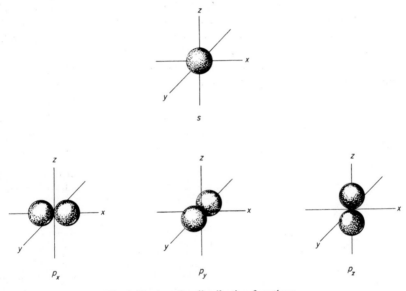

Fig. 1.12. Angular distribution functions

The quantity $A^2_{l,\,m}(\theta,\phi)$ gives the probability of finding an electron in the direction θ, ϕ. Thus the probability of finding an electron at a distance r from the nucleus and in the direction θ, ϕ is given by $R^2_{n,\,l}(r) . A^2_{l,\,m}(\theta,\phi)$ which is equal to the square of the wave function $\psi^2_{n,\,l,\,m}(r,\theta,\phi)$. The quantity $\psi_{n,\,l,\,m}(r,\theta,\phi)$ is the wave function of the system and for a particular value of n, l and m is referred to as an *atomic orbital*. An atomic orbital describes the spatial distribution of the energy and the angular momentum of an electron.

The importance of atomic orbitals in the present context is that the properties of electrons in molecules can be understood in terms of the combination of atomic orbitals to form molecular orbitals. For example the behaviour of the bonding electrons in the hydrogen molecule is represented by the molecular orbital formed by combination of the $1s$ orbitals of the two hydrogen atoms. The wave function, Ψ_g, of the molecular orbital is given by

$$\Psi_g = N_+(\psi_A + \psi_B) \tag{1.36}$$

where ψ_A, ψ_B are the wave functions for the $1s$ orbitals of each nucleus and N_+ is a constant.

The linear combination of the atomic orbitals ψ_A and ψ_B produces, in addition to the orbital Ψ_g, a second orbital Ψ_u which is represented by

$$\Psi_u = N_-(\psi_A - \psi_B) \tag{1.37}$$

where N_- is a constant. The subscripts g and u in the symbols for the molecular orbital functions indicate that these functions are symmetric and antisymmetric respectively with respect to inversion through the centre of the molecule.

The shapes of the atomic orbital functions ψ_A and ψ_B, the molecular orbital functions Ψ_g and Ψ_u, and the electron probability functions Ψ_g^2 and Ψ_u^2 when taken in section along the internuclear axis are shown in Figs. 1.13(a) and (b). It can be seen from these figures that for the orbital Ψ_g the probability, Ψ_g^2, of an electron being in the region between the nuclei is relatively high, while for the orbital Ψ_u this probability is relatively low. This means that for the orbital Ψ_g, bonding between the nuclei is favoured on account of the high electron density between the nuclei, while for orbital Ψ_u, the bonding is

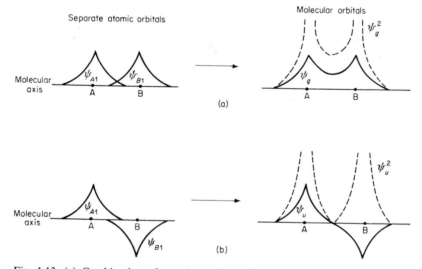

Fig. 1.13. (a) Combination of atomic orbitals to form a bonding molecular orbital. (b) Combination of atomic orbitals to form an antibonding molecular orbital

hampered on account of the repulsion of like charges on the nuclei. As a consequence the orbitals Ψ_g and Ψ_u are called bonding and anti-bonding orbitals respectively.

The spatial distribution and the energies of electrons in molecular orbitals determine the nature of the electronic transitions responsible for the absorption of radiation in the ultraviolet and visible regions of the spectrum. These factors are considered in the chapter on electronic spectroscopy.

C. TRANSITION MOMENTS

The intensity of an absorption band is determined by the probability of occurrence of the appropriate transition. It can be shown by the methods of wave mechanics (Eyring *et al.* 1944) that the spectral transition probability (transition moment) of an electronic, vibrational or rotational transition from a lower energy state i to an upper energy state j is given by

$$P_{i \to j} = \int_{-\infty}^{+\infty} \psi_i p \psi_j \, d\tau \qquad (1.38)$$

where ψ_i and ψ_j are the total wave functions of the two states, p is the electrical dipole moment of the molecule and $d\tau$ is a small element of volume. The subscripts i and j are quantum numbers which may relate to electronic energy ($i, j = n$), vibrational energy ($i, j = V$) or rotational energy ($i, j = J$). The quantum numbers n, V and J have integer values 0, 1, 2, 3, etc.

Since the dipole moment of a molecule consists of components p_x, p_y and p_z along the x, y and z axes, the integral $P_{i \to j}$ can be considered in terms of three component integrals P_x, P_y and P_z

$$P_x = \int_{-\infty}^{+\infty} \psi_i p_x \psi_j \, dx \qquad (1.39)$$

$$P_y = \int_{-\infty}^{+\infty} \psi_i p_y \psi_j \, dy \qquad (1.40)$$

$$P_z = \int_{-\infty}^{+\infty} \psi_i p_z \psi_j \, dz \qquad (1.41)$$

The intensity of a band in an electronic, vibrational or rotational spectrum is proportional to the square of the corresponding transition moment $P_{i \to j}$ and, obviously, the intensity of a band will only be finite if the value of at least one of the component integrals P_x, P_y and P_z is non-zero. Transition moments may be obtained in simple cases for diatomic molecules by substituting the appropriate wave function and dipole moment terms into the integrals and evaluating the integrals (Table 3.2). However, for most molecules the mathematics involved in calculating transition moments is formidable. When the value of the transition moment cannot be calculated, it is still possible to decide whether a transition is forbidden (zero transition moment) or permitted (non-zero transition moment) by considering the symmetry properties of the system.

IV. SYMMETRY CONCEPTS

Any object which has a characteristic shape (for example a molecule) possesses one or more *symmetry elements*. Each symmetry element has a *symmetry operation* associated with it. The relationship between these two concepts is that when the symmetry operation is performed on the object with respect to the appropriate symmetry element, an orientation is obtained which is indistinguishable from the initial orientation.

Consider a non-linear molecule of the type

Cartesian axes

It can be shown (Table 1.2) that the molecule possesses the following symmetry properties.

Table 1.2

Symmetry element	Symbol	Symmetry operation
Identity	E	Leave orientation unchanged
The two-fold rotation axis about z	$C_{2(z)}$	Rotate the molecule through 180° about the z axis passing through atom B
The yz plane	$\sigma_{(yz)}$	Reflect the molecule through the yz plane which lies in the plane of the molecule
The xz plane	$\sigma_{(xz)}$	Reflect the molecule through the xz plane which passes through atom B and is perpendicular to the plane of the molecule

The same symbol may be used to describe both the symmetry element and the symmetry operation, since these two concepts are closely related. The significance of these concepts can be demonstrated as follows.

Consider an observer who sees an object with a particular structure, before and after each of the following changes:

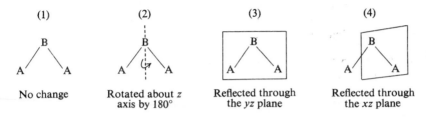

(1)	(2)	(3)	(4)
No change	Rotated about z axis by 180°	Reflected through the yz plane	Reflected through the xz plane

If neither A nor B have directional properties and the two A's are indistinguishable the observer has no means of knowing that the various operations 1, 2, 3 and 4 have been performed. The object is said to have the four *symmetry elements* which have been listed.

If all the symmetry elements possessed by various molecules are listed, then it is found that certain molecules, for example water, dichlorobenzene and

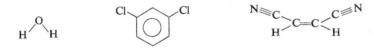

cis 1,2-dicyanoethylene, possess this same set of four symmetry elements. This indicates that these molecules belong to the same *point group*.

A. POINT GROUPS

The fact that a number of molecules possess the four elements of symmetry E, $C_{2(z)}$, $\sigma_{(yz)}$ and $\sigma_{(xz)}$ (and only these four) is represented by the statement that these molecules belong to the C_{2v} point group or, in short, they have C_{2v} symmetry. The terminology used here is the Schoenfliess notation. An alternative notation is the Hermann–Mauguin description in which the symbol 2 represents a two-fold axis, the symbol m represents a mirror plane and the point group is denoted by $2m$ in the example which has been considered.

Molecules which possess different sets of symmetry elements belong to different point groups. Molecules may possess the following symmetry elements in addition to those so far discussed (see Table 1.3).

The elements of symmetry possessed by any molecule may be established by inspection of the postulated shape of the molecule. The combination of elements of symmetry present defines the point group and all possible combinations are tabulated in texts, such as those of Jaffe and Orchin (1965), Cotton (1963), Herzberg (1945), Wilson *et al.* (1955) and others.

There are thirty-six point groups for all known molecules. A minority of molecules possess a large number of symmetry elements (for example methane,

Table 1.3

Symmetry element	Symbol	Symmetry operation
The n-fold rotation axis	C_n	Rotate the molecule about the axis by $360/n$
The n-fold alternating rotation–reflection axis	S_n	Rotate the molecule about the axis by $360/n$ and reflect the molecule through a plane perpendicular to the rotation axis
Centre of symmetry	i	Invert the molecule through a centre of symmetry

sulphur hexafluoride, ethylene and benzene), the vast majority of molecules possess very few elements of symmetry (for example propene, hydrogen peroxide and the majority of large organic molecules). All molecules possess at least the identity element E. The more elements of symmetry possessed by a molecule the higher symmetry it is said to have. However, the term "higher symmetry" is only of relative significance; the symmetry of any molecule is specified in a precise way by the point group to which that molecule belongs.

It is obvious that the symmetry of any molecule depends not only on its structure, but also on its configuration and conformation. Thus a particular compound may contain species with different shapes, each possibly having different symmetry.

Elements of symmetry conform to the properties of a formal mathematical group and are subject to the rules of that branch of mathematics known as *Group Theory*. However, the physical significance of symmetry considerations can be understood without recourse to the detailed mathematical methods of group theory. The properties of physical significance for each of the various point groups are contained in the *Character Table* of each point group.

B. CHARACTER TABLES

The character tables for each particular point group are listed in the various sources referred to in the previous section. The character table for the C_{2v} point group is reproduced in Table 1.4.

The symbols A_1, A_2, B_1 and B_2 in Table 1.4 are referred to as the *species*, *classes* or, more properly in group-theory terminology, the *irreducible representations* of the C_{2v} point group. The symbols specify a particular pattern of behaviour with respect to each symmetry operation. For each species there is a set of entries under each symmetry element given as $+1$ or -1. These entries are the *characters of the irreducible representations* of the C_{2v} point group. The symbol $+1$ conveys that the direction of some property is retained by the application of a particular symmetry operation (symmetric with respect to that operation), the symbol -1 conveys that the direction of some property is inverted by the application of a particular symmetry operation (antisymmetric with respect to that operation). Thus each of the four species in Table 1.4 has a different pattern of behaviour to the four symmetry operations and each pattern is specified by the four characters.

The characters take on the simple form $+1$ or -1 when the species are non-degenerate. However, the characters can take forms which are more difficult to relate to directional motions in the case of degenerate species.

Table 1.4. Character table for the C_{2v} point group

Species	\multicolumn{4}{c}{Symmetry elements}	\multicolumn{4}{c}{Cartesian coordinates of components of}
	E	$C_{2(z)}$	$\sigma_{(yz)}$	$\sigma_{(xz)}$	translation	rotation	dipole moment	polarizability
A_1	$+1$	$+1$	$+1$	$+1$	T_z		p_z	$\alpha_{xx}\ \alpha_{yy}\ \alpha_{zz}$
A_2	$+1$	$+1$	-1	-1		R_z		α_{xy}
B_1	$+1$	-1	-1	$+1$	T_x	R_y	p_x	α_{xz}
B_2	$+1$	-1	$+1$	-1	T_y	R_x	p_y	α_{yz}

Examples of directional properties are included in Table 1.4 in the form of cartesian coordinates of translations of molecules, rotations of molecules, displacement of electrical charge (dipole moment) and ease with which an electrical charge is displaced by an electric field (polarizability). The first three of these properties are vectors and have three cartesian coordinates, the last, polarizability, is a tensor property and requires six coordinates for its description (p. 130). The assignment of the components of the translational and rotational motions to a symmetry species as shown in Table 1.4 may easily be verified by inspection. Consider a translational motion along each of the three cartesian axes for any molecule with C_{2v} symmetry [Fig. 1.14(a)]. If the molecule is translating along the z axis and each of the four symmetry operations E, $C_{2(z)}$, $\sigma_{(yz)}$ and $\sigma_{(xz)}$ are performed in turn, a hypothetical observer would see no change in the direction of motion in each case. If the molecule is translating along the y axis, and the same four symmetry operations are performed, then to a hypothetical observer the direction of motion would be unchanged by the operations E and $\sigma_{(yz)}$ but would be reversed by the operations $C_{2(z)}$ and $\sigma_{(xz)}$. If the molecule is translating along the x axis the direction of motion would be unchanged by the operations E and $\sigma_{(xz)}$ but would be reversed by the operations $C_{2(z)}$ and $\sigma_{(xz)}$. Hence, the characters of each component under each symmetry operation are readily determined and the

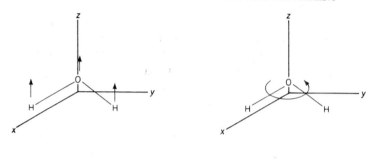

(a) Translation along z axis   (b) Rotation about z axis

Fig. 1.14. Molecular motion with respect to Cartesian axes

sets for translational motion in the z, y and x directions correspond to the A_1, B_2 and B_1 species respectively. Similarly for a rotational motion about each of the three cartesian axes [Fig. 1.14(b)] the character of the motion under each of the symmetry elements may be determined by considering the effect on the rotation as seen by a hypothetical observer looking along each particular axis when each symmetry operation is performed. The direction of the rotation is either unchanged ($+1$) or reversed (-1) and the assignments shown in Table 1.4 may be confirmed.

The symmetry properties of the components of the dipole moment are the same as those of the component of translation. This is because both are simple displacements, one electrical and the other mechanical, which may be represented by a vector with directional properties common to both. The symmetry of the components of polarizability α_{xx}, α_{xy}, α_{xz}, α_{yy}, α_{yz}, α_{zz} coincide with the symmetry of the products T_xT_x, T_xT_y, T_xT_z, T_yT_y, T_yT_z, T_zT_z. By using the rules of multiplication of symmetry operations the species of the components of polarizability may be determined. Their assignments in the character table of the C_{2v} point group may be readily confirmed.

C. MULTIPLICATION OF SYMMETRY OPERATIONS

Symmetry operations are defined as having been multiplied when each operation is performed in turn on a system which possesses the corresponding symmetry elements. One of the properties of a mathematical group is that the multiplication of any two symmetry operations in that group is equivalent to a single operation in that group. For example, the result of performing the operation $\sigma_{(yz)}$ followed by the operation $\sigma_{(xz)}$ on any of the translational or rotational components of a molecule with C_{2v} symmetry is equivalent to performing the single operation $C_{2(z)}$. This can be expressed by the equation

$$\sigma_{(yz)} \times \sigma_{(xz)} = C_{2(z)} \tag{1.42}$$

The products of every other possible combination of symmetry operations in the C_{2v} point group may be formed in a similar way and are summarized in Table 1.5.

Table 1.5. Multiplication table for the C_{2v} point group

	E	$C_{2(z)}$	$\sigma_{(yz)}$	$\sigma_{(xz)}$
E	E	$C_{2(z)}$	$\sigma_{(yz)}$	$\sigma_{(xz)}$
$C_{2(z)}$	$C_{2(z)}$	E	$\sigma_{(xz)}$	$\sigma_{(yz)}$
$\sigma_{(yz)}$	$\sigma_{(xz)}$	$\sigma_{(xz)}$	E	$C_{2(z)}$
$\sigma_{(xz)}$	$\sigma_{(xz)}$	$\sigma_{(yz)}$	$C_{2(z)}$	E

The multiplication of the characters of the operations $\sigma_{(yz)}$ and $\sigma_{(xz)}$ for the four species of the C_{2v} point group follows from equation 1.42. It can be seen from the equations below that the product in each case is the character of the $C_{2(z)}$ operation within each of the four species

$$A_1 \, (+1) \times (+1) = +1 \qquad\qquad (1.43a)$$

$$A_2 \, (-1) \times (-1) = +1 \qquad\qquad (1.43b)$$

$$B_1 \, (-1) \times (+1) = -1 \qquad\qquad (1.43c)$$

$$B_2 \, (+1) \times (-1) = -1 \qquad\qquad (1.43d)$$

This illustrates the general property of irreducible representations: they multiply in a way consistent with their respective symmetry operations.

D. SYMMETRY PROPERTIES OF TRANSITION MOMENTS

As discussed on page 23, a band will only be observed in an electronic, vibrational or rotational spectrum if at least one of the transition moment integrals P_x, P_y and P_z representing the transition has a non-zero value. A decision as to whether the values of the integrals P_x, P_y and P_z are zero or non-zero can be made by considering the symmetry of the individual terms in the integrals.

The integrals, like all observable properties of a molecule, must not be changed by performing a symmetry operation which simply takes the molecule from one orientation to another which is indistinguishable from the initial orientation. Considering the integral P_x; if this integral belonged to a symmetry species containing an operation for which the character was -1, then performing this operation on the integral would change the value of the integral, unless the value of the integral were zero. Thus, if the integral is to have a non-zero value then it must belong to a symmetry species in which all the characters are +1, i.e. the integral must belong to the totally symmetric species. The same considerations apply to the integrals P_y and P_z.

For a molecule belonging to the C_{2v} point group, one of the integrals P_x, P_y, P_z must belong to the A_1 species if the value of the transition moment $P_{i \to j}$ is to be non-zero. Now since the wave function in the ground state (ψ_i) is always totally symmetric (A_1 in the C_{2v} point group), the integrals P_x, P_y and P_z can only belong to the A_1 species if the respective products $p_x \psi_j$, $p_y \psi_j$ and $p_z \psi_j$ also belong to the A_1 species. The species to which the integrals P_x, P_y and P_z belong are derived by multiplication of the characters of the symmetry species to which ψ_i and $p_x \psi_j$, $p_y \psi_j$, $p_z \psi_j$ belong. The multiplication is illustrated for the product of ψ_i and $p_x \psi_j$ for each symmetry operation in Table 1.6. As can be seen the resultant characters are the same as those for the A_1 species.

Table 1.6. Multiplication of ψ_i and $p_x \psi_j$

	E	$C_{2(z)}$	$\sigma_{(yz)}$	$\sigma_{(xz)}$	Species
ψ_i	+1	+1	+1	+1	A_1
$p_x \psi_j$	+1	+1	+1	+1	A_1
$\psi_i p_x \psi_j$	+1	+1	+1	+1	A_1

The symmetry of the product $p_x \psi_j$ can only be A_1 if the terms p_x and ψ_j belong to the same symmetry species. For example, if both belong to the B_1 species then it may be shown (Table 1.7) that the product belongs to the A_1 species.

Table 1.7. Multiplication of p_x and ψ_j

	E	$C_{2(z)}$	$\sigma_{(yz)}$	$\sigma_{(xz)}$	Species
p_x	+1	−1	−1	+1	B_1
ψ_j	+1	−1	−1	+1	B_1
$p_x \psi_j$	+1	+1	+1	+1	A_1

Since the components p_x, p_y and p_z of the dipole moment p belong to the B_1, B_2 and A_1 species respectively in the C_{2v} point group, the transition moment can only be non-zero when the excited state ψ_j belongs to either the B_1, B_2 or A_1 species, since only then will the product of ψ_j and a component of p be of the A_1 species. If the excited state belongs to the A_2 species then the symmetry species of the product of ψ_j and each of the three components of p may be formed as shown in Table 1.8. Since in each case the product comes in a species other than the A_1 species it follows that the transition moment is zero.

Table 1.8. Multiplication of ψ_j and the components of p

	E	$C_{2(z)}$	$\sigma_{(yz)}$	$\sigma_{(xz)}$	Species
p_z	+1	+1	+1	+1	A_1
p_x	+1	−1	−1	+1	B_1
p_y	+1	−1	+1	−1	B_2
ψ_j	+1	+1	−1	−1	A_2
$p_z\psi_j$	+1	+1	−1	−1	A_2
$p_x\psi_j$	+1	−1	+1	−1	B_2
$p_y\psi_j$	+1	−1	−1	+1	B_1

It may be concluded that molecules with C_{2v} symmetry have permitted transitions to states belonging to the A_1, B_1 and B_2 species for all transitions associated with a dipole moment change—electronic, vibrational and rotational transitions. Rotational spectra will not be considered in any detail in this book. Transitions to levels belonging to the A_2 species are forbidden since no component of the dipole moment belongs to the A_2 species.

The above conclusions for the C_{2v} point group can be generalized to all point groups by the following statements.

(1) The transition moment $P_{i\rightarrow j}$ associated with a dipolar transition will have a non-zero value if at least one component of the transition moment P_x, P_y or P_z belongs to the totally symmetric species. It follows that electronic and vibrational transitions from the ground state to the first excited state are active if there is a component of the dipole moment which belongs to the same symmetry species as the excited state.

(2) The transition moment associated with a dipolar transition will have a zero value if none of the components of the transition moment belong to the totally symmetric species. It follows that electronic and vibrational transitions from the ground state to the first excited state are inactive if none of the components of the dipole moment belong to the same symmetry species as the excited state.

An alternative method for obtaining vibrational spectra (or rotational spectra) is by the Raman effect, details of which are given in Chapter 3. A Raman spectrum of a sample is observed as a shift in the frequency of monochromatic radiation when scattered by the sample. This shift is associated with a change between two vibrational energy levels (or rotational levels). The intensity of any Raman line is proportional to the square of the transition moment $P_{i \rightarrow j}$.

$$P_{i \rightarrow j} = \int \psi_i \, \alpha \psi_j \, d\tau \qquad (1.44)$$

where α is the polarizability of the molecule. By the same arguments used for electronic and vibrational transitions it can be shown that a transition from the ground state to the first excited vibrational state will be Raman active if the upper state belongs to the same symmetry species as at least one component of the polarizability function. Hence, for a molecule with C_{2v} symmetry all fundamental vibrations are Raman active because all species (Table 1.4) have at least one component of the polarizability in their symmetry class.

E. SYMMETRY PROPERTIES AND TRANSITIONS IN BENZENE

The activity of vibrational (infrared and Raman) and electronic transitions in benzene can be determined by consideration of the symmetry of the various excited levels of the molecule.

The structure of benzene is assumed to be a planar hexagon for which the following symbol is most frequently used.

It is obvious that this structure possesses many elements of symmetry. There are various six-fold rotation axes (C_6), various three-fold rotation axes (C_3), and various two-fold rotation axes (C_2), various planes of symmetry (σ), a centre of symmetry (i), and various n-fold alternating rotation–reflection axes (S_n). If all these elements of symmetry are listed it is found that they conform to the elements of symmetry for the D_{6h} point group. The character table of this point group is shown in Table 1.9.

There are twelve species within the D_{6h} point group of which those denoted by the symbols A_{2u} and E_{1u} contain components of the dipole moment function and those denoted by the symbols A_{1g}, E_{1g} and E_{2g} contain components of the polarizability function. The subscripts g (*gerade*) and u (*ungerade*) imply that the species are symmetric and anti-symmetric respectively with respect to an inversion through the centre of symmetry. The other symbols, A, B and E with an appropriate numerical subscript define the behaviour of the various species to the other symmetry operations. Symbols A and B are used for non-degenerate species and E is used for a doubly degenerate species.

Using the D_{6h} character table the following deductions may be made concerning the electronic and vibrational transitions in benzene (and all other

Table 1.9. Character table for the D_{6h} point group

Species	E	$2C_6$	$2C_3$	C_2	$3C_2'$	$3C_2''$	i	$2S_3$	$2S_6$	σ_h	$3\sigma_d$	$3\sigma_v$	translation	rotation	dipole moment	polarizability
A_{1g}	1	1	1	1	1	1	1	1	1	1	1	1				$\alpha_{xx}+\alpha_{yy},\ \alpha_{zz}$
A_{2g}	1	1	1	1	-1	-1	1	1	1	1	-1	-1		R_z		
B_{1g}	1	-1	1	-1	1	-1	1	-1	1	-1	1	-1				
B_{2g}	1	-1	1	-1	-1	1	1	-1	1	-1	-1	1				
E_{1g}	2	1	-1	-2	0	0	2	1	-1	-2	0	0		R_x, R_y		$\alpha_{yz},\ \alpha_{zx}$
E_{2g}	2	-1	-1	2	0	0	2	-1	-1	2	0	0				$\alpha_{xx}-\alpha_{yy},\ \alpha_{xy}$
A_{1u}	1	1	1	1	1	1	-1	-1	-1	-1	-1	-1				
A_{2u}	1	1	1	1	-1	-1	-1	-1	-1	-1	1	1	T_z		p_z	
B_{1u}	1	-1	1	-1	1	-1	-1	1	-1	1	-1	1				
B_{2u}	1	-1	1	-1	-1	1	-1	1	-1	1	1	-1				
E_{1u}	2	1	-1	-2	0	0	-2	-1	1	2	0	0	T_x, T_y		p_x, p_y	
E_{2u}	2	-1	-1	2	0	0	-2	1	1	-2	0	0				

Symmetry elements

Cartesian coordinates of components of

molecules with D_{6h} symmetry) for the transition from the ground state to the first excited electronic or vibrational energy level.

(1) Electronic transitions to levels which belong to either the A_{2u} or E_{1u} symmetry species are permitted. (These transitions are observed as absorption bands in the ultraviolet or visible regions of the spectrum.)

(2) Vibrational transitions to levels which belong to either the A_{2u} or E_{1u} symmetry species are permitted. (These transitions are observed as absorption bands in the infrared region of the spectrum.)

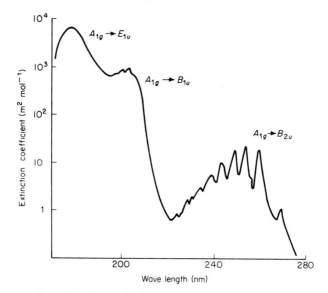

Fig. 1.15. Electronic absorption spectrum of benzene

(3) Vibrational transitions to levels which belong to either the A_{1g}, E_{1g} or E_{2g} symmetry species are permitted. (These transitions are observed as bands or lines in a Raman spectrum.)

The nature of the various upper electronic states of benzene have been the subject of wave mechanical calculations and it has been shown that the three lowest electronic excited states of benzene involve molecular orbitals belonging to the B_{1u}, B_{2u} and E_{1u} symmetry species. Of these, only the transition to the E_{1u} level is permitted by symmetry. Examination of the spectrum of benzene in the ultraviolet region (Fig. 1.15) shows three principal bands. The strongest band which appears at 180 nm is assigned to the transition to the E_{1u} level. Two weaker bands at 210 nm and 260 nm are assigned to transitions to the B_{1u} and B_{2u} levels respectively. These two bands have intensities less than the values expected for formally permitted transitions. They are forbidden by electronic symmetry factors but associated vibrational changes may momentarily change the symmetry of the molecule leading to the transition becoming less forbidden. The band at 210 nm is more intense than the band at 260 nm, probably because of borrowed intensity from the 180 nm band.

In the case of vibrational energy changes, methods exist firstly for the calculation of the number of fundamental modes of vibration in each symmetry species, and secondly for the calculation of the form and frequency of each of the fundamental modes of vibration on the basis of the classical models (Herzberg, 1945; Wilson et al. 1955). For example the 30 fundamental modes of vibration of benzene divide amongst the 12 species of the D_{6h} point group (Table 1.9) as follows: $2A_{1g} + 0A_{1u} + 1A_{2g} + 1A_{2u} + 0B_{1g} + 2B_{1u} + 2B_{2g} + 2B_{2u} + 1$ pair $E_{1g} + 3$ pairs $E_{1u} + 4$ pairs $E_{2g} + 2$ pairs E_{2u}. Six of the thirty fundamental modes of vibration will be associated with the stretching motion of the six C—H bonds. The stretching motions of these bonds couple together into modes of different symmetries. The forms and frequencies of these modes (calculated by Scherer and Overend, 1961) are shown in Fig. 1.16 together with the values of the observed infrared or Raman frequencies. The assignment of these modes to the respective symmetry species indicated in Fig. 1.16 may be

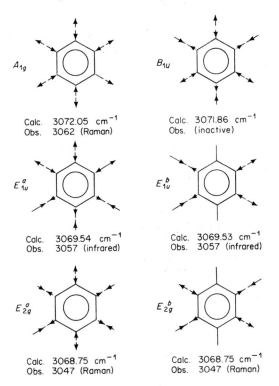

A_{1g}

Calc. 3072.05 cm^{-1}
Obs.   3062 (Raman)

B_{1u}

Calc. 3071.86 cm^{-1}
Obs.   (inactive)

E_{1u}^a

Calc. 3069.54 cm^{-1}
Obs.   3057 (infrared)

E_{1u}^b

Calc. 3069.53 cm^{-1}
Obs.   3057 (infrared)

E_{2g}^a

Calc. 3068.75 cm^{-1}
Obs.   3047 (Raman)

E_{2g}^b

Calc. 3068.75 cm^{-1}
Obs.   3047 (Raman)

Fig. 1.16. C—H stretching modes of benzene

readily confirmed by inspection. The forms and frequencies of the remaining 24 fundamental modes of vibration have also been reported by Scherer and Overend.

V. MASS SPECTROMETRY

The underlying principle of mass spectrometry is different from that of electronic, vibrational or nuclear magnetic resonance spectroscopy. In the last three methods information on molecular structure is obtained by analysing the effect of the interaction of electromagnetic radiation with a molecular system, whereas in mass spectrometry this information is normally obtained by interpretation of the effects of bombardment of the molecular systems with electrons of relatively high energy. The essential difference between mass spectrometry and other spectroscopic methods is that mass spectrometry involves production of an ionized species and interpretation of the fragmentation of this species, whereas other methods involve the interpretation of changes in energy states of the non-ionized molecule.

When a molecule absorbs energy upon impact by an electron the following changes can occur: (1) Excitation of the molecule to a higher electronic energy level and subsequent radiative or non-radiative return to the ground state. (2) Electron capture by the molecule to form a negative ion. (3) Ionization of the molecule to produce a positively charged molecular (or parent) ion. (4) Bond rupture within the molecular ion to produce fragment ions. The last two processes are those of most interest since they give most information upon molecular structure.

REFERENCES

Cotton, F. A. (1963). *In* "Chemical Applications of Group Theory". Interscience, New York.
Eyring, H., Walter, J., and Kimball, E. (1944). *In* "Quantum Chemistry". Wiley, New York.
Herzberg, G. (1945). *In* "Infrared and Raman Spectra of Polyatomic Molecules". Van Nostrand, London.
Jaffe, H. H., and Orchin, M. (1965). *In* "Symmetry in Chemistry". Wiley, New York.
Maccoll, A. (1947). *Q. Rev. chem. Soc.* **1**, 16.
Scherer, J. R., and Overend, J. (1961). *Spectrochim. Acta*, **17**, 719.
Wilson, E. B., Decius, J. C., and Cross, P. C. (1955). *In* "Molecular Vibrations". McGraw-Hill, New York.

Chapter 2

Electronic Spectroscopy

I. PRINCIPLES OF ELECTRONIC SPECTROSCOPY

Electronic spectroscopy is concerned with electronic transitions between orbitals of different energy. A common transition is one where an electron is promoted from the occupied orbital of highest energy to the vacant orbital of lowest energy. The energy difference between these orbitals in organic and inorganic molecules normally lies between $1\cdot5 \times 10^5$ and 6×10^5 J mol^{-1}, and the corresponding transition gives rise to absorption in the ultraviolet and visible regions of the spectrum.

Molecular orbitals may be described in terms of a linear combination of atomic orbitals. As considered in Chapter 1, the combination of the $1s$ orbitals of hydrogen atoms results in two molecular orbitals for the hydrogen molecule —one a bonding molecular orbital, Ψ_g, and the other an anti-bonding molecular orbital, Ψ_u. The orbitals are commonly denoted by the symbols σs and σs^* respectively. These molecular orbitals are represented in Fig. 2.1 along with the molecular orbitals formed by linear combination of atomic $2p$ orbitals. The bonding molecular orbitals formed by linear combination of $2p_x$ and $2p_y$ atomic orbitals are denoted by σp and πp respectively, while the corresponding anti-bonding orbitals are denoted by $\sigma^* p$ and $\pi^* p$. These symbols are commonly abbreviated to the forms σ, π, σ^* and π^*.

Another type of molecular orbital is the non-bonding orbital. An example is found in carbonyl compounds where the lone pair electrons on the oxygen

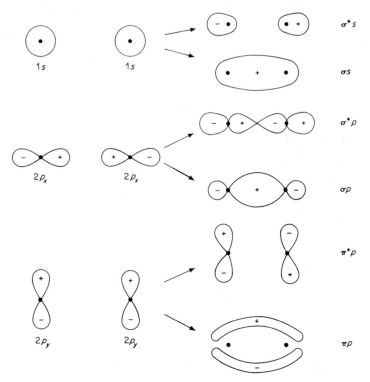

Fig. 2.1. Molecular orbitals formed by combination of atomic orbitals

atom are accommodated in the non-bonding $2p$ orbital localized on the oxygen atom. Electrons in non-bonding orbitals (n orbitals) can be promoted into σ^* and π^* orbitals of higher energy.

It should be noted that the orbitals shown in Fig. 2.1 do not individually define the probability distribution in space of the electrons in the molecule, although this probability distribution can be derived from the orbital system of the molecule by the methods of wave mechanics (Linnett, 1960). Since molecular orbitals are of primary importance in electronic spectroscopy, whereas the spatial distribution of electrons is to some extent of secondary importance, subsequent discussions in this chapter will relate to molecular orbitals only.

The relative positions of the energy levels corresponding to the different types of molecular orbital in organic molecules are shown in Fig. 2.2. The possible types of electronic transition between these orbitals are represented by vertical lines in the figure. The energy of a non-bonding orbital is generally greater than that of a σ or π orbital but lower than that of the anti-bonding orbitals. However, there are instances of molecules having a high degree of conjugation where the energy of the π orbital is greater than that of the n orbital. In such cases less energy is required to promote a $\pi \rightarrow \pi^*$ transition than to promote an $n \rightarrow \pi^*$ transition and the absorption band corresponding to the $\pi \rightarrow \pi^*$ transition will occur at a longer wavelength than that for the $n \rightarrow \pi^*$ transition.

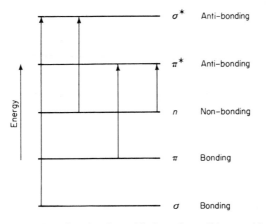

Fig. 2.2. Relative energies of molecular orbitals and possible transitions between the orbitals

A. DIATOMIC MOLECULES

The promotion of an electron from an orbital of lower energy to one of higher energy gives rise to an excited state of the molecule. The electronic configuration in the ground state and in the excited state of the hydrogen molecule may be depicted as follows.

$$\sigma^2(s) \quad \rightarrow \quad \sigma(s)\,\sigma^*(s)$$

Ground state   Excited state

If the molecule is stable in the excited state, i.e. it does not dissociate into atoms, then the properties in the excited state may be different to those in the ground state. For example, the dissociation energy, the equilibrium internuclear separation and the fundamental vibration frequency may differ in the two states. However the excited-state molecule will be essentially similar to the ground-state molecule and the potential energy function for the excited state will be similar to that of the ground state.

The potential energy function for a diatomic molecule can be represented by a two-dimensional curve, as is shown in Fig. 2.3(a) for a ground state and an excited state. Since the Franck–Condon principle stipulates that the internuclear separation of the atoms does not change during an electronic transition,

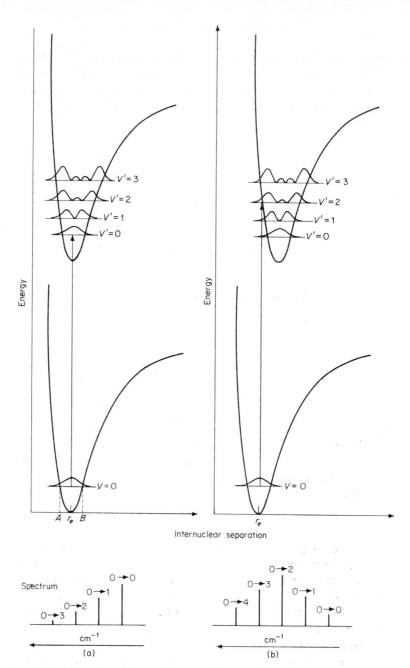

Fig. 2.3. Relationship between the potential-energy function, vibrational energy levels, vibrational probability functions and the absorption spectrum of a diatomic molecule. (The vibrational probability functions are represented by the curves above each vibrational energy level)

a transition from the ground state to an excited state can be denoted by a vertical line connecting the potential energy curves of the two states.

At room temperature the vibrational energy of nearly all the molecules corresponds to that of the lowest vibrational level ($V = 0$), and transitions from the ground electronic state can be considered as originating from this level. There are no general selection rules governing the change in vibrational quantum number in going from the ground state to an excited state, and transitions from the $V = 0$ level of the ground state to any level of the excited state are possible.

The intensity of an absorption band is dependent upon the value of the transition moment of the transition (p. 23) and this value will be high if the internuclear separation at the instant of the transition is such that the transition occurs between probable states of the molecule. The probability of a molecule having a particular internuclear separation is given by the square of the vibrational wave function of the system, and as can be seen from Fig. 2.3 the most probable internuclear separation for molecules in the $V = 0$ level of the ground state is close to the internuclear separation, r_e. If the probability function for a vibrational level in the excited state has a maximum at an internuclear separation corresponding to the equilibrium internuclear separation in the ground state, then the transition moment for a transition to this level will be large and an intense band will be observed in the spectrum. Examination of Fig. 2.3(a), where the ground state and excited states have the same internuclear separation, shows that the vertical line representing the $0 \rightarrow 0$ transition connects states of maximum probability. Thus the band in the absorption spectrum representing the $0 \rightarrow 0$ transition will be of maximum intensity.

As the vibrational quantum number, V', of the excited state increases so the probability function increases near the turning points (intersection of horizontal lines and potential energy curves) and decreases between these points. The maximum internuclear separation in the $V = 0$ level of the ground state, i.e. the distance AB, corresponds to the central region of the V' levels, and thus transitions from the $V = 0$ level to levels of increasing V' value will become increasingly less probable since the state of maximum probability in the higher V' levels will have internuclear separations which differ more and more from the equilibrium internuclear separation of the ground state. The absorption spectrum will therefore appear as a progression of bands of decreasing intensity.

If the equilibrium internuclear separation in the excited state is greater than that in the ground state, then the potential energy curves will be displaced relative to each other as in Fig. 2.3(b). In this particular case the vertical line from the $V = 0$ level intersects the potential energy curve for the excited state at the turning point for the $V' = 2$ level. From the classical viewpoint, the probability of the molecule having the internuclear separation corresponding to that at the turning point is a maximum since the molecule spends most time at the turning point in the course of a vibration. The line representing the transition from the $V = 0$ level to the $V' = 2$ level thus connects states of maximum probability and the corresponding band in the spectrum will be the most intense. The probability of transitions to other levels in the excited state will decrease as the vibrational quantum number, V', increases or decreases from the value of 2, and the spectrum will appear as a progression of bands of decreasing intensity on either side of the band representing the $0 \to 2$ transition.

Information on the physical parameters of diatomic molecules in the excited state can be obtained by analysis of such band spectra (see Expt. E3).

B. ORGANIC POLYATOMIC MOLECULES

The electronic configuration of the ground state of a polyatomic molecule may have electrons in σ, π or n orbitals, dependent upon the structure of the particular molecule. The possible electronic transitions in polyatomic molecules are $n \to \pi^*$, $\sigma \to \sigma^*$, $n \to \sigma^*$ and $\pi \to \pi^*$. The main features of these four types of transition are discussed in the following sections.

I. $n \to \pi^*$ Transitions

(a) Symmetry Properties and Band Intensities

Unsaturated molecules which contain atoms such as oxygen, nitrogen and sulphur often exhibit a weak band in their absorption spectrum which can be assigned to an $n \to \pi^*$ transition. For aldehydes and ketones this band arises from excitation of a lone-pair electron on the oxygen atom into the antibonding π^* orbital of the carbonyl function. The alternative ways of representing this transition are shown in Fig. 2.4.

The bonding in the carbonyl group consists of a σ bond formed by overlap of the $2p_x$ orbitals on the carbon and oxygen atoms and a π bond formed by

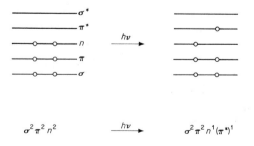

Fig. 2.4. Representations of an $n \to \pi^*$ transition

overlap of the $2p_z$ orbitals. There are two electrons in the $2p_y$ orbital of the oxygen atom which are not involved in bonding. On absorption of radiation of the appropriate energy a non-bonding electron in the $2p_y$ orbital is promoted to the anti-bonding π^* orbital.

The observation that the bands corresponding to $n \rightarrow \pi^*$ transitions are weak suggests that the electronic transition is forbidden. The probability of an $n \rightarrow \pi^*$ transition and hence the intensity of the corresponding absorption band can be predicted by considering the symmetry properties of the ground and excited states. For ketones which belong to the C_{2v} point group, the symmetry species of the ground and excited states will be one of the four species A_1, A_2, B_1 and B_2 of this point group (p. 27). The electronic configurations for the ground state and for the excited state of a carbonyl compound after an $n \rightarrow \pi^*$ transition can be represented by:

Ground state   [Inner electrons]$\sigma^2 \pi^2 n^2$

Excited state   [Inner electrons]$\sigma^2 \pi^2 n^1 (\pi^*)^1$

The symmetry species to which each of these states belong can be readily assigned if it is assumed that the symmetry species for the total wave functions of these states is equal to the product of the symmetry species for the wave functions describing the behaviour of the individual electrons. The symmetry species for the individual electrons are obviously the same as the symmetry species for the orbitals in which the electrons are accommodated, and thus the symmetry species for the total wave functions is derived from the product of the symmetry species for the orbitals containing the electrons.

The multiplication is simplified somewhat since the symmetry species of an orbital containing two electrons must be totally symmetric. For example, it can be seen from Fig. 2.5 that the π orbital associated with the carbonyl function belongs to the B_1 species, and since there are two electrons in this orbital the symmetry species of the combined electronic system is given by the product $B_1 \times B_1$

$$
\begin{array}{lcccc}
B_1 & +1 & -1 & -1 & +1 \\
B_1 & +1 & -1 & -1 & +1 \\
\hline
B_1 \times B_1 & +1 & +1 & +1 & +1 \equiv A_1
\end{array}
$$

All the characters for the A_1 species are $+1$ and multiplication of a species such as B_1 by A_1 does not alter the sign of any of the characters of the B_1 species. Thus the species for the product will be B_1. In general terms, the result of multiplication of any species, or any product of species, by the A_1 species will be the original species or the original product of species.

Since all the "inner" electrons in the electronic configurations of the ground and excited state are accommodated in filled orbitals, the symmetry species corresponding to each orbital will be A_1. Also the electrons in the σ, π and n orbitals of the ground state and in the σ and π orbitals of the excited state are in filled orbitals and the species for each of these systems will be A_1. The symmetry species for the single electrons in the n and π^* orbitals are B_2 and B_1 respectively (Fig. 2.5). The symmetry species for the total wave functions of the ground and excited states are derived from the products of the symmetry species of the individual electronic systems.

Ground state   [Inner electrons] $\sigma^2 \pi^2 n^2$
$\qquad\qquad [A_1 \times A_1 \ldots] A_1 A_1 A_1 \equiv A_1$

Excited state   [Inner electrons] $\sigma^2 \pi^2 n^1 (\pi^*)^1$
$\qquad\qquad [A_1 \times A_1 \ldots] A_1 A_1 B_2 B_1 \equiv B_2 \times B_1 \equiv A_2$

It may be recalled from Chapter 1, page 29, that the intensity of an electronic transition is related to the symmetry species to which the transition moment belongs. If, for a molecule with C_{2v} symmetry, the transition moment belongs to the A_1 species then the value of the transition moment is non-zero and the transition is allowed, whereas if the transition moment belongs to a species

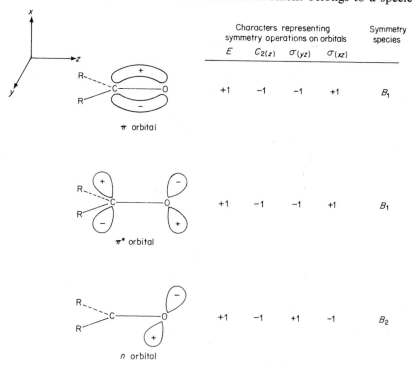

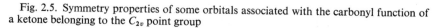

Fig. 2.5. Symmetry properties of some orbitals associated with the carbonyl function of a ketone belonging to the C_{2v} point group

other than A_1 the value is zero and the transition is forbidden. It may also be recalled that the symmetry species of the transition moment is only A_1 if the wave function for the excited state belongs to the same symmetry species as one of the dipole moment components p_x, p_y and p_z. Inspection of the character table for the C_{2v} point group given on page 27 shows that the components p_x, p_y and p_z belong to the B_1, B_2 and A_1 species respectively. Since the wave function of the excited state after an $n \rightarrow \pi^*$ transition belongs to the A_2 species and not the B_1, B_2 or A_1 species, it may be concluded that the transition moment for the $n \rightarrow \pi^*$ transition is zero and that the $n \rightarrow \pi^*$ transition is forbidden. It is thus to be expected that the corresponding absorption band will be of low intensity.

The intensity of an absorption band at any particular wavelength is proportional to a constant termed the "extinction coefficient". The extinction coefficient, ϵ, can be derived experimentally from absorbance measurements (p. 68) and can be used as a guide as to whether a transition is allowed or forbidden. Fully allowed transitions will have ϵ values greater than about 1000 m^2 mol^{-1} while transitions of low probability will have ϵ values below 100 m^2 mol^{-1}. The extinction coefficients for $n \rightarrow \pi^*$ transitions are generally in the range 1–10 m^2 mol^{-1}.

(b) General Features of $n \rightarrow \pi^*$ Transitions

Absorption bands arising from $n \rightarrow \pi^*$ transitions are typified by those observed in the spectra of carbonyl compounds. In saturated aldehydes and ketones, the band due to the $n \rightarrow \pi^*$ transition generally occurs in the range 270–300 nm and the value of the extinction coefficient at the peak maximum lies between 1 and 2 m^2 mol^{-1}. In unsaturated carbonyl compounds where the double bonds are separated by two or more single bonds the absorption spectrum is normally a composite of the ethylenic and carbonyl chromophore absorptions. If the carbonyl group forms part of a conjugated system the band due to the $n \rightarrow \pi^*$ transition generally appears in the range 300–350 nm and has an extinction coefficient of the order of 10 m^2 mol^{-1}.

The nature of the substituents on the carbonyl group affects the position of the absorption band due to the $n \rightarrow \pi^*$ transition (see Expt. E4). The substituent can affect the carbonyl group by the inductive effect or by the mesomeric (resonance interaction) effect. The inductive effect alters the energies of the n orbital and the π^* orbital to approximately the same extent and there is no marked change in the position of the band on account of this effect. Consequently changes in the position of the band have to be explained in terms of the mesomeric effect. As a result of this effect substituents such as OH, Cl, NH$_2$ and OAlk, which possess lone-pair electrons, act as electron donors to the π system of the carbonyl group. This causes an increase in energy of all the π orbitals, bonding and anti-bonding, but does not affect the energy of the n orbital. Thus the energy difference between the n and π^* orbitals is increased in compounds containing such substituents and the corresponding absorption band will be shifted to shorter wavelength.

Replacement of the hydrogen atom in an aldehydic function by a methyl group results in a shift of the band due to the $n \rightarrow \pi^*$ transition to shorter wavelength. The methyl group acts as an electron donor by virtue of the hyperconjugation effect, and has the same effect on the relative energies of the n and π^* orbitals as substituents with lone-pair electrons. Increasing the chain length and degree of branching of the alkyl substituent reduces the electron donor power and, as would be expected, this is observed as a shift of the band to progressively longer wavelengths.

Examination of the effect of solvents on the position of a band in a spectrum can give information as to the type of transition responsible for that band. One of the characteristic features of bands due to $n \rightarrow \pi^*$ transitions is that the bands shift to shorter wavelength on changing from a hydrocarbon to a hydroxylic solvent (see Expt. E4). This behaviour is explained in terms of solute–solvent interactions. If a carbonyl compound is dissolved in a hydroxylic solvent then hydrogen bonding will occur between the carbonyl function and the hydrogen atom of the hydroxylic group. The formation of the hydrogen bond stabilizes the ground state of the carbonyl compound. When an $n \rightarrow \pi^*$ transition occurs an electron is removed from the n orbital, leaving a formal positive charge on the oxygen atom and the hydrogen bond is broken. The net effect of changing from a hydrocarbon to a hydroxylic solvent is therefore to lower the energy of the ground state relative to that of the excited state and thus to increase the energy required for an $n \rightarrow \pi^*$ transition.

Another method of characterizing $n \rightarrow \pi^*$ transitions is to observe the spectrum in acidic solutions. Bands due to $n \rightarrow \pi^*$ transitions generally disappear in acidic solution because a bond is formed between the lone-pair electrons on the heteroatom and the acidic proton. For example, the band due to the $n \rightarrow \pi^*$ transition in pyridine disappears in acidic solution owing to the reaction

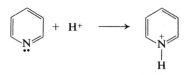

Where feasible an $n \rightarrow \pi^*$ transition may be assigned by comparison of the spectrum of the compound containing the heteroatom with a similar compound not containing a heteroatom. For instance the spectrum of pyrazine exhibits a band with λ_{max} at about 300 nm which is not present in the spectrum of benzene. This band arises from the promotion of a lone-pair electron on the nitrogen atom into an anti-bonding π^* orbital of the aromatic system.

2. σ → σ * Transitions

σ → σ* transitions are the only type of transition which can occur in compounds in which all the electrons are involved in single bonds, e.g. saturated hydrocarbons. The energy required for such transitions is large and the absorption bands lie in the vacuum ultraviolet region of the spectrum. For instance, the longest wavelength band of ethane is found at 135 nm. Since commercial spectrophotometers do not generally operate at wavelengths less than 180–200 nm, σ → σ* transitions cannot normally be observed.

3. n → σ * Transitions

n → σ* transitions can occur in compounds containing atoms with lone-pair electrons. The energy required for an n → σ* transition is generally less than that required for a σ → σ* transition and the corresponding absorption bands appear at longer wavelengths. The wavelength of the peak maximum in the n → σ* band of compounds such as ammonia, alkyl halides, aliphatic alcohols and aliphatic amines, is less than 200 nm and a gradual increase in absorption is all that is observed on scanning the spectra of such compounds from higher wavelength down to 200 nm. Since compounds such as aliphatic alcohols and alkyl halides only start to absorb at about 260 nm they are commonly used as solvents when absorption measurements are made in the ultraviolet region. However, it is not possible to use these solvents when making measurements in the region 200–260 nm and saturated hydrocarbons which only give rise to σ → σ* transitions must be used. Saturated hydrocarbons suffer from the disadvantage that they are poor solvating agents.

4. π → π * Transitions

(a) Symmetry Properties and Band Intensities

A π → π* transition corresponds to the promotion of an electron from a bonding π orbital to an anti-bonding π* orbital and can in principle occur in any molecule containing a π-electron system. However, selection rules, such as those based on symmetry concepts, determine whether a transition to a particular π* orbital is allowed or forbidden. In molecules where there is a number of π orbitals constituting the π-electron system there are several possibilities for transitions from bonding to anti-bonding orbitals. The extent to which transitions are forbidden or allowed varies and spectra may exhibit bands with a range of intensities.

ETHYLENE. There are twelve electrons in the ethylene molecule and in the ground state of the molecule ten electrons are accommodated in the five σ-bonding orbitals and two in the π-bonding orbital. The electronic transitions in ethylene which give rise to absorption bands in the ultraviolet region involve the electrons in the π orbital, and in the excited state of the molecule

Table 2.1. Character table for the D_{2h} point group

Species	E	$C_{2(z)}$	$C_{2(y)}$	$C_{2(x)}$	i	$\sigma_{(xy)}$	$\sigma_{(xz)}$	$\sigma_{(yz)}$	Dipole moment components
A_g	+1	+1	+1	+1	+1	+1	+1	+1	
B_{1g}	+1	+1	−1	−1	+1	+1	−1	−1	
B_{2g}	+1	−1	+1	−1	+1	−1	+1	−1	
B_{3g}	+1	−1	−1	+1	+1	−1	−1	+1	
A_u	+1	+1	+1	+1	−1	−1	−1	−1	
B_{1u}	+1	+1	−1	−1	−1	−1	+1	+1	p_z
B_{2u}	+1	−1	+1	−1	−1	+1	−1	+1	p_y
B_{3u}	+1	−1	−1	+1	−1	+1	+1	−1	p_x

the promoted electron is localized in the π^* anti-bonding orbital. The electronic configurations of the ground and excited states of ethylene may be given as:

Ground state   [Inner electrons] π^2

Excited state   [Inner electrons] $\pi^1(\pi^*)^1$

Ethylene belongs to the point group D_{2h} (Table 2.1) and possesses the following elements of symmetry: identity element E; two-fold rotation axes $C_{2(x)}$, $C_{2(y)}$, $C_{2(z)}$; centre of symmetry i; planes of symmetry $\sigma_{(xy)}$, $\sigma_{(xz)}$, $\sigma_{(yz)}$. The characters representing the effect of the operations corresponding to these symmetry elements on the π and π^* orbitals are given in Fig. 2.6. It can be

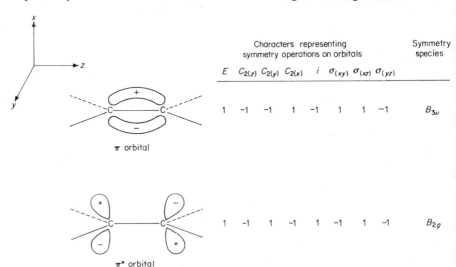

Fig. 2.6. Symmetry properties of bonding and anti-bonding π orbitals of ethylene

seen from this figure that the characters for the π and π^* orbitals are the same as those for the B_{3u} and B_{2g} species respectively. Thus the symmetry species for unpaired electrons in these orbitals will be B_{3u} and B_{2g}. As considered previously, the symmetry species for electrons in a filled orbital must be the totally symmetric species, i.e. the A_g species for the D_{2h} point group.

The symmetry species for the total wave functions of the ground and excited states of ethylene can be deduced by multiplying the symmetry species for the individual electronic systems.

Ground state   [Inner electrons] π^2
$$[A_g \times A_g \ldots] A_g \equiv A_g$$
Excited state   [Inner electrons] $\pi^1 (\pi^*)^1$
$$[A_g \times A_g \times \ldots] B_{3u} B_{2g} \equiv B_{3u} \times B_{2g} \equiv B_{1u}$$

All the electrons are in filled orbitals in the ground state and consequently the ground state belongs to the totally symmetric A_g species. The symmetry species of the excited state is derived by multiplying the characters, for each symmetry operation, for the B_{3u} and B_{2g} species, whence it is found that the characters for the product are the same as those for the B_{1u} species.

Since the excited state belongs to the B_{1u} species and since the component of the dipole moment in the z direction, p_z, also belongs to the B_{1u} species (Table 2.1), the transition moment for the $A_g \to B_{1u}$ transition will have a non-zero value (p. 31). $A_g \to B_{1u}$ is thus an allowed transition and the band corresponding to this transition will be intense. The ultraviolet absorption spectrum of ethylene exhibits an intense band at 174 nm ($\epsilon \sim 1700$ m^2 mol^{-1}) and this band has been assigned as arising from the $A_g \to B_{1u}$ transition. The spectrum also exhibits a weak band at 200 nm ($\epsilon < 1$ m^2 mol^{-1}). This latter band has been accounted for in terms of a transition in which both electrons in the π orbital are promoted to the π^* orbital. The electronic configuration and symmetry species of the resulting excited state will be

$$[Inner\ electrons]\ \pi^0 (\pi^*)^2$$
$$[A_g \times A_g \ldots] (B_{2g} \times B_{2g}) \equiv B_{2g} \times B_{2g} \equiv A_g$$

Since the excited state in this case belongs to a symmetry species, viz. A_g, to which none of the dipole moment components p_x, p_y and p_z belong, the value of the transition moment for the $A_g \to A_g$ transition will be zero and the band corresponding to this forbidden transition would be expected to be weak.

BENZENE. Combination of the p_z orbitals on the six carbon atoms in benzene leads to the formation of six molecular orbitals. The shapes of the molecular orbitals are represented in Fig. 2.7(a) and their relative energies in Fig. 2.7(b). In the ground state of the molecule the six π electrons are accommodated in the three molecular orbitals of lowest energy. These orbitals belong to the A_{2u} and E_{1g} symmetry species of the D_{6h} point group (p. 33), and thus in the ground state of benzene there will be two electrons in the orbital of A_{2u} symmetry and four in the orbitals of E_{1g} symmetry. Since all the occupied orbitals in the ground state are completely filled, the ground state of benzene belongs to the totally symmetric species A_{1g}. Absorption of ultraviolet radiation results in the promotion of an electron from the E_{1g} degenerate

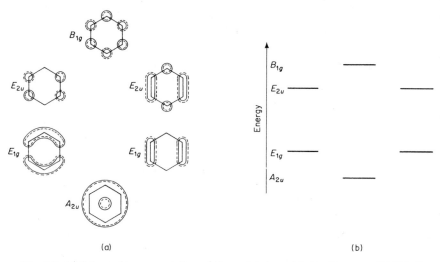

(a)   (b)

Fig. 2.7. (a) Schematic representation of the molecular orbitals of benzene. (b) Relative energies of the molecular orbitals of benzene. (Benzene belongs to the D_{6h} point group and the species to which the individual molecular orbitals belong are given. Solid contours outline the positive parts of wave function, dotted contours the negative parts)

pair of orbitals into the E_{2u} degenerate pair of orbitals; and thus in the excited molecule there will be two electrons in the orbital of A_{2u} symmetry, three in the orbitals of E_{1g} symmetry and one in the orbitals of E_{2u} symmetry. It may be shown by the methods of group theory that this configuration gives rise to three states belonging to the B_{1u}, B_{2u} and E_{1u} symmetry species. Thus theory predicts that in the spectrum of benzene there will be three absorption bands, corresponding to the transitions $A_{1g} \rightarrow B_{1u}$, $A_{1g} \rightarrow B_{2u}$ and $A_{1g} \rightarrow E_{1u}$. Three bands are in fact observed on the spectrum (Fig. 1.15). Of the three transitions expected, $A_{1g} \rightarrow B_{1u}$ and $A_{1g} \rightarrow B_{2u}$ are forbidden, and $A_{1g} \rightarrow E_{1u}$ is allowed (p. 34).

(b) General Features of $\pi \rightarrow \pi$ * Transitions

ETHYLENIC COMPOUNDS. As mentioned previously the spectrum of ethylene exhibits an intense band at 174 nm and a weak band at 200 nm, both of which are due to $\pi \to \pi^*$ transitions. The following discussion will be concerned only with the band at 174 nm representing an allowed transition.

Substitution of an atom containing lone-pair electrons for a hydrogen atom in ethylene results in a shift of the band to longer wavelength (red shift). This shift can be related to the effect of resonance interaction, i.e. the effect of migration of unshared electrons into the π-electron system, on the energy levels of the π orbitals. Combination of the bonding and anti-bonding orbitals of the ethylenic linkage with the n orbital of the X group in the structure C=C—X results in the formation of three molecular orbitals π_1, π_2 and π_3^*.

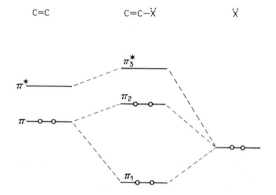

Fig. 2.8. Changes in orbital energies accompanying bond formation

The energy levels of these orbitals relative to those of the orbitals of the separate components C=C and X are shown in Fig. 2.8. The energy separation between the π_2 and π_3^* orbitals in the combined structure is less than that between the π and π^* orbitals in ethylene, and consequently the $\pi \to \pi^*$ transition occurs at longer wavelength in the substituted compounds. The shifts in wavelength brought about by certain substituents are given below:

Substituent	Cl	CH$_3$	OH	NH$_2$
Shift, nm	5	5	30	40

An inductive effect, as well as a resonance effect, is introduced on substitution of an atom having lone-pair electrons. However, the inductive effect causes the π and π^* orbital energies to be lowered by about the same amount and so there is no marked change in the energy difference between the π_2 and π_3^* orbitals. Consequently the position of the band due to the $\pi \to \pi^*$ transition is not appreciably altered by the inductive effect.

Alkyl substitution in the ethylene chromophore causes a lowering in the energy required for the $\pi \to \pi^*$ transition. It has been postulated that the interaction of the electrons in the C—H bonds with the π electrons gives alternative resonance structures for the molecule (hyperconjugative effect). For propylene the structures are

$$H_3 \equiv C - CH = CH_2 \longleftrightarrow \overset{+}{H_3} = C = CH - \bar{C}H_2$$

The methyl group is acting as an electron donor and therefore has the same effect on the relative positions on the π orbitals as atoms with lone-pair electrons.

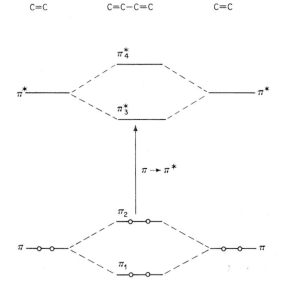

Fig. 2.9. Changes in molecular orbital energy levels of ethylene on conjugation

If the ethylenic link is part of a conjugated system then the band due to the $\pi \to \pi^*$ transition moves to longer wavelengths; the shift increasing with the extent of conjugation. In the case of butadiene the combination of two ethylenic units results in the formation of two filled π orbitals and two unoccupied π^* orbitals. The energies of these orbitals relative to the energies of the orbitals in ethylene are shown in Fig. 2.9. It can be seen that the energy required for the transition of lowest energy is less than in ethylene and consequently the corresponding absorption band is displaced to the red. If the extent of conjugation is increased the band is displaced further and further to the red. This effect is illustrated in Table 2.2 where the longest wavelength maxima for some conjugated polyenes are listed.

The approximate position of the longest wavelength band in conjugated

Table 2.2. Maxima of longest wavelength bands in the spectra of some polyenes

Compound	$n = 1$	2	3	4	Solvent
		λ_{max}, nm			
CH_3—$(CH{=}CH)_n$—COOH	204	254	294	327	Ethanol
CH_3—$(CH{=}CH)_n$—CHO	220	271	315	353	Benzene
Ph—$(CH{=}CH)_n$—Ph	306	334	358	384	Ethanol

systems can be calculated with reasonable accuracy in cases where there is free electron movement along the chain and all the bonds in the chain are of the same bond order. The cyanine dyes provide an example of a system where there is reasonable agreement between calculated and experimental values of the longest wavelength band (Expt. E5). In the case of a carbon chain of conjugated double bonds the bond order is not the same for all the bonds along the chain and poor agreement is obtained between the calculated and observed value for the longest wavelength band.

The absorption characteristics of substituted dienes have been extensively investigated. As in the case of ethylene, substitution of an alkyl group or a group containing lone-pair electrons in place of a hydrogen atom results in a red shift of the allowed $\pi \to \pi^*$ band at longest wavelength. It has been found empirically that for diene systems the shifts caused by substituents are additive. A set of rules have been drawn up to enable the prediction of the position of the bands due to $\pi \to \pi^*$ transitions in open-chain dienes and dienes in six-membered rings. These rules are given in Table 2.3.

Table 2.3. Woodward's rules for prediction of diene absorption (Woodward, 1941, 1942)

	nm
Wavelength assigned to parent heteroannular or open chain diene	214
Wavelength assigned to parent homoannular diene	253
Add increments for:	
—O Acyl	0
—Cl, Br	5
—Alkyl substituent or ring residue	5
—Exocyclic double bond	5
—O Alkyl	6
—Double bond extending conjugation	30
—S Alkyl	30
—N Alkyl$_2$	60
	____
Calculated λ_{max} = Total	

3

The spectra of α, β unsaturated aldehydes and ketones exhibit bands due to both $n \rightarrow \pi^*$ and $\pi \rightarrow \pi^*$ transitions. The weak $n \rightarrow \pi^*$ band lies at longer wavelength than the intense $\pi \rightarrow \pi^*$ bands. The longest wavelength $\pi \rightarrow \pi^*$ band has been the subject of much study and the position of this band can also be predicted by use of a set of empirically derived rules. These rules are given in Table 2.4.

Table 2.4. Woodward, Fieser, Scott rules for prediction of absorption of α, β unsaturated aldehydes and ketones (Scott, 1964)

$$C_\delta{=}C_\gamma{-}C_\beta{=}C_\alpha{-}C{=}O$$

				nm
Wavelength assigned to cyclopentenone				202
Wavelength assigned to parent α, β-unsaturated six ring or acyclic ketone				215
Wavelength assigned to parent α, β-unsaturated aldehyde				207
Add increments for:	α	β	γ	δ
—O Ac	6	6	6	6
—Alkyl or ring residue	10	12	18	18
—O Alkyl	35	30	17	31
—Br	25	30	—	—
—OH	35	30	—	50
Add increments for:				
—Exocyclic double bond				5
—Double bond extending conjugation				30
—Homodiene component				39
			Calculated λ_{max} in EtOH =	Total

AROMATIC COMPOUNDS. The ultraviolet absorption spectrum of benzene shows three bands—two intense bands at 180 and 200 nm, and a weak band at 260 nm. The effect of substituents on the position of the band at 260 nm has been the subject of a great deal of study. If an electron-donating substituent, e.g. —NH$_2$, —OH, —Cl, is introduced into the ring then the band shifts to the red and is intensified. If the substituent group has non-bonded or lone-pair electrons then resonance interaction between these electrons and the π-electron system leads to a partial migration of the unshared electrons into the conjugated system. The resonance interaction and hence the red shift increases as the electron donor power of the substituent increases. As discussed for substituted ethylenes (p. 51), the effect of the resonance interaction is to raise the energy of the highest filled orbital relative to that of the lowest unfilled orbital and hence to cause a lowering of the energy required for the transition between these orbitals. Since the unfilled orbital is localized on the phenyl ring, the effect of the transition is to transfer charge from the substituent to the phenyl ring. Such a transition is termed an intramolecular charge transfer transition, and can be represented:

Electron-withdrawing substituents, e.g. $-NO_2$, $-CHO$, $-COOH$, also give rise to a red shift of the band appearing at 260 nm in the spectrum of benzene. Here the resonance interaction between the π-electron system and the electrons of the substituent is in the reverse direction to that for electron releasing substituents, and in this case the absorption of ultraviolet radiation causes a transfer of charge from the phenyl ring to the substituent. This type of intramolecular charge transfer transition is represented:

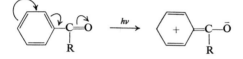

For monosubstituted derivatives of benzene, an intramolecular charge transfer transition normally appears as a band in the region 230–290 nm. The extinction coefficient for the peak maximum of the band is generally in the region 20–150 $m^2 mol^{-1}$. Empirical rules have been drawn up for predicting the wavelength maximum of the charge transfer band in compounds of the general formula (Table 2.5)

$$X \overline{} \langle \text{benzene ring} \rangle \overline{} -\underset{R}{\overset{}{C}}=O$$

Application of these rules to the prediction of the wavelength of the intramolecular charge transfer band in aromatic carbonyl compounds is considered in Expt. E7.

Increasing the extent of conjugation on aromatic compounds causes the absorption bands to shift to longer wavelengths. For instance, in the series of linear condensed-ring hydrocarbons—naphthalene, anthracene, naphthacene, pentacene—the λ_{max} of the longest wavelength bands are 314, 380, 480 and 580 nm respectively.

The effect of solvents on the position of bands due to $\pi \rightarrow \pi^*$ transitions is opposite to that found for $n \rightarrow \pi^*$ transitions. Whereas $n \rightarrow \pi^*$ bands move to shorter wavelengths (blue shift) upon changing from a hydrocarbon to a hydroxylic solvent, $\pi \rightarrow \pi^*$ bands move to longer wavelength (red shift). This red shift of $\pi \rightarrow \pi^*$ bands can be understood in terms of the polarities of the ground and excited states. The distribution of π electrons in the ground state of benzene is symmetrical and the molecule is non-polar. Excitation of an electron into a π^* orbital destroys the symmetric distribution of the π-electron system and consequently the excited state will be more polar than the ground state. The excited state will therefore be stabilized relative to the ground state in a polar solvent. Thus for solvents of increasing polarity the energy of the excited state will decrease relative to that of the ground state

Table 2.5. Scott's Rules for prediction of absorption in substituted benzene derivatives X—C$_6$H$_4$—COR (Scott, 1961)

			nm
Wavelength assigned to parent chromophore:			
X = OH or O Alkyl			230
R = Alkyl or ring residue			246
= H			250
Add increments for substituent X:			
	o	*m*	*p*
—Cl	0	0	10
—Br	2	2	15
—Alkyl or ring residue	3	3	10
—OH or O Alkyl	7	7	25
—O$^-$	11	20	78
—NH$_2$	15	15	58
—NHAc	20	20	45
—NHMe	—	—	73
—NMe$_2$	20	20	85
Calculated λ_{max} in EtOH =		Total	

and the $\pi \to \pi^*$ band will be successively shifted to the red. For solutions in which hydrogen bonds are formed between the solute and solvent, e.g. phenol–dioxane, benzophenone–ethanol, a $\pi \to \pi^*$ transition will not affect the hydrogen bond (unlike an $n \to \pi^*$ transition as discussed on p. 46) and a red shift of the $\pi \to \pi^*$ band will still occur if the polarity of the medium is increased. It is to be noted, however, that the formation of a hydrogen bond in a solute–solvent system will change the absorption spectrum of the solute. For example, the absorption spectrum of β-naphthol in a mixture of dioxane and *n*-hexane as compared to the spectrum in *n*-hexane shows additional bands due to absorption by hydrogen-bonded species. Spectral analysis of such systems can give information on the equilibrium constant for the reaction leading to hydrogen bond formation (see Expt. E12).

C. CHARGE TRANSFER TRANSITIONS

The absorption spectrum of mixtures of two components, one of which can act as an electron donor, D, and the other as an electron acceptor, A, often exhibit broad featureless bands which are not observable in the spectra of the individual components (Fig. 2.10). Such bands arise from the absorption of radiation by a molecular complex formed between the components

$$D + A \; \rightleftharpoons \; (DA)_{complex} \; \overset{h\nu}{\rightleftharpoons} \; (DA)^{*}_{complex}$$

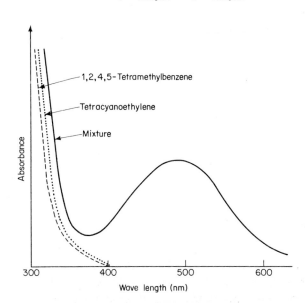

Fig. 2.10. Charge transfer absorption of 1,2,4,5-tetramethylbenzene–tetracyanoethylene complex

The new bands are described as intermolecular charge transfer bands since the absorption of radiation by the complex is associated with the transfer of an electron from a molecular orbital of the donor to a molecular orbital of the acceptor. The electronic transition of lowest energy in the complex will correspond to the promotion of an electron from the highest occupied molecular orbital of the donor to the lowest vacant molecular orbital of the acceptor (Fig. 2.11). Promotion of an electron to higher unoccupied orbitals of the acceptor can occur and additional charge transfer absorption bands may be observed. However, these latter bands are often obscured by the absorption bands of the "free" donor and acceptor.

The molecular complex formed by the interaction of donor and acceptor can be considered as a resonance hybrid of two structures: one a no-bond structure (D, A) in which the two molecular species are bound together by van der Waals forces, and the other an ionic structure $(D^+—A^-)$ in which the two species are bound by a covalent bond. The no-bond structure (D, A) is the main contributing structure to the ground state, and the wave function of the ground state, Ψ_g, may be represented as the sum of the wave functions of the no-bond state (D, A) and the ionic state $(D^+—A^-)$, with the contribution from the no-bond state predominating

$$\Psi_g = a\Psi_{(D,\,A)} + b\Psi_{(D^+-A^-)} \tag{2.1}$$

where a and b are constants and $a \gg b$.

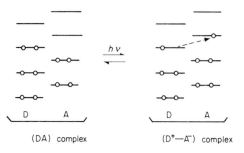

Fig. 2.11. Charge transfer transition of lowest energy in a molecular complex

In the formation of the excited state an electron is transferred from donor to acceptor and consequently the excited state is predominantly ionic $(D^+—A^-)$ in character, with only a small contribution from the no-bond structure (D, A). The wave function of the excited state, Ψ_e, is described by

$$\Psi_e = a^* \Psi_{(D^+-A^-)} - b^* \Psi_{(D,\,A)} \tag{2.2}$$

where $a^* \gg b^*$.

Charge transfer bands are generally very broad and show no fine structure. This results from the loose nature of the bonding between the donor and acceptor in the ground state of the complex. As a consequence of the weak bonding, the components of the complex can adopt different orientations with respect to each other, and since each complex with a particular orientation of the components will absorb radiation at a slightly different wavelength to every other complex there will be a wide range over which absorption of energy occurs. The wavelength of maximum absorption corresponds to the energy required to excite the electron from the donor to acceptor in the most probable ground state configuration of the complex.

Since a charge transfer absorption band arises from the transfer of an electron from a donor to an acceptor, the energy required for the process depends upon the ionization potential of the donor (i.e. the energy necessary to remove an electron from the donor) and the electron affinity of the acceptor (i.e. the energy necessary to place an electron on the acceptor). It has been found empirically that the energy, hv_{CT}, of electron transfer from donor to acceptor is related to the ionization potential, I, of the donor and the electron affinity, EA, of the acceptor by the expression

$$hv_{CT} = I - EA + \Delta \qquad (2.3)$$

where Δ is a constant for a related series of donors.

For a series of donors with the same acceptor the position of the charge transfer absorption band will move to the red (lower energy) as the ionization potential of the donor decreases (see Expt. E8). Quantitative analysis of the charge transfer band can yield a value for the equilibrium constant for complex formation (see Expt. E12).

There are two main categories of donor compound. The first category includes compounds in which the electrons available for donation to the acceptor are located in π molecular orbitals, i.e. aromatic hydrocarbons, alkenes, alkynes. They are said to form π complexes and are termed "π donors". The second category includes compounds in which the electrons available for donation are in non-bonding orbitals, as in alcohols, amines, ethers, etc.

There are many types of acceptor compound. Among the inorganic acceptors are the halogens, metal halides and silver salts of copper(I) and mercury(II). The range of organic acceptors include such compounds as tetracyanoethylene, trinitrobenzene, tetrachlorophthalic anhydride and others which contain highly electronegative substituents.

The compounds of transition elements often exhibit absorption bands which can be attributed to charge transfer transitions. For example in the halide containing complexes $[Co(NH_3)_5X]^{3+}$ the absorption spectra show a strong band which shifts progressively to longer wavelength as X is changed along the series F, Cl, Br, I. The observed band results from an electronic transition in which an electron is removed from the halide ligand and transferred to the metal ion. As would be expected the energy required for the transition is decreased as the electron attracting power of the halide is decreased. The above complexes provide an example in which intramolecular charge transfer occurs. As discussed previously, intramolecular charge transfer can also occur in derivatives of benzene.

D. $d \rightarrow d$ TRANSITIONS

The spectra of complexes of transition metal ions often exhibit absorption bands in the visible region of the spectrum and the complexes appear highly coloured. The absorption bands are caused by the promotion of an electron from a lower to a higher energy d orbital of the metal ion, and are labelled as $d \to d$ bands. The discussion which follows is limited to $d \to d$ transitions in certain octahedral complexes and for further information on this type of transition the reader is referred to texts such as Figgis (1966) and Sutton (1968).

The relative orientation of the five d orbitals $d_{x^2-y^2}$, d_{z^2}, d_{xy}, d_{yz}, and d_{xz} are shown in Fig. 2.12. The $d_{x^2-y^2}$ and d_{z^2} orbitals are labelled thus because they lie along the x and y and z axes respectively, while the d_{xy}, d_{yz}, and d_{xz} are labelled in terms of the planes in which they lie.

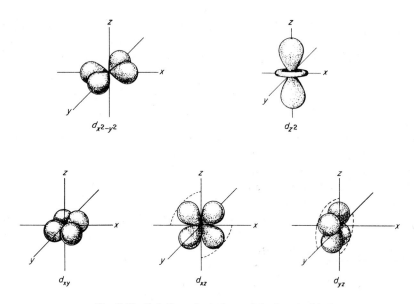

Fig. 2.12. Relative orientations of the five d orbitals

In a free, gaseous transition-metal ion the five d orbitals have the same energy, i.e. they are degenerate. If it is supposed that the transition-metal ion is surrounded by a spherical field of negative charge, then the energy of the five d orbitals will be increased because of repulsion between the electron density on the metal ion and the field of negative charge (Fig. 2.13b).

In an octahedral complex the ligands, which are negatively charged, are located at the corners of an octahedron [Fig. 2.13(a)] and the metal ion located at the centre of the complex will experience a non-spherical field. The electron density of the ligands is orientated directly along the x, y and z axes and the electron repulsion between the electrons of the metal ion and the electrons of the ligand is greater for the electrons in the $d_{x^2-y^2}$ and d_{z^2} orbitals than for those in the d_{xy}, d_{yz} and d_{xz} orbitals. As a consequence the five d orbitals are split into two sets of different energy: a doubly degenerate set (e_g) comprising

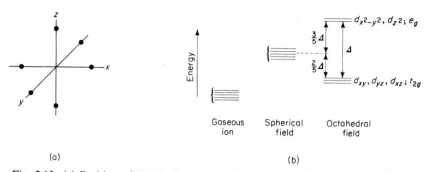

(a)   (b)

Fig. 2.13. (a) Position of ligands in an octahedral complex (b) Orbital splitting in an octahedral field

the $d_{x^2-y^2}$ and d_{z^2} orbitals and a triply degenerate set (t_{2g}) of lower energy comprising the d_{xy}, d_{yz} and d_{xz} orbitals [Fig. 2.13(b)]. The difference in energy between the sets of orbitals is termed the ligand field splitting and is represented by the symbol Δ. When an octahedral complex is formed the energy of each of the e_g orbitals is raised by 3/5 Δ relative to their energy in a spherical field while the corresponding decrease in energy for each of the t_{2g} orbitals is 2/5 Δ. Thus the average energy of the d orbitals in the complex is equal to that of the orbitals in a spherical field.

Octahedral complexes with only one electron in the d orbitals (d^1 complex) exhibit relatively simple spectra. For instance, $[Ti(H_2O)_6]^{3+}$ has a single absorption band in the visible region with an absorption maximum at about 500 nm. For d^1 complexes the single electron is located in the t_{2g} set of orbitals and the absorption band corresponds to the promotion of the electron to the e_g set of orbitals, i.e. it is a $t_{2g} \rightarrow e_g$ transition. The difference in energy between the t_{2g} and e_g orbitals changes with ligand field splitting as illustrated in Fig. 2.14. For $[Ti(H_2O)_6]^{3+}$ the absorption band maximum is at 500 nm and the

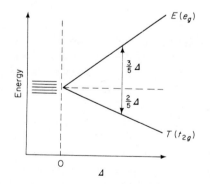

Fig. 2.14. Energy level diagram for a d^1 complex (T represents triply degenerate, E represents doubly degenerate)

difference in energy, $\Delta E_{(500\,nm)}$, when represented in wave number units is 20 000 cm^{-1}. Thus the ligand field splitting, Δ, for water towards Ti^{3+} is 20 000 cm^{-1}.

Octahedral complexes with nine electrons in the five d orbitals (d^9 complex) also have relatively simple absorption spectra in the visible region. Here the nine d electrons are divided so that the t_{2g} set of orbitals are filled with six electrons and the remaining three electrons are sited in the e_g set of orbitals. The ground state electronic configuration is then $(t_{2g})^6 (e_g)^3$. The spectra of octahedral d^9 complexes exhibit a single $d \rightarrow d$ absorption band arising from promotion of an electron from the t_{2g} set of orbitals to the e_g set of orbitals.

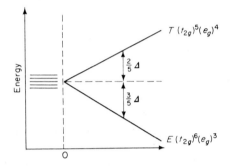

Fig. 2.15. Energy level diagram for a d^9 complex

The excited state configuration is $(t_{2g})^5 (e_g)^4$. The ground state is doubly degenerate since there are two configurations for $(t_{2g})^6 (e_g)^3$ which give states of the same energy. These two configurations are $(t_{2g})^6 (d_{x^2-y^2})^1 (d_{z^2})^2$ and $(t_{2g})^6 (d_{x^2-y^2})^2 (d_{z^2})^1$. The excited state is triply degenerate since the configuration $(t_{2g})^5 (e_g)^4$ can be made up three ways, viz. $(d_{xy})^2 (d_{yz})^2 (d_{xz})^1 (e_g)^4$, $(d_{xy})^1 (d_{yz})^2 (d_{xz})^2 (e_g)^4$ and $(d_{xy})^2 (d_{yz})^1 (d_{xz})^2 (e_g)^4$. Thus the energy level diagram for a d^9 complex (Fig. 2.15) is inverted relative to that of the d^1 complex where the ground state is triply degenerate and the excited state doubly degenerate.

The nature of the ligands in the complex affect the energy of the $d \rightarrow d$ transitions for a particular transition-metal ion. It has been found that the following ligands can be arranged in a series I$^-$, Br$^-$, SCN$^-$, Cl$^-$, NO$_3^-$, F$^-$, OH$^-$, H$_2$O, NH$_3$, CN$^-$, where on moving from left to right along the series the effect on the absorption maximum of the $d \rightarrow d$ band is to move it to longer wavelength, i.e. the Δ value of the ligand increases on moving from left to right. This effect can be illustrated for the ligands NH$_3$ and H$_2$O by substituting NH$_3$ for H$_2$O in the complex $[Cu(H_2O)_6]^{3+}$ whence it is found that the absorption maximum of the $d \rightarrow d$ band in the complex $[Cu(H_2O)_{6-n}(NH_3)_n]^{3+}$ shifts progressively to the blue as n is changed from 0 to 4 (see Expt. E9).

E. RADIATIVE AND NON-RADIATIVE TRANSITIONS

The absorption of ultraviolet and visible radiation by a molecule results in the formation of states of high energy. For example, the absorption of radiation of wavelength of 500 nm corresponds to an energy uptake of 2.38×10^5 J mol^{-1}, while absorption of radiation of wavelength of 200 nm corresponds to an uptake of 5.98×10^5 J mol^{-1}. Naturally such high-energy states are relatively short lived and the molecule quickly returns to the ground state by loss of the excess energy. There are two types of energy-dissipation process by which molecules can lose the absorbed energy. One type is termed a "radiative" process since the energy is lost by the emission of radiant energy, while the other is classified as a "non-radiative" process since no radiation is emitted during the conversion from the excited state to the ground state.

I. Radiative Processes

The multiplicity of a state of a molecule is equal to $2I + 1$ where I is the algebraic sum of the spin quantum numbers of the electrons in the molecule. For a state in which all the electrons in the molecule are paired, I will equal zero and the multiplicity of the state will therefore be unity. Such a state is termed a "singlet state" and assigned the symbol S. In the majority of organic molecules, the electrons are all paired in the ground state of the molecule and the ground state is symbolized as the S_0 state.

The absorption of ultraviolet or visible radiation by a molecule is generally accompanied by the promotion of a single electron to an unoccupied orbital of higher energy. The spin orientation of the electron in the higher orbital may be retained or it may be reversed after excitation. In the former case the excited state of the molecule has all the electrons paired in orbitals, except for two, which are in different orbitals but have opposite spin. Here the multiplicity S is still unity since the sum of the spin quantum numbers is still zero and the state is termed a "singlet excited state" $(S_1, S_2, \ldots)$. If the spin orientation of the promoted electron is reversed after excitation then the multiplicity is three, since I is now unity, and the state is termed an "excited triplet state" $(T_1, T_2, \ldots)$. For each excited singlet state there is always a corresponding triplet state of lower energy. The lower energy of the triplet state is a consequence of the Pauli principle which demands that electrons with parallel spin in different orbitals cannot occupy the same position in space simultaneously. Since this does not apply to the singlet state there is a greater electron repulsion in the singlet state than in the triplet state, so that the former state has the higher energy. The relative energies of singlet and triplet states of a typical molecule are shown in the energy-level diagram of Fig. 2.16.

A molecule can be raised directly from the ground state S_0 to an excited singlet state by absorption of radiation of the appropriate wavelength but can only generally be raised to an excited triplet state by initial promotion to the corresponding singlet state followed by reversal of the spin orientation of the excited electron. The transfer from a singlet state to a triplet state is termed "intersystem crossing". The lifetimes of molecules in upper excited singlet states $S_2, S_3, \ldots$ and upper excited triplet states $T_2, T_3, \ldots$ are extremely short and within a period of about 10^{-13} s there is a radiationless transfer of energy from the molecule to its environment and the molecule is deactivated to the lowest excited singlet state S_1, or lowest triplet state T_1. Once the molecule is in either of these states the excess energy may be dissipated by the emission

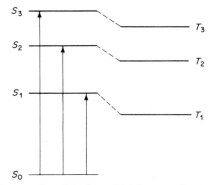

Fig. 2.16. Relative energies of singlet and triplet states in a typical molecule

of radiation. The radiation emitted is termed "fluorescence" when the transition occurs between states of the same multiplicity, e.g. $S_1 \rightarrow S_0$, and termed "phosphorescence" when the transition is between states of different multiplicity, e.g. $T_1 \rightarrow S_0$.

The relationship of the absorption spectrum of a molecule with a singlet ground state to the fluorescence and phosphorescence spectra of the molecule is shown in Fig. 2.17. On absorption of radiation the molecule may be excited to an upper vibrational level of the first excited singlet state or end up in such a level after deactivation from an upper excited singlet state. The excess vibrational energy is rapidly lost by collisional deactivation and the molecule finishes up in the lowest vibrational level before fluorescence can be emitted. The emission of fluorescence is due to the transition from the lowest vibrational level of the first excited singlet state to various vibrational levels of the ground state. It can be readily seen from Fig. 2.17 that there is an approximate "mirror-image" relationship between the absorption spectrum and the fluorescence spectrum. The spacing between the bands in the absorption spectrum is equal to the difference in energy between the vibrational levels of the excited state while that between the bands in the fluorescence spectrum is equal to the difference in energy between the vibrational levels of the ground state.

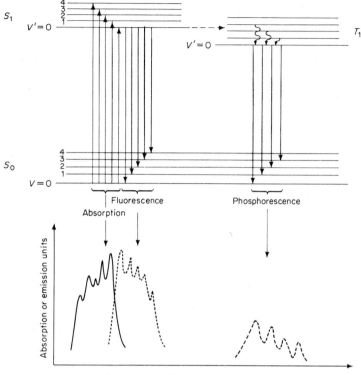

Fig. 2.17. Relationship between the absorption spectrum, fluorescence spectrum and phosphorescence spectrum of a molecule

Phosphorescence results from the radiative transition from the lowest vibrational level of the first triplet state to the various vibrational levels of the ground state. Since the energy of the lowest triplet state is lower than that of the lowest excited singlet state the phosphorescence band occurs at longer wavelength than the fluorescence band. Transitions between states of different multiplicity are forbidden by selection rules and consequently phosphorescence is generally weak. On the other hand fluorescence is generally intense since it corresponds to a transition between states of the same multiplicity. Apart from the difference in intensity, fluorescence differs from phosphorescence in that the fluorescence is generally emitted within 10^{-6} s of the exciting light being cut off while phosphorescence may be emitted for periods lasting up to several seconds after the exciting light is cut off.

The fluorescence or phosphorescence of a system may be quenched by the presence of a second molecule which can act as an energy quencher. For example, irradiation of a dilute solution of naphthacene in solid anthracene with radiation which is mainly absorbed by the anthracene results in the green fluorescence of naphthacene being emitted and no fluorescence from anthracene can be observed. This is explained on the basis that the lowest singlet state of naphthacene lies below that of anthracene in energy, and that the excitation energy is transferred from the excited anthracene molecule to naphthacene before the anthracene can emit fluorescence. Similarly, energy transfer can occur between the triplet states of molecules and phosphorescence can be quenched.

Phosphorescence can only be observed for those molecules in which the rate of transfer from the first excited singlet state S_1 to the associated triplet state T_1 is comparable to, or greater than, the rate of deactivation from the singlet state S_1 to the ground state S_0. The rate of the transfer $S_1 \rightarrow T_1$ is rapid in aromatic ketones such as 2-acetonaphthone and 1-naphthaldehyde and the phosphorescence of these compounds can be readily observed.

2. Non-radiative Transitions

Most compounds do not exhibit fluorescent or phosphorescent properties and for those that do the energy re-emitted by either fluorescence or phosphorescence is generally less than the energy absorbed. Obviously there are other processes whereby the absorption energy may be lost and the excited molecule deactivated. Such processes are discussed below.

On absorption of radiation a molecule may be excited from the lowest vibrational level of the ground state to a higher vibrational level of the excited state. For molecules in solution, molecular collisions with the solvent leads to rapid vibrational deactivation of the excited state, probably by one vibrational level at a time, with the vibrational energy being transferred to the solvent as kinetic energy. This process is represented for a diatomic molecule by the wavy arrows in Fig. 2.18(a). If another potential-energy curve crosses that of the excited state then during the vibrational deactivation process the molecule will in the course of a vibration pass through the point of intersection of the curves. At the cross-over point the molecule may change over into the configuration represented by the potential-energy curve S', or it may stay in the configuration represented by the potential-energy curve S_1. If the molecule crosses over to the former configuration then the vibrational deactivation process will continue until another cross-over point is reached when the molecule may again change configuration. The change from one excited state to another is known as "internal conversion". For polyatomic molecules there may be a large number of potential-energy surfaces crossing each other, and if one such surface crosses the ground-state potential-energy surface then the molecule may eventually return to the ground state by a succession of internal conversions. Consequently the molecule may reach the ground state before any emission of radiation can occur. If a molecule is initially excited to a higher electronic state than the first excited state then it will almost certainly return to the first excited state before emission of radiation has time to occur. However, once in the first excited state the remainder of the excess energy may be lost by either a radiative or a non-radiative process.

The potential-energy curve representing the excited state may be crossed by a repulsive state, as represented by state S' in Fig. 2.18(b), and if the molecule crosses over to state S' dissociation will occur. This phenomenon is named "predissociation". The absorption energy may also be dissipated in causing the decomposition of the molecule by the processes as shown in Figs. 2.18(c) and (d). In the former case the transition is to an unstable state while in the latter case the transition is to a vibrational level of greater energy than the dissociation energy, D, of the excited state.

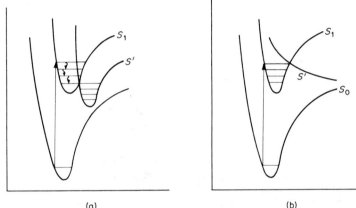

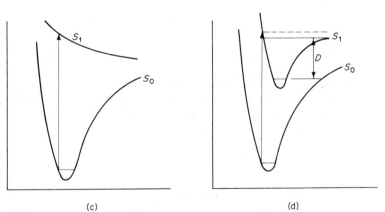

Fig. 2.18. Modes of dissipation of electronic energy

F. APPLICATIONS OF ELECTRONIC SPECTROSCOPY

I. Quantitative Analysis

The concentration of a species in a system may be determined by measurement of the absorption of radiation by the species. There are two laws governing the absorption of radiation: the Lambert law and the Beer law. The Lambert law states that the fraction of incident radiation absorbed by a transparent medium is independent of the intensity of the incident radiation, and that each successive layer of the medium absorbs an equal fraction of the incident radiation. The Beer law states that the amount of radiation absorbed is proportional to the number of molecules absorbing the radiation. From the Beer law, the absorption of a species in a solution will be proportional to the concentration of the species if it is dissolved in a transparent medium. The two laws may be combined to give the Beer–Lambert law which can be expressed in the form

$$\log_{10}\frac{I_0}{I} = \epsilon c l \tag{2.4}$$

where I_0 and I are the intensities of the incident and transmitted radiation respectively, c is the molar concentration of the absorbing species, l is the thickness of the absorbing layer and ϵ the extinction coefficient. The term $\log_{10} I_0/I$ is called the absorbance and is represented by the symbol A.

The SI units for c, l and ϵ in equation 2.4 are mol dm^{-3}, mm and m^2 mol^{-1} respectively, whereas the units which have been used in the literature to date are mole litre^{-1}, cm and litre mole^{-1} cm^{-1} respectively. The extinction coefficient, ϵ, is a constant at any given wavelength for a given species and is independent of the concentration of the absorbing species. It has been the convention to date not to express the units when values for extinction coefficients are quoted. However, if the units of ϵ are the SI units then the values will be a factor of ten less than the corresponding values expressed in the non-SI units given above. SI units are used in this book and the units of the extinction coefficient are quoted along with the numerical value in order to avoid confusion.

If the molecular weight of the absorbing species is not known, then the expression $E_{1\%}^{10\text{mm}}$ may be quoted instead of the extinction coefficient. This expression relates to an absorbing layer of thickness 10 mm of a 1% solution of the species, and is related to the extinction coefficient by the equation

$$E_{1\%}^{10\text{mm}} = \frac{\epsilon \times 10}{\text{molecular weight}} \tag{2.5}$$

Although the Beer–Lambert law holds for all systems, apparent deviations from the law can occur on account of physical factors such as fluorescence of the solute, stray radiation from the source, non-linear response of the detector, and on account of chemical factors such as molecular association of the solute and ionization of the solute. If either of the first two physical factors are operative, then the intensity of the radiation reaching the detector will be different from the intensity of the radiation transmitted by the sample and the measured absorbance will not be the same as the true absorbance. The effect of the two chemical factors given is to change the effective concentration of the absorbing species, and the measured absorbance will not be the same as that which would correspond to the initial concentration of the absorbing species.

It is possible to check whether any deviation from the Beer–Lambert law is caused by physical or chemical effects. A plot of absorbance against thickness of absorbing layer, at constant concentration of absorbing solute, should give a straight line if chemical effects are causing the deviation, since under these conditions the factor causing the deviation is held constant. If physical effects are causing the deviation then the plot will not give a straight line.

Since deviations from the Beer–Lambert law can occur, it is always advisable to check the validity of the law for any system under study. This may be done by plotting absorbance against concentration of solute over a range of concentrations, whence a straight line should be obtained if the law holds. This plot may be used as a calibration for determining unknown concentrations of solute (see Expt. E10). If deviations from the law occur then the plot of absorbance against concentration will not give a straight line. Nevertheless, the concentration of solute in a solution of unknown concentration may be determined from the plot, provided the unknown concentration falls within the range of concentrations used to draw up the plot.

If there are several absorbing species present then the measured absorbance at any wavelength will be the sum of the absorbances of the individual species, provided that the species do not interact with each other. For a system of n components the absorbance at wavelengths $\lambda_1, \lambda_2, \ldots \lambda_n$ will be given by

$$\lambda_1 \quad A_1 = \epsilon_1^a c^a l + \epsilon_1^b c^b l + \ldots \epsilon_1^n c^n l \qquad \text{2.6(a)}$$

$$\lambda_2 \quad A_2 = \epsilon_2^a c^a l + \epsilon_2^b c^b l + \ldots \epsilon_2^n c^n l \qquad \text{2.6(b)}$$

$$\vdots \quad \vdots \quad \vdots \quad \vdots \quad \vdots$$

$$\lambda_n \quad A_n = \epsilon_n^a c^a l + \epsilon_n^b c^b l + \ldots \epsilon_n^n c^n l \qquad \text{2.6(c)}$$

where $A_1, A_2, \ldots A_n$ represent the absorbance of the system at wavelengths $\lambda_1, \lambda_2, \ldots \lambda_n$; $\epsilon_1^a, \epsilon_2^a, \ldots \epsilon_n^a$ the extinction coefficients of the component A at wavelengths $\lambda_1, \lambda_2, \ldots \lambda_n$; $c^a, c^b, \ldots c^n$ the concentrations of the components; and l the thickness of the absorbing layer.

If the absorbance of the system is measured at n different wavelengths, and if the extinction coefficients of the individual species are determined at each of these wavelengths, then a set of n simultaneous equations may be set up containing n unknowns, viz. the concentrations of the species A, B,...N. Thus in principle the concentration of any number of absorbing species in a mixture may be determined. In practice this method of quantitative analysis is generally only applied to two or three component systems (see Expt. E11).

2. Determination of Molecular Weights

Molecular weights of compounds can be determined spectrophotometrically if suitable derivatives of the compounds can be prepared. The method is dependent upon the formation of a derivative, such as a picrate, in which the extinction coefficient associated with a chromophoric grouping of the reagent is unaltered upon forming the derivative. The principle of the method is outlined in Expt. E13.

3. Kinetic Measurements

In order to determine the kinetics of a reaction the change in concentration of either a reactant or product with time is measured. Since absorbance is directly proportional to concentration, absorption spectrophotometry provides a convenient method for following the course of a reaction. The method is dependent upon one of the reactants or products exhibiting suitable absorption in the ultraviolet or visible regions which is not overlapped by absorption due to other species present. The reaction of triiodide ion with arsenious acid provides an example of a system for which kinetics may be determined from spectrophotometric measurements (see Expt. E14).

4. Dissociation Constants of Acids and Bases

The dissociation of an acid, HA, in water may be written as

$$HA + H_2O \rightarrow H_3O^+ + A^-$$

where A^- is the conjugate base. The acid dissociation constant, K_a, is given by the expression

$$K_a = \frac{a_{H_3O^+} \times a_{A^-}}{a_{HA}} \qquad (2.7)$$

where a represents the activity of the species at equilibrium.

For dilute solutions the activity terms may be represented by concentrations of the species at equilibrium and the expression for the acid dissociation constant becomes

$$K_a = \frac{[H_3O^+][A^-]}{[HA]}$$

Taking negative logarithms on both sides

$$-\log K_a = -\log [H_3O^+] - \log [A^-]/[HA]$$

Therefore

$$pK_a = pH + \log[HA]/[A^-]$$

Thus the pK_a value for an acid can be determined from the measurement of pH and the ratio $[HA]/[A^-]$ at that pH. The ratio $[HA]/[A^-]$ can be readily determined spectrophotometrically (see Expt. E15).

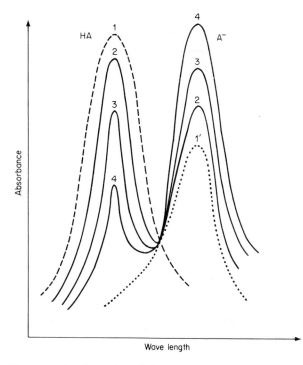

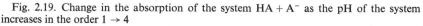

Fig. 2.19. Change in the absorption of the system $HA + A^-$ as the pH of the system increases in the order $1 \rightarrow 4$

For the dissociation $HA \rightleftharpoons H^+ + A^-$ the position of equilibrium will depend upon the pH of the medium. In strong alkaline solution the equilibrium will lie to the right and the absorption spectrum of the system will be that of the anion A^-, while in strong acid solution the equilibrium will lie to the left and the absorption spectrum will be that of HA. The absorption spectra of A^- and HA are represented diagrammatically as curves $1'$ and 1 of Fig. 2.19. If the pH of the system is decreased then the concentration of HA, and hence the absorbance due to HA, will increase while the concentration of A^-, and hence the absorbance due to A^-, will decrease. The overall absorption of the system will change with pH as indicated by the successive curves 2, 3 and 4 of Fig. 2.19. The extinction coefficients for HA and A⁻ will vary from a maximum value at the wavelengths corresponding to the peak maxima to zero at wavelengths where there is no absorption. If there is a wavelength at which the extinction coefficients of the two species are equal, then at this wavelength the absorbance of the system will remain constant since the total concentration of solute, acid and conjugate base, is constant. The point at which the absorbance remains invariant is termed an isosbestic point (see Expt. E15).

II. INSTRUMENTATION AND SAMPLE HANDLING PROCEDURES

A. DESCRIPTION OF A SPECTROPHOTOMETER

Spectrometers for measurement of absorption in the ultraviolet, visible and infrared regions of the spectrum are similar in that they consist of the same basic units, viz. source of radiation, monochromator unit, sample compartment and detector system. The basic layout of a double beam recording spectrophotometer for use in the ultraviolet and visible is shown in Fig. 2.20. One difference in design from that of an infrared spectrometer is that the sample compartment is sited after the monochromator. In this arrangement only a

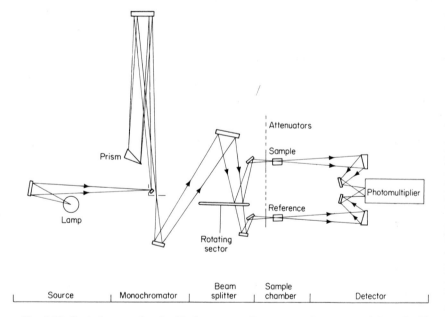

Fig. 2.20. Basic layout of a double beam recording spectrophotometer. Adapted with permission from Pye Unicam Ltd.

small percentage of the radiation from the source traverses the sample at any one time and photodecomposition of the sample is minimized.

In an ultraviolet/visible spectrophotometer radiation from the source is passed via a mirror system to the monochromator unit where the radiation is dispersed into different wavelengths by means of a prism. The prism/mirror system in the monochromator is arranged so that only a narrow range of wavelengths passes through the exit slit. The wavelength of the radiation passing through the exit slit is varied by rotating the prism. In double-beam operation the monochromatic light is split into two beams by means of a rotating sector. The radiation from the monochromator is either reflected off the front surface of the sector and passed through the reference cell or passes through the cut out section of the sector to be reflected through the sample cell. Thus the beam is alternately passed through the sample and reference cells at a frequency controlled by the speed of rotation of the sector. After passing through the sample and reference cells the light beams are focused onto the detector. The output of the detector is connected to a phase-sensitive amplifier which responds to any change in transmission through the sample cell by feeding out a signal which causes the withdrawal or insertion of wedge-shaped combs (attenuators) into the sample or reference beams. If, for instance, the transmission of the sample relative to the reference decreases, then the attenuator in the sample beam is withdrawn while that in the reference beam is inserted. In this way equality of transmission is obtained in the two beams. The movement of the attenuators is transmitted to the recorder and is followed by movement of the pen on the chart. The chart drive is coupled to the rotation of the prism and thus the absorbance or transmittance of the sample is recorded as a function of wavelength.

1. Source

The commonly used sources in commercial ultraviolet and visible spectro-photometers are tungsten-filament lamps for the visible region and hydrogen or deuterium lamps for the ultraviolet region. Tungsten-filament lamps can be usefully used over a range 800–300 nm while the hydrogen or deuterium lamps cover the range 350–120 nm.

2. Monochromator

The monochromator unit consists of entrance and exit slits, a mirror system and a prism. In commercial instruments the slit widths are generally pro-grammed to take account of the change in energy output of the source and the change in response of the detector as the wavelength changes. In this way the output from the detector is kept reasonably constant over the wavelength range of the instrument. The actual slit width at any wavelength is a com-promise between the width required to give maximum resolution consistent with sufficient energy reaching the detector. One prism mounting which is frequently used is the Littrow system in which the back surface of the prism is silvered and the beam which passes through the prism is reflected off the silvered surface and returned through the prism. The prism material may be of glass for the visible and very near ultraviolet region, but fused silica must be used to cover the remainder of the range down to 200 nm. Calcium or lithium fluoride prisms are used for the range 200–125 nm. The mirrors in the optical system are front surfaced since glass starts to absorb in the near ultraviolet region.

3. Detector

Photoelectric detectors are generally used in commercial spectrophotometers. The radiation from the sample and reference beams is incident upon a photosensitive metal surface housed in an evacuated glass or silica envelope, and the electrons ejected from the metal surface are collected on a positively charged plate. The plate current which is measured is proportional to the number of electrons incident upon the plate and hence proportional to the intensity of radiation striking the active surface.

4. Presentation of Spectra

Commercial instruments record either the absorbance or the transmittance of the sample at a particular wavelength. The transmittance, T, is equal to the ratio I/I_0 and is related to the absorbance by the expression $A = -\log_{10} T$. The absorbance is preferred to the transmittance as a measure of absorption characteristics and in the presentation of a spectrum it is normal practice for the absorbance to be plotted as ordinate against the wavelength in nm or against the wave number (cm^{-1}). The latter presentation is often preferred since a wave number scale is linearly related to the energy of the radiation. It is convenient for many purposes to replot the spectrum using either ϵ or $\log \epsilon$ as the ordinate. Plotting $\log \epsilon$ as ordinate has the advantage that intense bands can be shown on the same scale as weak bands. Two different methods of presentation of the absorption spectrum of the same absorbing species are shown in Fig. 2.21.

When tabulating absorption data the convention used is to record the positions of the peak maxima (λ_{max}) and the extinction coefficients at the maxima (ϵ_{max}). Before presenting results the wavelength calibration of the spectrophotometer should always be checked. This may be done by comparing the λ_{max} values for peaks in a selected compound with the values quoted in standard tables, or, for example, by measuring the λ_{max} values for the absorption bands in didymium and holmium glass filters. The spectral transmission and the resolution of the instrument should also be checked from time to time (see Expt. E1).

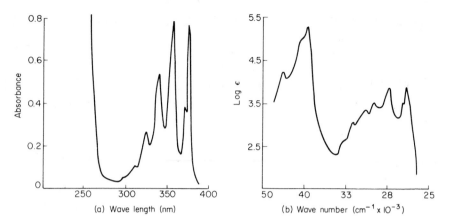

Fig. 2.21. Different presentation of the ultraviolet spectrum of anthracene. (a) Solution containing 2×10^{-4} mol dm^{-3} of anthracene in cyclohexane. (b) Solution in cyclohexane

B. SAMPLE HANDLING PROCEDURES

I. Samples and Sample Cells

Absorption spectra may be recorded when the absorbing species is in the gas phase, in solution or in the solid. Gas-phase spectra are taken in enclosed cells with a similar evacuated enclosed cell in the reference beam. Solution spectra are generally run in rectangular or cylindrical cells which are fitted with caps or stoppers to prevent evaporation of the solute or solvent. The spectra of solids may be examined in various ways. The solid may be deposited as a thin film on a transparent medium, such as an optically-flat silica plate, by casting, by evaporation from solution or by painting on directly. The medium is then mounted in the cell compartment and the spectrum recorded. Plastic or polymeric samples may also be mounted onto a supporting medium while the spectrum is scanned. The spectra of powdered solids may be taken by pressing the solid in a KBr or NaCl disc as in infrared spectroscopy. Another technique used for examining the spectra of solids is that of diffuse reflectance where the reflectance off the solid surface is measured instead of the transmittance through the solid.

The sample cells may be of glass if measurements are to be made in the visible region or very near ultraviolet (down to ~350 nm) but if the measurements are to be extended to shorter wavelengths then silica cells must be used. The path length of the cell depends upon the system being investigated. In most cases 10 mm cells are used but standard cells of path lengths from 100 mm down to 1 mm are commercially available. If the absorption of the solute is too high in any region for the path length used then it may be reduced by appropriate dilution of the solution. Cells used in gas-phase work are often of long path length since the concentration of absorbing species in the gas phase is likely to be low.

2. Solvents

Any solvent used should be reasonably transparent in the region in which the solute absorbs. It may be permissible for the solvent to have a low absorption in the solute region provided the solvent absorption is cancelled out by having a cell containing solvent in the reference beam. This technique can only be

Table 2.6. Cut-off points of some common solvents

Solvent	Approximate cut-off point (nm)	Solvent	Approximate cut-off point (nm)
Diethylether	210	Methylene dichloride	238
Ethanol	210	Chloroform	252
Isopropanol	222	Carbon tetrachloride	268
Methanol	215	Benzene	285
Cyclohexane	220	Xylene	300
Iso-octane	215	Pyridine	312
Dioxane	226	Acetone	335

used when the solvent absorption is low, otherwise the intensity of the light reaching the detector may be too low to enable readings to be made. The cut-off points (i.e. the wavelength below which solvents have high absorbance) for some common solvents are given in Table 2.6. The absorbance of solvents should always be checked before use since the presence of small amounts of impurities may give rise to appreciable absorption in the range of absorption of the sample. Spectroscopically pure solvents, i.e. solvents free of impurities which absorb in the region where the solvent transmits, may be purchased commercially.

III. EXPERIMENTS

A. PERFORMANCE OF INSTRUMENTS AND EXPERIMENTAL TECHNIQUES

The experiments in this section are designed to give familiarity with the operation of an ultraviolet/visible spectrophotometer and with various aspects of experimental technique.

EI. Performance of an Ultraviolet and Visible Spectrophotometer

Object

To check the performance of a spectrophotometer in relation to wavelength accuracy and reproducibility, absorbance accuracy and reproducibility, resolution and beam balance.

Theory

A basic prism spectrophotometer has been described on page 72. As there may be a gradual deterioration in the performance of a spectrophotometer the test specifications for the instrument should be checked regularly and any necessary adjustments made to the preset controls. The test specifications and procedures for correcting faults are often given in the manufacturer's handbook. If the manufacturer's specifications are not available the procedures outlined below may be carried out in order to give an indication of the performance of an instrument. Naturally the performance of any make of instrument varies from any other and the performance limits for any particular test on one instrument may be greater or less than that of another instrument. This experiment outlines suggested tests for determining the performance of an instrument.

Apparatus

Spectrophotometer (200–800 nm); 10-mm silica cells; holmium and didymium glass filters; benzene, potassium dichromate, sulphuric acid.

Procedure

WAVELENGTH ACCURACY. Record the absorption spectrum of a standard reference material and compare the measured values of the peak maxima with the literature values. (The standards generally used are holmium glass for the range 190–460 nm and didymium glass for the range 530–800 nm. Typical tolerance limits are ± 0.4 nm at 200 nm, ± 0.7 nm at 350 nm, ± 1.9 nm at 500 nm, ± 4.8 nm at 700 nm and ± 7 nm at 800 nm.)

WAVELENGTH REPRODUCIBILITY. Superimpose repeat scans of the spectrum of the standard reference on the same chart and measure the wavelength differences between the peak maxima for the different scans. (Typical values for the maximum allowable difference between the greatest and least of a set of scans are 0.2 nm at 200 nm, 0.65 nm at 350 nm, 1.9 nm at 500 nm, 4.8 nm at 700 nm and 7 nm at 800 nm.)

ABSORBANCE ACCURACY. Measure the absorbance in a 10-mm cell of a solution containing 120 ± 0.5 mg dm^{-3} of potassium dichromate in 5×10^{-3} mol dm^{-3} ($N/100$) sulphuric acid at 235, 257, 313 and 350 nm. Correct the absorbance values for the absorption of the solvent at these wavelengths. Compare the experimentally observed absorbance values with the literature values which for the above wavelengths are 1·495, 1·738, 0·586 and 1·288 respectively. (The wavelengths quoted correspond to the peaks and troughs in the spectrum of potassium dichromate.)

ABSORBANCE REPRODUCIBILITY. Record the spectrum of the standard solution of potassium dichromate three or four times. Measure the difference between the greatest and least absorbance reading at each peak and trough. (The difference should not normally be greater than 0·005 absorbance units.)

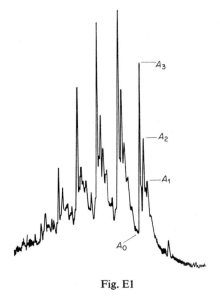

Fig. E1

RESOLUTION. If possible, set the control for the energy of the source to a minimum. Record the absorption spectrum of benzene in the vapour phase by placing a drop of benzene in the bottom of a stoppered cell, pouring the drop out of the cell down one of the ground glass sides, replacing the stopper and measuring the absorption spectrum of the vapour trapped inside over the range 220–330 nm. Repeat this procedure until the absorbance of the peak A_3 (see Fig. E1) at 258·9 nm has a value between 1·3 and 2. Calculate the resolution of the instrument using the equation

$$R = \frac{A_2 - A_1}{A_1 - A_0}$$

where the A terms represent the absorbance values at the points indicated in the figure. (The value of R should be greater than 0·50 for a spectrophotometer with reasonably good resolution.)

BEAM BALANCE. Scan a spectrum without cells in the sample and reference cell holders and measure the difference between the greatest and least reading on the zero absorbance line so drawn. (This difference should not normally exceed 0·02 absorbance units.)

Results and Discussion

Compare the experimental data on the performance of the instrument with the typical specifications given for each test or, if available, with the manufacturer's specifications. Comment on the performance of the instrument.

E2. The Cut-off Point of Solvents and the Calibration of Absorption Cells

Object

To measure the relative difference in transmission between a pair of absorption cells and to test the purity of a selection of solvents.

Theory

The difference in transmission between a matched pair of absorption cells supplied commercially is normally very small but for the greatest accuracy in obtaining absorbance readings this difference should be determined.

Small amounts of impurities in solvents can often give rise to strong absorption bands and thus the "spectroscopic" purity of solvents, i.e. freedom from absorption in the wavelength region of interest, should always be checked before use. The presence of extraneous absorption may render a particular batch of solvents useless for measurements in a particular wavelength range and it may be necessary to purify the solvent.

Apparatus

Spectrophotometer (200–800 nm); a pair of 10-mm silica cells; range of solvents (see Table 2.6).

Procedure

Mark the absorption cells so as to distinguish one from another. Fill each cell with distilled water and measure the transmission of each over the desired wavelength range. Choose the cell with the highest transmission over this range as the reference cell and measure the transmission of the other cell relative to this reference.

Select a number of solvents and measure the absorption spectra of these solvents in the ultraviolet region using 10-mm silica cells. Compare the measured cut-off points, i.e. the wavelength at which the absorbance becomes greater than 1·5, with those given in Table 2.6. If the cut-off points differ markedly, then attempt to purify the solvents by passing them down a chromatographic column containing either silica gel or alumina. Measure the absorption spectra of the purified solvents and again determine the cut-off points.

Results and Discussion

(1) Calculate the correction in transmission to be applied to the sample cell for any observed transmission.

(2) Tabulate the observed cut-off points of the unpurified solvents and of any purified solvents along with the specified values which are given in Table 2.6. Suggest the likely impurities in those solvents which were observed to contain impurities and comment on the ease of purification.

(3) It is considered that solvent purity is generally more critical in ultraviolet than in infrared absorption studies. Why should this be so?

B. FUNDAMENTAL ASPECTS OF ELECTRONIC TRANSITIONS AND RELATIONSHIP TO STRUCTURE

Electronic transitions can be classified according to the type of transition involved, e.g. $n \to \pi^*$, $\pi \to \pi^*$, $d \to d$. Such transitions give rise to absorption bands of various intensity throughout the ultraviolet and visible regions of the spectrum; the position and intensity of the bands depending on the nature of the transition and on the structure of the absorbing species. Examples of various types of electronic transitions and of the effects of structure on transitions are given in the following experiments.

E3. Absorption Spectrum of Iodine Vapour

Object

To calculate thermodynamic parameters for iodine by analysis of the absorption spectrum of iodine in the visible region.

Theory

The absorption spectrum of iodine in the visible region shows a series of bands at the red end which are depicted, together with the transitions to which they are assigned, in Fig. E3. As can be seen from the figure, the separation between bands in the spectrum gives the difference in energy between adjacent vibrational levels in the excited state. Since the band separation can be conveniently measured in units of cm^{-1}, the expressions for energy used in this experiment are given in these units. The expressions can be converted to the SI unit of energy, the Joule, on multiplication by the factor hc.

The energy, $\tilde{E}_{V'}$, of a vibrational level in the *excited* state is given, in units of cm^{-1}, by the expression

$$\tilde{E}_{V'} = \tilde{v}'_e(V' + \tfrac{1}{2}) - \tilde{v}'_e x'_e(V' + \tfrac{1}{2})^2 \tag{E3.1}$$

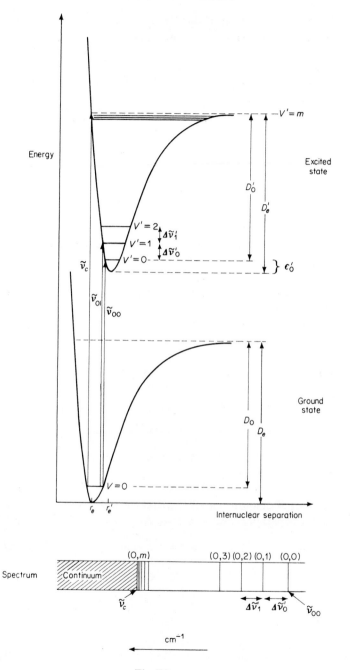

Fig. E3

where $\tilde{v}'_e$ is the wave number of the equilibrium vibration in the excited state, x'_e is the anharmonicity constant in the excited state and V' is the vibrational quantum number of the level.

The difference in energy, $\Delta\tilde{v}'$, between adjacent levels in the excited state is $\tilde{E}_{V'+1} - \tilde{E}_{V'}$ where V' is the vibrational quantum number of the lower level and $\tilde{E}_{V'+1}$ and $\tilde{E}_{V'}$ are given by

$$\tilde{E}_{V'+1} = \tilde{v}'_e(V' + \tfrac{3}{2}) - \tilde{v}'_e x'_e(V' + \tfrac{3}{2})^2 \qquad (E3.2)$$

$$\tilde{E}_{V'} = \tilde{v}'_e(V' + \tfrac{1}{2}) - \tilde{v}'_e x'_e(V' + \tfrac{1}{2})^2 \qquad (E3.3)$$

Thus the expression for $\Delta\tilde{v}'$ is

$$\Delta\tilde{v}' = \tilde{v}'_e - 2\tilde{v}'_e x'_e(V' + 1) \qquad (E3.4)$$

The relationship given in equation E3.4 suggests that a plot of $\Delta\tilde{v}'$ against the corresponding values of $(V' + 1)$ will yield a straight line from which the values of $\tilde{v}'_e$ and x'_e can be obtained.

The values of $\tilde{v}'_e$ and x'_e are used in the following calculations.

DISSOCIATION ENERGY IN THE EXCITED STATE (D'_0). The dissociation energy in the excited state is the energy required to raise the molecule from the lowest vibrational level of the excited state to the vibrational level corresponding to $V' = m$ (see Fig. E3). This energy can be calculated in two ways:

(a) The energy separation between the $V' = 0$ level and the $V' = m$ level is the sum of the individual energy separations of the adjacent levels between $V' = 0$ and $V' = m$. This may be expressed as

$$D'_0 = \sum_{V'=0}^{V'=m} \Delta\tilde{v}' \qquad (E3.5)$$

Thus the dissociation energy D'_0 is equal to the area under the plot of $\Delta\tilde{v}'$ against $(V' + 1)$.

(b) The separation of the vibrational levels in any electronic state decreases as the vibrational quantum number increases and at the level corresponding to $V' = m$ the separation becomes zero. At this point equation E3.4 becomes

$$0 = \tilde{v}'_e - 2\tilde{v}'_e x'_e m \qquad (E3.6)$$

Therefore

$$m = \frac{1}{2x'_e} \qquad (E3.7)$$

Now

$$D'_0 = \tilde{E}_m - \tilde{E}_o \qquad (E3.8)$$

where $\tilde{E}_m$ is the energy of the $V' = m$ level and $\tilde{E}_0$ is the energy of the $V' = 0$ level.

Substituting for $\tilde{E}_m$ and $\tilde{E}_0$ in equation E3.8 gives

$$D'_0 = [\tilde{\nu}'_e(m + \tfrac{1}{2}) - \tilde{\nu}'_e x'_e(m + \tfrac{1}{2})^2] - [\tilde{\nu}'_e(\tfrac{1}{2}) - \tilde{\nu}'_e x'_e(\tfrac{1}{2})^2]$$
$$= \tilde{\nu}'_e m - \tilde{\nu}'_e x'_e m^2 - \tilde{\nu}'_e x'_e m \qquad \text{(E3.9)}$$

Now substituting the value for m in equation E3.9 gives the following expression for D'_0

$$D'_0 = \frac{\tilde{\nu}'_e}{4}\left[\frac{1}{x'_e} - 2\right] \qquad \text{(E3.10)}$$

It can be seen from equation E3.10 that the value of D'_0 can be calculated from the experimentally derived values of $\tilde{\nu}'_e$ and x'_e.

CONVERGENCE LIMIT ($\tilde{\nu}_c$). The convergence limit is defined as the wave number value at which the band system in the absorption spectrum merges into a continuum. The convergence limit is equal to $D'_0 + \tilde{\nu}_{00}$ where $\tilde{\nu}_{00}$ is the wave number for the energy of the transition from the lowest vibrational level of the ground state to the lowest vibrational level of the excited state. The energy represented by the convergence limit is the energy required for the process

$$I_2 \rightarrow I + I^*$$

where I* represents an excited iodine atom.

DISSOCIATION ENERGY IN THE GROUND STATE (D_0). The dissociation energy in the ground state is the energy required for the process

$$I_2 \rightarrow I + I$$

The relationship between this energy and the energy at the convergence limit is

$$D_0 = \tilde{\nu}_c - \Delta\tilde{\nu}$$

where $\Delta\tilde{\nu}$ is the difference between the energy of an unexcited iodine atom I and the energy of an excited iodine atom I* in units of cm^{-1}. This difference in energy is $4\cdot55 \times 10^{27}$ cm^{-1} mol^{-1} ($\equiv 9\cdot03 \times 10^4$ J mol^{-1}), and since $\tilde{\nu}_c$ can be derived from the absorption spectrum the value of D_0 can also be determined.

FORCE CONSTANT IN THE EXCITED STATE (f'). The force constant or restoring force per unit displacement of the nuclei is related to the wave number of the equilibrium vibration, $\tilde{\nu}'_e$, by the classical expression for a harmonic oscillator

$$\tilde{\nu}'_e = \frac{1}{2\pi c}\left(\frac{f'}{\mu}\right)^{1/2} \qquad \text{(E3.11)}$$

where c is the velocity of light and μ is the reduced mass [$1/\mu = (1/m_a) + (1/m_b)$ where m_a and m_b are the masses of the atoms].

MORSE POTENTIAL-ENERGY FUNCTION. Morse proposed the following function to represent the potential energy of a diatomic molecule

$$E_{(r-r_e)} = D_e[1 - e^{-\alpha(r-r_e)}]^2 \qquad (E3.12)$$

where $E_{(r-r_e)}$ is the potential energy as a function of the difference between the internuclear separation, r, and the equilibrium internuclear separation, r_e, D_e is the dissociation energy of the molecule measured from the minimum of the potential-energy curve (Fig. E3), and α is a constant. The terms D_e and α are represented by D'_e and α' for the excited state.

D'_e is related to the dissociation energy, D'_0, in the excited state by the expression

$$D'_e = D'_0 + E'_0 \qquad (E3.13)$$

where E'_0 is the zero-point energy of the excited state and is equal to $\frac{1}{2}\tilde{\nu}'_e$.
The constant α' is given by the expression

$$\alpha' = 0.121\ 77\tilde{\nu}'_e \left(\frac{\mu}{D'_e}\right)^{1/2} \qquad (E3.14)$$

Thus if the values of D'_e and α' are obtained from equations E3.13 and E3.14 respectively using the experimental values for D'_0 and $\tilde{\nu}'_e$, then the potential-energy function for the excited state can be derived if suitable values of r and r_e are chosen.

Apparatus

Spectrograph fitted with photographic detection system; photographic plates sensitive to visible light; source of continuous radiation in visible region (e.g. tungsten-filament lamp); condensing lens; iron arc; standard chart of iron emission spectrum; travelling microscope and plateholder; iodine.

Procedure

Set up the source and lens system so that the source is focused onto the entrance slit of the spectrograph. Take a test-tube containing a few crystals of iodine and clamp this immediately in front of the slit. Remove the tube and heat carefully with a bunsen burner or electrical heating element until a strong violet colour is obtained, insert a cork and replace in front of the slit. Take a series of exposures at definite time intervals, e.g. one-second exposure at 30 s intervals may be a convenient condition.

Set up the iron arc so that the emission from the arc enters the entrance slit of the spectrograph and add an iron-arc emission spectrum alongside each iodine absorption spectrum. Develop, fix, wash and dry the plate. If the plate has been over or under exposed repeat the experiment using different exposure conditions. If the absorption of iodine vapour is too high or too low repeat the exposure using different concentrations of iodine vapour. Note: The concentration of iodine vapour in the light path will fall with time and by taking a series of exposures at different time intervals an exposure may be taken when the concentration of iodine vapour is at the optimum for obtaining an absorption spectrum showing clearly defined bands.

By comparison of the iron spectrum on the plate with that on the standard chart pick out the iron line at 550·678 nm on the plate. This line may be used as the reference point since the line is almost coincident with the head of the (0,24) band in one of the band systems in the iodine absorption spectrum. Measure the distance from the iron reference line of each band of the series, (0,24), (0,25) and so on, for about the first twenty bands. Convert the distance from the reference line into a value for the wavelength of each band by using the calibration equation for the plate. The method of deriving this equation is outlined on page 169.

Results

Complete Table E3 from the experimental data:

Table E3

Band	Distance from reference line	Wavelength nm	Wave number $\tilde{\nu}$ cm^{-1}	$\Delta\tilde{\nu}'$ cm^{-1}
0, 24	x	550·7868	18 155·8	
				84·7
0, 25	y	548·2290	18 240·5	
0, 26				
.				
.				

Plot $\Delta\tilde{\nu}'$ against $(V'+1)$ and calculate $\tilde{\nu}'_e$ and x'_e. Calculate values for the dissociation energy of iodine in the excited state (D'_0) in J mol^{-1} by methods (a) and (b).

Given that the origin of the band system, $\tilde{\nu}_{00}$, is at 641·016 nm, calculate the energy of the convergence limit in cm^{-1} and in J mol^{-1}. Using the latter value calculate the dissociation energy of the iodine molecule in the ground state.

Using the Morse potential energy function calculate values of $E_{(r-r_e)}$ in cm^{-1} for the excited state taking the following values of r: 0·25, 0·26, 0·27, 0·28, 0·29, 0·30, 0·3015, 0·32, 0·34, 0·36, 0·40, 0·50 and 0·60 nm. Take the value of r_e to be 0·3015 nm. Plot a potential-energy curve from the above for iodine in the excited state; calculating energy in cm^{-1} units. Calculate the energy in

4

cm^{-1} units corresponding to the vibrational levels $V' = 0, 1, 2, 3, 10, 20, 30,$ 40 and draw in these levels on the potential-energy curve.

Discussion

(1) Comment on any differences between the values for the dissociation energy in the excited state calculated by methods (a) and (b). Give possible reasons which could account for any difference.

(2) How can the wave number for the equilibrium vibration of iodine in the ground state be obtained?

(3) Is there a significant departure from linearity in the plot of $\Delta\bar{v}'$ against $(\bar{v}' + 1)$? Suggest reasons for any deviation from linearity.

E4. $n \rightarrow \pi^*$ Transitions of the Carbonyl Chromophore

Object

To examine solvent and substituent effects on the $n \rightarrow \pi^*$ transition of a carbonyl group.

Theory

The effect of substituents and of solvents on the energy of the $n \rightarrow \pi^*$ transition in the carbonyl group is discussed on page 45. Substitution of electron-donor groups adjacent to the chromophore and a change from hydrocarbon to hydroxylic type solvents are factors responsible for blue shifts of bands arising from $n \rightarrow \pi^*$ transitions.

Apparatus

Spectrophotometer (200–400 nm); 10-mm silica cells; Solutes: acetone, acetaldehyde, acetylchloride, acetophenone and o- and m-hydroxyaceto-phenone; Solvents: n-heptane, chloroform, acetonitrile, ethanol and water.

Procedure

Record the ultraviolet absorption spectra of the solvents and check that there are no impurities present which absorb in the region of the $n \rightarrow \pi^*$ band in acetone, i.e. 250–290 nm, and in the aromatic carbonyls, i.e. 270–350 nm.

Prepare approximately 0·04 mol dm^{-3} solutions of acetone in each of the solvents.

Record the spectra of the solutions and measure the λ_{max} value for the band representing the $n \rightarrow \pi^*$ transition (the transition of lowest energy) in each of the solvents.

Prepare solutions of the solutes in n-heptane having the following con-centrations: acetaldehyde 5×10^{-2} mol dm^{-3}, acetyl chloride $2\cdot5 \times 10^{-2}$ mol dm^{-3}, acetophenone $2\cdot5 \times 10^{-2}$ mol dm^{-3}, o-hydroxyacetophenone

5×10^{-5} mol dm^{-3}, m-hydroxyacetophenone sat. solution. Record the spectrum of each solution and measure the λ_{max} value for the $n \rightarrow \pi^*$ transition in each solute.

Results

Tabulate the results as shown in Table E4.

Table E4

Solvent	Acetone $n \rightarrow \pi^*$ band (nm)	Solute O ∥ CH$_3$—C—R	$n \rightarrow \pi^*$ band of solute (nm)
CH$_3$(CH$_2$)$_5$CH$_3$		R = CH$_3$	
CHCl$_3$		H	
CH$_3$CN		Cl	
C$_2$H$_5$OH		OH	
H$_2$O		OH ⬡—	
		OH ⬡—	

Discussion

(1) Rationalize the variation of λ_{max} of the $n \rightarrow \pi^*$ band of acetone with solvent.

(2) Relate the values of λ_{max} for the $n \rightarrow \pi^*$ band of CH$_3$COR in n-heptane solution to the electron-donor or electron-withdrawing properties of the substituent R.

E5. Absorption Spectrum of a Conjugated Dye

Object

To compare the experimental value for the absorption maximum of the longest wavelength band in the spectrum of a conjugated dye with the theoretical value calculated on the basis of a free electron model.

Theory

Electrons in the π-electron system of a conjugated aromatic compound are not restricted to specific nuclei but are free to move throughout the system. In a linear conjugated system the potential energy of the electrons will vary along the chain, being lowest near the nuclei and highest between them. However, it can be assumed that, to a first approximation, the potential energy of the electrons is constant along the length of the chain and increases to infinity one bond distance beyond the end atoms of the chain. The electrons can be envisaged as being contained in a one-dimensional potential well, of length l, and occupying orbitals of different energy within the well. The wave functions representing the orbitals and the energy levels of the orbitals can be obtained by solving the Schrödinger equation for the system (p. 15). The form of the Schrödinger equation for a one-dimensional system is

$$\frac{\partial^2 \psi(x)}{\partial x^2} + \frac{8\pi^2 m}{h^2}(E - V)\psi(x) = 0 \tag{E5.1}$$

Since the potential energy V is a constant for the particular system under consideration it may conveniently be set equal to zero for purposes of deriving solutions of the equation. The Schrödinger equation can then be written as

$$\frac{-h^2}{8\pi^2 m}\frac{\partial^2 \psi(x)}{\partial x^2} = E\psi(x) \tag{E5.2}$$

It is now necessary to find suitable functions $\psi(x)$ which are solutions for equation E5.2 and which satisfy the conditions stated on page 15. Inspection shows that functions of the following type satisfy the requirements

$$\psi(x) = A \sin\frac{n\pi x}{l} \tag{E5.3}$$

where A is a constant and $n = 1, 2, 3,\ldots$ etc.

Substitution of the function $\psi(x)$ given in equation E5.3 into the left-hand and right-hand sides of equation E5.2 shows that the function is a solution of the Schrödinger equation.

$$\text{Left-hand side} = -\frac{h^2}{8\pi^2 m}\left(-\frac{n^2 \pi^2}{l^2}\right)A \sin\frac{n\pi x}{l}$$

$$= \frac{n^2 h^2}{8ml^2}\left(A \sin\frac{n\pi x}{l}\right)$$

$$\text{Right-hand side} = E\left(A \sin\frac{n\pi x}{l}\right)$$

The two sides of the equation are of equivalent form and the function $\psi(x)$ is a solution of the Schrödinger equation when the energy E of the system has the values

$$E = \frac{n^2 h^2}{8ml^2}$$

(E5.4)

The allowed energies, E, for the system are represented in Fig. E5.

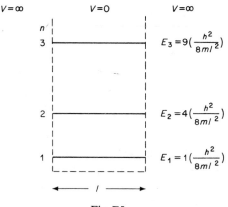

$$V = \infty \qquad\qquad V = 0 \qquad\qquad V = \infty$$

n

3 ——————————— $E_3 = 9\left(\frac{h^2}{8ml^2}\right)$

2 ——————————— $E_2 = 4\left(\frac{h^2}{8ml^2}\right)$

1 ——————————— $E_1 = 1\left(\frac{h^2}{8ml^2}\right)$

$\longleftarrow \quad l \quad \longrightarrow$

Fig. E5

The ground state configuration for the conjugated system is derived by allocating two electrons to each of the energy levels. A system with N electrons forming π bonds will have the $N/2$ lowest energy levels filled and all the higher energy levels empty. The first electronically excited state of the molecule is attained by promotion of an electron from the highest filled orbital ($n = N/2$) to the lowest empty orbital [$n = (N/2) + 1$]. The energy, ΔE, required for this transition is

$$\Delta E = E_{N/2+1} - E_{N/2}$$

(E5.5)

$$= \frac{(N/2 + 1)^2 h^2}{8ml^2} - \frac{(N/2)^2 h^2}{8ml^2}$$

(E5.6)

$$= \frac{(N + 1) h^2}{8ml^2}$$

(E5.7)

The wavelength corresponding to the energy for this transition is

$$\lambda = \frac{hc}{\Delta E} = \frac{8ml^2 c}{h(N + 1)}$$

(E5.8)

where c is the velocity of light.

The wavelength maximum of the band representing the transition of lowest energy in a π electron system can be calculated using equation E5.8.

The free-electron model has been applied to carbocyanine dyes by Kuhn (1948a, 1948b, 1949). Carbocyanine dyes may be represented by the two limiting structures

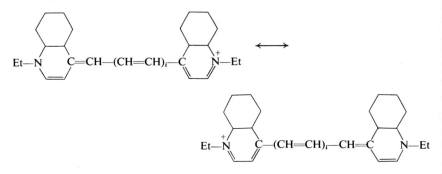

Resonance between the two limiting structures results in a constant bond length along the chain. The extra valence electrons on the carbon atoms and the three electrons from the two nitrogen atoms are delocalized along the length of the chain. Since the number of carbon atoms in the polymethine chain is $2i + 7$, the number of π electrons, N, is $2i + 10$.

The length of the chain between the nitrogen atoms is $(2i + 8)a$ where a is the bond length between atoms in the chain. The total length, l, of the potential well is the chain length plus a bond length at either end and is therefore $(2i + 10)a$.

Substituting for N and l in equation E5.8 gives

$$\lambda = \frac{8mc(2i + 10)^2 a^2}{h(2i + 11)} \tag{E5.9}$$

For a conjugated polyene the average distance a between the atoms may be taken as 0·139 nm. Thus the absorption maximum of the longest wavelength transition in carbocyanine dyes may be calculated by substituting for the constants in equation E5.9 and converting the resulting expression to nm units.

Apparatus

Spectrophotometer (400–750 nm); 10-mm cells; carbocyanine dye, e.g. 1,1^1-diethyl-4,4^1-carbocyanine iodide (kryptocyanine, $i = 1$).

Procedure

Prepare an approx. 10^{-4} mol dm^{-3} solution of the dye in methanol. Record the absorption spectrum of the dye in the visible region. Dilute the initial solution until a spectrum is obtained with an absorbance reading of about one at the peak. Note: Since the dyes dimerize in solution the band shape will change with concentration. At the lower concentrations the monomer peak will be predominant and the dimer peak will either disappear or be present as a shoulder on the low wavelength side of the monomer peak.

Results and Discussion

(1) Calculate the theoretical wavelength for the longest wavelength absorption maximum and compare with the observed value. Comment on the reasons for any difference between these two values.

(2) All-*trans*-β-carotene, which has the following structure

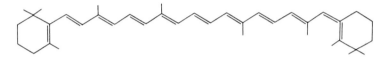

has an absorption band centred at 451 nm. This is the longest wavelength at which it has an absorption peak. Calculate an approximate length for the molecule.

E6. Absorption Spectra of Aromatic Compounds

Object

To examine solvent and substituent effects on $\pi \rightarrow \pi^*$ transitions in benzene and phenol.

Theory

The two bands which occur at approximately 200 and 260 nm in the absorption spectrum of benzene arise from transitions from a π orbital to π^* orbitals of differing symmetry. The symmetry of the excited states for the bands at 200 and 260 nm are such that the transitions are "forbidden" (p. 34). However, perturbing influences from vibrational interaction effects result in some breakdown of the selection rules and the transition at 200 nm appears as a relatively strong band, and that at 260 nm as a band of low intensity. The absorption spectra of substituted benzenes also exhibit two bands in the region of 200 and 260 nm. However, the symmetries of the ground and excited electronic states are reduced in substituted benzenes and the transition at 260 nm becomes less "forbidden" and therefore stronger. If the substituent is a polar group having non-bonding electrons, then resonance interaction between the non-bonding electrons and the π-electron system of the phenyl ring results in a raising of the energy of the ground state relative to that of the excited state and hence to a red shift in the position of the $\pi \rightarrow \pi^*$ bands.

The positions of the bands in the spectrum are also affected by the solvent. Increasing the polarity of the solvent causes a red shift in $\pi \to \pi^*$ transitions. Promotion of an electron from a π orbital to a π^* orbital results in a decreased "electron symmetry" in the system with the result that the excited state is more polar than the ground state. Thus the excited state is stabilized, i.e. the energy of the state is lowered, relative to the ground state as the polarity of the solvent is increased.

In non-polar solvents the band at 260 nm exhibits vibrational fine structure. The fine structure is not observed for substituted benzenes in polar solvents. This is because dipolar interaction between the solvent and the solute in the excited state results in a merging of the energies of the different vibrational levels of the excited state and transitions to distinct vibrational levels of the excited state cannot be observed.

Apparatus

Spectrophotometer (200–300 nm); 10-mm silica cells; benzene; phenol; "Spectrosol" grade *n*-hexane; "Spectrosol" grade ethanol.

Procedure

Dilute 0·5 cm^3 benzene to 25 cm^3 with *n*-hexane. Take 0·5 cm^3 of the solution and make up to a further 25 cm^3 in *n*-hexane. Record the spectrum of the solution over the range 200–300 nm using *n*-hexane in the reference cell. Reduce the absorbance of the band near 200 nm to a measurable value by either diluting the sample by a known amount or by using a shorter path-length cell. Record the spectrum of the band near 200 nm. Repeat the procedure using ethanol as solvent.

Make up accurately 50 cm^3 of an approximately 10^{-2} mol dm^{-3} solution of phenol in *n*-hexane. Take 1·0 cm^3 of this solution and dilute to 25 cm^3 in *n*-hexane. Repeat the procedure using ethanol as solvent. Repeat the spectrum of phenol in ethanol after adding to the sample 1 drop of a 1 mol dm^{-3} NaOH solution.

Results

Tabulate the extinction coefficients (in units of m^2 mol^{-1}) and the wavelengths of the main absorption bands in the spectra of benzene and phenol in the different solvent systems. Note: Density of benzene $= 0·88$ g cm^{-3}.

Discussion

Comment on the following:

(1) The comparison between the values of λ_{max} and ϵ for the two principal absorption peaks near 200 and 260 nm in the spectra of benzene and phenol in *n*-hexane.

(2) The comparison between the values of λ_{max} and ϵ for the band at 260 nm in spectra of phenol in n-hexane and in ethanol.

(3) The change in the absorption spectrum of phenol on adding alkali.

(4) The vibrational structure of the 260 nm band comparing

> benzene in ethanol
> benzene in n-hexane
> phenol in ethanol
> phenol in n-hexane

(5) Assuming the minimum measurable absorbance is 0·01, what is the limit of detection of (a) phenol (b) benzene in g dm^{-3} in n-hexane at 10 mm path length using the band near 260 nm for the analysis?

E7. Intramolecular Charge Transfer Transitions

Object

To measure the effect of substituents on the energy of intramolecular charge transfer transitions in aromatic carbonyl compounds.

Theory

The electronic absorption spectrum of benzene exhibits bands at 180, 200 and 260 nm, the band at 260 nm being relatively weak. When an electron-withdrawing substituent, e.g. —COR, —NO$_2$, is substituted into the benzene ring, then a relatively intense band appears at about 260 nm. This band is due to an intramolecular charge transfer transition (p. 55). Such a transition is represented below

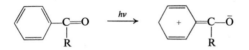

The effect of substituents in the benzene ring on the absorption spectra of aromatic carbonyl compounds of the type

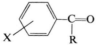

has been analysed by Scott (1961), and it has been found that the changes in the energy of the charge transfer transitions on introducing substituents are additive. Thus knowing the increments to be allotted for the substituent, the position of the charge transfer band in the spectra of carbonyl compounds can be predicted.

Apparatus

Spectrophotometer (200–400 nm); 10-mm silica cells; "spectroscopically pure" ethanol, a selection of substituted acetophenone and benzoic acid derivatives such as: acetophenone; o-, m- and p-hydroxyacetophenone; 2,6-dihydroxyacetophenone; benzoic acid; o-, m- and p-methoxybenzoic acid; 2,4- and 3,4-dimethoxybenzoic acid.

Procedure

Make up solutions of the solutes in ethanol in the cell either by successively adding solute until the concentration is such that the band at 260 nm can be observed, or by successively diluting concentrated solutions of the solutes until the band at 260 nm has a measurable value. Note: A band may be present on the long wavelength side of the 260 nm band. This band will arise from an $n \rightarrow \pi^*$ transition in the carbonyl group.

Results

Tabulate the λ_{max} values for the band at 260 nm in the various compounds as shown in Table E7.

$$\lambda_{max} \text{ parent: acetophenone} = \quad \text{nm}$$
$$\text{benzoic acid} = \quad \text{nm}$$

Table E7

Acetophenone			Benzoic acid		
substituent	λ_{max} (nm)	$\Delta\lambda_{max}$ (nm)	substituent	λ_{max} (nm)	$\Delta\lambda_{max}$ (nm)
o-OH			o-OCH$_3$		
m-OH			m-OCH$_3$		
p-OH			p-OCH$_3$		
2,6-diOH			2,4-diOCH$_3$		
			3,4-diOCH$_3$		

Discussion

(1) Using your results, predict the λ_{max} value for the electron transfer band in 2,6-dihydroxyacetophenone and 2,4- and 3,4-dimethoxybenzoic acid. Compare these predicted values with the experimental values and comment on any differences.

(2) Comment on the accuracy of Scott's rules for predicting the position of electron transfer bands in carbonyl compounds.

(3) Compare your results for the incremental values $\Delta\lambda_{max}$ with those given in Table 2.5 on page 56. Discuss any significant differences between your results and those given in Table 2.5.

(4) Using the values given in Table 2.5 predict the position of the charge transfer band for the compounds having the following structures

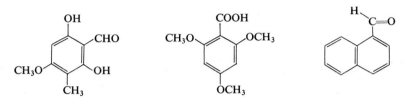

E8. Intermolecular Charge Transfer Transitions

Object

To demonstrate the relationship between the ionization potential of benzene derivatives and the energy of the charge transfer transition in complexes formed with tetracyanoethylene.

Theory

Complexes are often formed in solutions which contain a component capable of donating electrons and one capable of accepting electrons. One of these components may be the solvent. For a solution containing a donor component, D, and an acceptor component, A, a loosely bound complex, DA, may be formed

$$D + A \rightleftharpoons (DA)_{complex}$$

When the complex absorbs radiation, an electron is promoted from an orbital of the donor to a higher energy orbital of the acceptor; the process being termed a "charge transfer transition". The energy required for this transfer process frequently falls in the visible region of the spectrum and solutions containing the complexes are often coloured.

Since an electron is transferred from the donor D to the acceptor A the energy required for the transfer process is dependent upon the ionization potential of D and on the electron affinity of A. It has been demonstrated that the following empirical relationship holds for many charge transfer transitions (McConnell *et al.*, 1953; Merrifield and Phillips, 1958).

$$(h\nu)_{CT} = I_D - E_A + \Delta \qquad (E8.1)$$

where $(h\nu)_{CT}$ is the energy of the charge transfer transition, I_D the ionization potential of the donor, E_A the electron affinity of the acceptor and Δ a constant.

Thus for a series of different donors with the same acceptor a plot of $(h\nu)_{CT}$ against I_D should be a straight line.

Apparatus

Spectrophotometer (300–600 nm); 10-mm glass cells; a saturated solution of tetracyanoethylene in chloroform; solutions of benzene (2×10^{-1} mol dm^{-3}, toluene (2×10^{-1} mol dm^{-3}), o-xylene (1×10^{-1} mol dm^{-3}), m-xylene (1×10^{-1} mol dm^{-3}), mesitylene (5×10^{-2} mol dm^{-3}), anisole (3×10^{-1} mol dm^{-3}) and chlorobenzene (4×10^{-1} mol dm^{-3}) in chloroform. (The concentrations given in brackets refer to approximate concentrations which will give measurable charge transfer absorption bands.)

Procedure

Mix 5 cm^3 aliquots of the solution of tetracyanoethylene (acceptor solution) with a 5 cm^3 aliquot of each of the solutions of the benzenoid compounds (donor solutions). Record the absorption spectra of (a) the mixed solutions, (b) the individual donor solutions and (c) the solution of tetracyanoethylene over the range 300–600 nm using chloroform in the reference cell.

Results

From your results complete Table E8. Plot a graph of transition energy against ionization potential and determine the equation of the line which gives the best fit to the points.

Table E8

Donor	Ionization potential J mol^{-1}	Charge-transfer transition λ_{max}, nm	Charge-transfer transition energy J mol^{-1}
Benzene	$8 \cdot 78 \times 10^{-5}$		
Toluene	$8 \cdot 49$		
o-Xylene	$8 \cdot 30$		
m-Xylene	$8 \cdot 30$		
Mesitylene	$8 \cdot 10$		
Anisole	$7 \cdot 91$		
Chlorobenzene	$8 \cdot 73$		

Discussion

Comment on the shape of the absorption bands due to the complexes and suggest a reason for the appearance of more than one band in the spectra of some of the mixed solutions.

E9. $d \rightarrow d$ Transitions in an Octahedral Complex

Object

To demonstrate the effect of ligand-field strength on the energy of a $d \rightarrow d$ transition in a Cu(II) complex.

Theory

Electronic transitions in the d orbitals of transition-metal ions often occur on absorption of visible radiation (p. 59). The energies of the five d orbitals are determined by the environment of the ion and in an octahedral complex the orbitals are split into two sets of different energy: the lower energy set

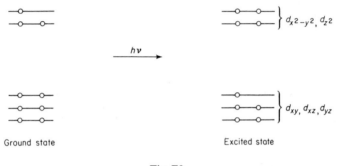

Fig. E9

comprising the d_{xy}, d_{xz} and d_{yz} orbitals, and the higher energy set comprising the $d_{x^2-y^2}$ and d_{z^2} orbitals [see Fig. 2.13(b)]. The splitting of the orbital energies is termed the "ligand-field splitting" and is dependent upon the nature of the coordinating ligand.

The Cu^{2+} ion has nine electrons in the five d orbitals. Promotion of an electron from the lower energy set of orbitals to the higher energy set is represented in Fig. E9. This transition gives rise to a single broad absorption band in the visible region of the spectrum. The energy required for the transition increases in the series of complexes $[Cu(H_2O)_n(NH_3)_{6-n}]^{2+}$ as n is decreased from 4 to 0. This is reflected by the change in colour in going from aqueous solutions of the ion which are pale blue to ammoniacal solutions which are blue-violet.

Apparatus

Spectrophotometer (500–800 nm); 10-mm cells; stock solutions of (1) exactly $1 \cdot 0$ mol dm^{-3} $Cu(NO_3)_2$ $3H_2O$ in water, (2) $2 \cdot 0$ mol dm^{-3} ammonium nitrate and (3) exactly $1 \cdot 0$, $2 \cdot 0$ and $3 \cdot 0$ mol dm^{-3} solutions of ammonium hydroxide.

Procedure

Prepare aqueous solutions of the ions $[Cu(H_2O)_n(NH_3)_{6-n}]^{2+}$ as follows:

$n = 6$: Dilute 0.5 cm^3 of the stock copper solution to 25 cm^3 with water.

$n = 5$: Saturate 5 cm^3 of copper solution with a large excess of ammonium nitrate. Add 5 cm^3 of 1.0 mol dm^{-3} ammonium hydroxide slowly and with stirring. Re-saturate with ammonium nitrate if necessary and then dilute 2 cm^3 of the solution to 25 cm^3 with 2.0 mol dm^{-3} ammonium nitrate.

$n = 4$: Repeat the above procedure using 2.0 mol dm^{-3} ammonium hydroxide.

$n = 3$: Repeat the above procedure using 3.0 mol dm^{-3} ammonium hydroxide.

$n = 2$: Add 1 cm^3 of 0.880 ammonia solution to 1 cm^3 of copper solution and then dilute to 50 cm^3 with water.

$n = 1$: Dilute 0.5 cm^3 of copper solution to 25 cm^3 with 0.880 ammonia solution.

Record the absorption spectra of the ions over the range 500–800 nm. Measure the λ_{max} value for the $d \rightarrow d$ transition in solutions 2–6.

Discussion

Comment on the following:

(1) The correlation between the position of the peak maxima and the number of ammonia molecules in the complex.

(2) The band maximum for the complex $[Cu(H_2O)(NH_3)_5]^{2+}$ lies at longer wavelength than that for the complex $[Cu(H_2O)_2(NH_3)_4]^{2+}$.

(3) The region in which the band maximum for the complex $[Cu(NH_3)_6]^{2+}$ will lie.

C. APPLICATIONS OF THE BEER–LAMBERT LAW

The Beer–Lambert law relates the absorbance of a species to its concentration and to the thickness of the absorbing layer of the species. If the thickness of the absorbing layer is kept constant throughout a series of measurements, then the absorbance will be proportional to the concentration of the absorbing species. Thus the concentration of species present in a system can often be determined directly by absorption spectrophotometry. Examples of systems in which measurement of the concentration of species present yields quantitative information on the system are given in the following experiments.

E10. Quantitative Analysis of Benzene

Object

To determine the concentration of benzene in a test solution in cyclohexane.

Theory

If the absorption spectrum of a solute–solvent system exhibits a band or bands arising from the solute in a region where the solvent does not absorb, then it is possible to analyse quantitatively for the solute in this system. An example of a solute–solvent system of this type is that of benzene–cyclohexane. Here the absorption of benzene gives rise to a band in the region 230–270 nm, whereas there is no absorption from cyclohexane in this region. This difference arises because benzene possesses π electrons which are involved in the absorption of radiation in this region.

From the Beer–Lambert law (p. 68), the absorbance of a solution of benzene in cyclohexane at any wavelength in the region 230–270 nm will be proportional to the concentration of benzene in the mixture, and a plot of absorbance at any wavelength against concentration should be a straight line provided there are no factors causing deviations from the law. The calibration plot can be used to find the concentration of benzene in a test solution.

Apparatus

Spectrophotometer (200–300 nm) suitable for quantitative measurements; 10-mm silica cells; benzene, cyclohexane, test solution.

Procedure

Record the spectrum of cyclohexane over the range 230–270 nm to test that there are no impurities present which absorb in this region. Make up a stock calibration solution in cyclohexane by making 0.5 cm^3 benzene up to 25 cm^3 and, taking 0.5 cm^3 of this solution and again making up to 25 cm^3.

Prepare the following solutions in 10 cm^3 graduated flasks:

Solution (1) 8 cm^3 stock solution, make up to 10 cm^3
      (2) 6 cm^3 stock solution, make up to 10 cm^3
      (3) 4 cm^3 stock solution, make up to 10 cm^3
      (4) 2 cm^3 stock solution, make up to 10 cm^3
      (5) 1 cm^3 stock solution, make up to 10 cm^3

Record the spectrum of the test solution over the range 230–270 nm and note the absorbance value for the strongest absorption. If this value exceeds 0.7 then dilute the solution by an appropriate amount so that the absorbance at the peak maximum of the strongest absorption band lies in the range 0.3–0.7. Note: (a) The dilution factor must be known accurately. (b) The accuracy of absorbance values is generally highest in the absorbance range 0.3–0.7.

Measure the absorbance values, with cyclohexane in the reference cell, of solutions 1–6 and the test solution at the wavelength corresponding to the peak maximum of the strongest absorption band. Repeat the measurements at the wavelength corresponding to the peak of the next strongest absorption band. Note: The preliminary spectrum of the test solution may be run on a recording spectrophotometer but the measurements of the absorbance values should be taken on a manual spectrophotometer unless the recording instrument is suitable for quantitative measurements, i.e. capable of giving a reading of absorbance to three decimal places.

Results

Calculate the concentration of benzene in mol dm^{-3} in each of the solutions 1–6 and draw up a plot of absorbance against concentration for each of the wavelengths at which measurements were made. Note: Density of benzene = 0·88 g cm^{-3}.

Determine the concentration of benzene in the test solution using the absorbance values recorded at the two chosen wavelengths.

Calculate the extinction coefficients of benzene at the two wavelengths.

Discussion

(1) Estimate the error in the values obtained for the concentration of benzene, and indicate what experimental factors could contribute to this error.

(2) The extinction coefficient for benzene at the wavelength for the strongest band in the infrared spectrum is 40 m^2 mol^{-1}. Since cyclohexane also absorbs at this wavelength the maximum cell path length that can be used for making quantitative measurements is 0·1 mm. (For normal practical purposes the maximum cell path length that is used in ultraviolet spectroscopy is 100 mm.) What then is the ratio of the relative limits of detection of benzene in cyclohexane by infrared compared with ultraviolet spectroscopy?

(3) In view of the shape of the peaks on which absorbance measurements were made, what experimental precautions would be necessary if accurate analysis of benzene–cyclohexane mixtures were to be made on a day-to-day basis?

EII. Quantitative Analysis of a Two-component Mixture
Object

To determine the concentration of cobalt and of chromium in a mixture of cobalt(II) and chromium(III) nitrate.

Theory

If the absorption of radiation by two components X and Y in a mixture is additive then the absorbance, A, at wavelengths λ_1 and λ_2 is given by

$$\lambda_1: \quad A_1 = \epsilon_1^X c^X l + \epsilon_1^Y c^Y l$$
$$\lambda_2: \quad A_2 = \epsilon_2^X c^X l + \epsilon_2^Y c^Y l$$

where ϵ_1^X, ϵ_2^X, ϵ_1^Y and ϵ_2^Y are the extinction coefficients of the components at wavelengths λ_1 and λ_2, c^X and c^Y are the concentrations of the components in the mixture and l is the path length of the cell. The units of concentration and path length are mol dm^{-3} and mm respectively. The extinction coefficients can be determined at λ_1 and λ_2 from Beer–Lambert law plots for pure components X and Y at these wavelengths. If the extinction coefficients are known, the above two simultaneous equations have only two unknowns, viz. c^X and c^Y. Thus, by measuring the absorbance of a mixture of two components at two wavelengths at which the extinction coefficients of each component is known, the composition of the mixture can be determined.

Apparatus

Spectrophotometer (350–650 nm) suitable for quantitative measurements; 10-mm cells; aqueous solutions of 0·1880 mol dm^{-3} cobalt(II) nitrate and 0·0500 mol dm^{-3} chromium(III) nitrate.

Procedure

(a) From the 0·1880 mol dm^{-3} cobalt(II) nitrate stock solution make up solutions 0·0376, 0·0752 and 0·1504 mol dm^{-3} in cobalt. Record the absorption spectrum of the 0·0752 mol dm^{-3} cobalt solution over the range 350–650 nm.

(b) From the 0·500 mol dm^{-3} chromium(III) nitrate stock solution make up solutions 0·0100, 0·0200, 0·0300 and 0·0400 mol dm^{-3} in chromium. Record the absorption spectrum of the 0·0200 mol dm^{-3} chromium solution over the range 350–650 nm.

(c) Make up a 50:50 mixture of the cobalt and chromium stock solutions.

By reference to the absorption spectra of the cobalt and chromium solutions find two suitable wavelengths at which to carry out the analysis of the cobalt–chromium mixture. Select wavelengths where the absorption of one component is low while that for the other is high, and where the rate of change of absorbance with wavelength is low. Measure the absorbances at the chosen wavelengths, of all the solutions prepared under sections (a)–(c) above.

Results

Draw up Beer's law plots (absorbance versus concentration) for cobalt nitrate and chromium nitrate at the chosen wavelengths. Calculate the extinction coefficients for the cobalt and chromium ions at these wavelengths.

Calculate the concentrations of cobalt and chromium in solution (c).

Discussion

(1) Compare the experimentally derived values for the cobalt and chromium concentrations in the mixed solution with the correct values and comment on the accuracy of the method for determining the concentrations of components in a mixture.

(2) Discuss any limitations to the method and point out any assumptions that are made in analysing a mixture by this method.

E12. Molecular Complexes: (a) Charge-transfer Complexes; (b) Hydrogen-bonded Complexes

Object

To evaluate the equilibrium constant for complex formation between either (a) durene and chloranil or (b) β-naphthol and p-dioxane.

Theory

Many molecules which are not capable of reacting "chemically" often interact "physically" and are bound together in the form of a molecular complex by weak forces of coordination. For example, compounds which can act as electron donors, D, can interact with compounds which can act as electron acceptors, A, to form complexes which are in equilibrium with the donor and acceptor components

$$D + A \;\rightleftharpoons\; (DA)_{complex} \tag{E12.1}$$

In the case of the interaction between durene and chloranil, the π-electron system of durene acts as the donor site while the electron deficient π-electron system of chloranil (effect of electronegative chlorine atoms) acts as acceptor site. The absorption spectrum of the mixture exhibits a band in the visible region corresponding to the transition of an electron from a filled orbital of anthracene to a higher energy-vacant orbital of chloranil (Foster *et al.* 1956).

In the case of the interaction between p-dioxane and β-naphthol, the lone-pair electrons on the oxygen atom in p-dioxane act as donor site while the hydrogen atom in the hydroxylic function of β-naphthol acts as acceptor site, i.e. a hydrogen bond is formed between the components. The formation of a hydrogen bond causes a change in the energy levels of the ground and excited states of the phenol and consequently there is a difference between the electronic absorption spectrum of "free" phenol and "complexed" phenol (Baba and Suzuki, 1961).

For dilute solutions, the equilibrium constant, K_a, for complex formation is given by

$$K_a = \frac{[DA]}{[D][A]} \tag{E12.2}$$

where square brackets represent concentrations at equilibrium.

The absorbance, A_{obs}, of a solution at any wavelength will be the sum of the absorbances of the species D, A and DA, and is given by

$$A_{obs} = \epsilon_A[A]l + \epsilon_D[D]l + \epsilon_{DA}[DA]l \tag{E12.3}$$

where ϵ represents extinction coefficients and l the thickness of the absorbing layer.

If the initial concentration of the donor is D_i, the [D] will equal $D_i - [DA]$, and if it is arranged that D is in excess as compared to A, then [D] will be approximately equal to D_i.

If the initial concentration of the acceptor equals A_i, then [A] will equal $A_i - [DA]$.

Substituting for [D] and [A] as given above in equation E12.2 gives

$$K_a = \frac{[DA]}{D_i(A_i - [DA])} \tag{E12.4}$$

If only the complex absorbs at the wavelength at which absorbance measurements are made, then equation E12.3 reduces to

$$A_{obs} = \epsilon_{DA}[DA]l \tag{E12.5}$$

Substitution for [DA] from equation E12.5 into equation E12.4 and rearranging gives

$$\frac{A_i l}{A_{obs}} = \frac{1}{K\epsilon_{DA}} \cdot \frac{1}{D_i} + \frac{1}{\epsilon_{DA}} \tag{E12.6}$$

Equation E12.6 can be applied to the durene–chloranil system since neither of the components absorb at the wavelength of maximum absorption of the complex. Thus if the initial acceptor concentration A_i is kept constant and the initial donor concentration D_i varied, a plot of A_i/A_{obs} against $1/D_i$ should give a straight line. The values of K_a and ϵ_{DA} may be calculated from the slope and the intercept of the line on the y axis.

Equation E12.6 cannot be used for the p-dioxane–β-naphthol system since the absorption of the "free" β-naphthol and the "hydrogen-bonded" β-naphthol overlap considerably. p-Dioxane does not absorb in the region of the longest wavelength band in the spectrum of β-naphthol and thus for the p-dioxane–β-naphthol system in this region equation E12.3 can be reduced to

$$A_{obs} = \epsilon_A[A]l + \epsilon_{DA}[DA]l \qquad \text{(E12.7)}$$

The "apparent" extinction coefficient, ϵ_a, of a solution of p-dioxane and β-naphthol at a wavelength where only β-naphthol absorbs is defined by the equation

$$A_{obs} = \epsilon_a([A] + [DA])l \qquad \text{(E12.8)}$$

Equation E12.9 can be derived from equations E12.7 and E12.8 and the definitions of D_i and A_i

$$\frac{1}{\epsilon_a - \epsilon_A} = \frac{1}{K_a(\epsilon_{DA} - \epsilon_A)} \frac{1}{D_i} + \frac{1}{(\epsilon_{DA} - \epsilon_A)} \qquad \text{(E12.9)}$$

If the initial concentration, A_i, of β-naphthol is kept constant then the term $([A] + [DA])$ in equation E12.8 will be a constant, equal to A_i, and equation E12.8 can be written as

$$A_{obs} = \epsilon_a A_i l \qquad \text{(E12.10)}$$

Now for a concentration A_i of β-naphthol in a non-interacting system, the absorbance of A_{obs}^f of "free" β-naphthol will be given by

$$A_{obs}^f = \epsilon_A A_i l \qquad \text{(E12.11)}$$

Subtracting equation E12.11 from E12.10 and rearrangement gives

$$\frac{1}{\epsilon_a - \epsilon_A} = \frac{A_i l}{A_{obs} - A_{obs}^f} \qquad \text{(E12.12)}$$

Substituting for the left-hand side of equation E12.12 into equation E12.9 gives

$$\frac{A_i l}{A_{obs} - A_{obs}^f} = \frac{1}{K_a(\epsilon_{DA} - \epsilon_A)} \frac{1}{D_i} + \frac{1}{(\epsilon_{DA} - \epsilon_A)} \qquad \text{(E12.13)}$$

Thus if the left-hand side of equation E12.13 is plotted against $1/D_i$ for the p-dioxane–β-naphthol system, a straight line should be obtained. The value of K_a may be calculated from the slope and the intercept of the line on the y axis.

Apparatus

Spectrophotometer (300–700 nm) suitable for quantitative measurements; 10-mm silica cells; durene (1,2,4,5-tetramethylbenzene), chloranil, p-dioxane, β-naphthol, dichloromethane, iso-octane.

Procedure

DURENE–CHLORANIL SYSTEM. Make the following stock solutions up accurately in 50 cm^3 graduated flasks:

Solution A: Approximately 0·26 mol dm^{-3} durene in dichloromethane
Solution B: Approximately 6×10^{-3} mol dm^{-3} chloranil in dichloromethane

Make up the following solutions from stock solutions A and B:

Solution (1) 5 cm^3 Solution A + 5 cm^3 dichloromethane
     (2) 5 cm^3 Solution B + 5 cm^3 dichloromethane
     (3) 5 cm^3 Solution A + 5 cm^3 Solution B

Measure the absorption spectra of solutions 1–3 in the region 300–700 nm using dichloromethane as reference, and hence determine the wavelength of maximum absorbance of the complex.

Make up the following solutions in 10 cm^3 graduated flasks and measure the absorbance of the solutions at λ_{max} for the complex using solution 1 as the blank:

Solution (4) 5 cm^3 B + 5 cm^3 A
     (5) 5 cm^3 B + 4 cm^3 A, make up to 10 cm^3
     (6) 5 cm^3 B + 3 cm^3 A, make up to 10 cm^3
     (7) 5 cm^3 B + 2 cm^3 A, make up to 10 cm^3
     (8) 5 cm^3 B + 1·5 cm^3 A, make up to 10 cm^3
     (9) 5 cm^3 B + 1 cm^3 A, make up to 10 cm^3

p-DIOXANE–β-NAPHTHOL SYSTEM. Make the following stock solutions up accurately in 50 cm^3 graduated flasks:

Solution A: Approximately 0·2 mol dm^{-3} p-dioxane in iso-octane
Solution B: Approximately 6×10^{-4} mol dm^{-3} β-naphthol in iso-octane

Make up the following solutions in 10 cm^3 graduated flasks:

Solution (1) 5 cm^3 B + 5 cm^3 iso-octane
     (2) 5 cm^3 B + 5 cm^3 A
     (3) 5 cm^3 B + 2·5 cm^3 A, make up to 10 cm^3
     (4) 5 cm^3 B + 2·0 cm^3 A, make up to 10 cm^3
     (5) 5 cm^3 B + 1·0 cm^3 A, make up to 10 cm^3
     (6) 5 cm^3 B + 0·5 cm^3 A, make up to 10 cm^3

Record the spectra of solution 1–6 over the range 300–450 nm.
Measure the absorbance of the solutions at 332 nm.

Results and Discussion

DURENE–CHLORANIL SYSTEM. Plot $A_i l / A_{obs}$ against $1/D_i$ and hence obtain a value for the equilibrium constant K_a and for the extinction coefficient ϵ_{DA} of the complex. Estimate the error in the value of K and ϵ_{DA}.

Indicate how the value of K_a would be expected to change as the electron-donor ability of the donor molecule is increased.

p-DIOXANE–β-NAPHTHOL SYSTEM. Plot $A_i l / A_{obs} - A^f_{obs}$ against $1/D_i$ and hence obtain a value for the equilibrium constant K_a. Estimate the error in the value of K_a.

Comment on the effect of hydrogen bonding on the longest wavelength transition of phenol.

E13. Determination of Molecular Weights

Object

To determine the molecular weight of an amine.

Theory

The absorption spectrum of a compound depends upon the chromophoric groupings present. If a derivative is formed via a functional group of the compound remote from the chromophoric system then the absorption arising from the chromophoric system will remain unaltered provided the new group does not absorb in the same region. Thus the extinction coefficient, ϵ, for wavelengths within the region of the chromophoric absorption will remain constant. However, the absorbance, A, will change since this value is dependent upon the molecular weight of the absorbing species:

$$A = \epsilon c l$$

$$= \frac{\epsilon c^1 l}{M} \tag{E13.1}$$

where M is the molecular weight of the absorbing species, c^1 the concentration of the absorbing species in g dm^{-3} and l the thickness of the absorbing layer in mm. It can be seen from equation E13.1 that the molecular weight of the absorbing species can be determined by measuring the absorbance of a solution containing a known weight of the species in a known volume.

This method can be used for determining the molecular weights of compounds which form picrates, e.g. amines (Cunningham *et al.* 1951).

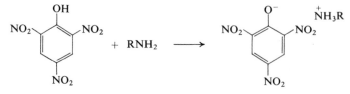

Picric acid exhibits an absorption band centred at 360 nm arising from the —NO_2 chromophoric groupings. This band can be utilized for determining the molecular weight of the picrate provided the group R does not absorb at the wavelength at which measurements are made.

Apparatus
Spectrophotometer (300–400 nm); 10-mm silica cells; picrates of aniline, N-methylaniline, N,N-dimethylaniline and pyridine.

Procedure
Record the spectra between 300–400 nm of approximately 10^{-3} mol dm^{-3} solutions of aniline, N-methylaniline, N,N-dimethylaniline, pyridine and picric acid in benzene and determine a wavelength suitable for measurement on the picrates. Prepare accurately solutions of the picrates containing about 5×10^{-2} g picrate per 50 cm^3 of methanol and dilute 1 cm^3 of the solution to 100 cm^3 in methanol. Measure the absorbance of these solutions at the chosen wavelength.

Results
Calculate the extinction coefficient of each picrate at the chosen wavelength of measurement. Determine the mean extinction coefficient for the picrates of aniline, N,N-dimethylaniline and pyridine. Using this value calculate an experimental value for the molecular weight of the N-methylaniline picrate.

Discussion
(1) Compare the experimental value for the molecular weight of N-methyl-aniline with the theoretical value and give reasons for any difference.

(2) Comment on the accuracy of this method for determining molecular weights.

E14. Reaction Kinetics

Object
To determine the form of the rate equation and the rate constant for the reaction of the triiodide ion with arsenious acid.

Theory
The reaction of the triiodide ion with arsenious acid in aqueous solution produces arsenic acid, iodide ions and hydrogen ions

$$H_3AsO_3 + I_3^- + H_2O \rightleftharpoons H_3AsO_4 + 3I^- + 2H^+$$

It has been shown that the rate law for the forward reaction is of the form

$$-\frac{d[I_3^-]}{dt} = k_f[H_3AsO_3]^a [I_3^-]^b [I^-]^c [H^+]^d \qquad \text{(E14.1)}$$

where k_f is the rate constant for the forward reaction and a, b, c, d are integers which may be negative, zero or positive.

The triiodide ion absorbs in the visible region of the spectrum and since none of the other species present absorb in this region the disappearance of the triiodide ion during the reaction can be followed spectrophotometrically. If the concentration of H_3AsO_3, I^- and H^+ are high compared with that of the triiodide ion, then only the concentration of the triiodide ion will change appreciably in the course of the reaction. From the Beer law the absorbance of the triiodide ion is proportional to the concentration and thus by following the change of absorbance of the triiodide ion with time the overall kinetics of the reaction with respect to triiodide may be established. Also by varying the concentration of one of the species H_3AsO_3, I^-, H^+ while keeping that of the other two constant and measuring the rate of disappearance of the triiodide ion the values of a, c and d may be determined. Once a, b, c and d are known the value of k_f can be calculated.

Apparatus

Spectrophotometer (400–700 nm) suitable for quantitative measurements; 10-mm glass cells; 60% $HClO_4$ solution; NaI, $NaClO_4$, As_2O_3, I_2.

Procedure

Prepare the stock solutions listed below. Note: The concentrations of the solutions are given to the number of significant figures to which the concentrations should be known.

(1) ~1·000 mol dm^{-3} $HClO_4$
    Make up 16·75 g of 60% $HClO_4$ to 100 cm^3 with water.
(2) ~0·050 mol dm^{-3} NaI_3 plus ~0·300 mol dm^{-3} NaI in ~0·60 mol dm^{-3} $NaClO_4$
    Weigh out accurately about 4·0 g NaI, 7·3 g $NaClO_4$ and 1·4 g I_2. Mix the NaI and $NaClO_4$, dissolve in water, add the I_2 and after dissolution make up to 100 cm^3 in water.
(3) 1·00 mol dm^{-3} NaI
(4) 0·200 mol dm^{-3} H_3AsO_3 in 1·00 mol dm^{-3} $NaClO_4$
    Weigh out accurately about 2·0 g of As_2O_3 and 12·2 g $NaClO_4$. Dissolve the As_2O_3 in 10 cm^3 of a solution containing 4 mol dm^{-3} NaOH, add HCl until the solution is just acid, then add the $NaClO_4$ and after dissolution make up to 100 cm^3 with water.

Record the absorption spectra of the following solutions: (a) 2 cm³ stock solution 2 plus 20 cm³ water, and (b) 4 cm³ stock solution 2 plus 18 cm³ water. Observe the wavelength at which the absorbance of each of these solutions is 0·5. Note: Subsequent measurements are made at one or other of these wavelengths dependent upon whether the reactant solutions contains 2 or 4 cm³ of triiodide stock solution.

Prepare reactant solution 1 using the volumes of stock solution given in Table E14 adding the triiodide solution last and noting the time at which approximately half of the triiodide solution is added. Immediately measure the absorbance of the reactant solution at one of the wavelengths selected

Table E14

Reactant solution	Volume of stock solution (cm³)					Final volume (cm³)
	H_3AsO_3	I^-	H^+	H_2O	I_3^-	
1	10	6	2	2	2	22
2	5	6	2	7	2	22
3	10	3	2	5	2	22
4	10	6	4	—	2	22
5	10	6	2	—	4	22
6	5	3	4	6	4	22

above and note the time after mixing at which the absorbance reading is made. Measure the absorbance of the solution at approximately two-minute intervals for up to about 16 minutes. Repeat the procedure for reactant solutions 2–6. Note: Reaction is rapid in reactant solution 3 and readings should be taken about every 30 seconds.

Results

Using the results for reactant solution 1, plot log absorbance, 1/absorbance and 1/(absorbance)² against time and hence determine whether the reaction is first order, second order or third order with respect to triiodide ion. When the order with respect to triiodide ion has been determined, plot the appropriate function for absorbance against time for solutions 1–6 on the same graph. By inspection of the slope of the plots and knowing the relative concentrations of the reactants in the different solutions, determine the values of the constants a, c and d in equation E14.1. Calculate a value for k_f for each of the six solutions from the slopes of the plots and the values of the initial concentrations.

Discussion

(1) Comment on the agreement between the values of k_f obtained from the different plots.

(2) Suggest how the rate law for the back reaction may be determined knowing that it is of the form

$$\frac{d[I_3^-]}{dt} = k_b[H_3AsO_4]^x [I^-]^y [H^+]^z$$

where x, y and z can be assumed to be integers.

E15. Determination of Dissociation Constants: Isosbestic Point

Object

To determine the dissociation constant of p-nitrophenol and to demonstrate the existence of an isosbestic point in the p-nitrophenol–p-nitrophenate anion system.

Theory

If two or more species are in equilibrium in solution and if there is an overlap of the absorption spectra of two of the species, then there will be a point, termed the "isosbestic point", at which the absorbance of the solution will remain constant regardless of the ratio of the concentration of the two species (p. 71).

The equilibrium between p-nitrophenol and p-nitrophenate anion has been studied by Biggs (1954). The equilibrium can be represented as

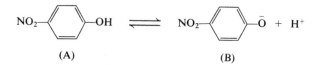

$$(A) \qquad\qquad\qquad (B)$$

If the total concentration of solute is x mol m^{-3} and the degree of dissociation of the phenol is y, the concentrations of the species at equilibrium are $[A] = (x - y)$, $[B] = y$, $[H^+] = y$ mol m^{-3}. If both components obey Beer's law and if absorption due to the hydrogen ion is neglected then the absorbance A at any wavelength is given by

$$A = \epsilon_A[A]l + \epsilon_B[B]l \qquad (E15.1)$$

or

$$A = \epsilon_A(x - y)l + \epsilon_B yl \qquad (E15.2)$$

where ϵ_A, ϵ_B represent the extinction coefficients of species A and B at the wavelength of measurement and l the thickness of the absorbing layer. When $\epsilon_A = \epsilon_B$, equation E15.2 can be written as

$$A = \epsilon_A xl$$

Since x is constant the absorbance at the point where $\epsilon_A = \epsilon_B$ will be constant. The dissociation constant, K_a, for p-nitrophenol is given by the expression

$$K_a = \frac{[H^+][B]}{[A]} \qquad (E15.3)$$

where square brackets represent concentration of the species at equilibrium. Equation E15.3 can be written in the form

$$pK_a = pH + \log\frac{[A]}{[B]} \qquad (E15.4)$$

If the ionic strength of the solution is taken into account equation E15.4 becomes

$$pK = pH + \log\frac{[A]}{[B]} + \frac{0{\cdot}51\sqrt{I}}{1 + 1{\cdot}25\sqrt{I}} \qquad (E15.5)$$

where I is the ionic strength of the solution and equals $\frac{1}{2}\sum_i c_i Z_i^2$ where c_i is the concentration of the ith ion, Z_i is its charge, and the summation extends over all the ions in the solution.

Equation E15.1 gives

$$A = \epsilon xl = \epsilon_A[A]\,l + \epsilon_B[B]\,l \qquad (E15.6)$$

where x is the total concentration of the solute and ϵ the molar extinction coefficient of the mixed solution.

Since $x = [B] + [A]$ equation E15.6 may be expressed as

$$\epsilon([A] + [B])\,l = \epsilon_A[A]\,l + \epsilon_B[B]\,l \qquad (E15.7)$$

Rearranging equation E15.7 gives

$$\frac{[A]}{[B]} = \frac{(\epsilon - \epsilon_B)}{(\epsilon_A - \epsilon)} \qquad (E15.8)$$

Substituting for $[A]/[B]$ in equation E15.5 gives

$$pK = pH + \log\frac{\epsilon - \epsilon_B}{\epsilon_A - \epsilon} + \frac{0{\cdot}51\sqrt{I}}{1 + 1{\cdot}25\sqrt{I}} \qquad (E15.9)$$

Thus if I, ϵ_A and ϵ_B are known and the extinction coefficient ϵ can be determined at a known pH, then the pK_a value and hence the K_a value for the ionization of the species A can be evaluated.

Apparatus

Spectrophotometer (280–440 nm); 10-mm silica cells; p-nitrophenol; sodium chloride; buffer solutions (pH range 4·0–9·5).

Procedure

Make up accurately an aqueous solution (250 ml) containing approx. 8×10^{-5} mol dm^{-3} of *p*-nitrophenol. Make up buffer solutions of accurately known pH within the range pH 4–9·5. (Suggested values are pH $\approx$ 4·0, 6·0, 6·5, 6·7, 7·0, 9·0, 10·0.) Mix 40 cm^3 of phenol solution with 10 cm^3 of buffer solution and add a weighed amount (approx. 0·263 g) of sodium chloride. Repeat with other buffer solutions. Record the spectra of the solutions on the same chart between 280 and 450 nm. Measure the absorbance of the solutions at the peak maxima.

Results

Calculate the concentration of sodium chloride in each solution and hence the ionic strength of each solution. Assuming that in the strongest acid solution the only absorbing species present is the acid form, A, calculate the values of ϵ_A at the wavelengths corresponding to the peak maxima. Similarly, calculate the values of ϵ_B at these wavelengths from the absorbance measurements made on the strongest base solution. Calculate the extinction coefficients, ϵ, for the solutions containing both species A and B and hence calculate the corresponding pK_a and K_a values. Determine the mean value for pK_a and K_a for *p*-nitrophenol. Quote the estimated error in these values.

Record the wavelength of the isosbestic point.

Discussion

Explain why more than one isosbestic point might be observed in a set of spectra for an equilibrium mixture if the spectra are recorded over a wide wavelength range.

REFERENCES

Baba, H., and Suzuki, S. (1961). *J. Chem. Phys.* **35**, 1118.
Biggs, A. I. (1954). *Trans. Faraday Soc.* **50**, 800.
Cunningham, K. G., Dawson, W., and Spring, F. S. (1951). *J. Chem. Soc.* 2305.
Figgis, B. N. (1966). *In* "Introduction to Ligand Fields". Wiley, London.
Foster, R., Hammick, D. L., and Parsons, B. W. (1956). *J. Chem. Soc.* 555.
Kuhn, H. (1948a). *Helv. Chim. Acta*, **31**, 1441.
Kuhn, H. (1948b). *J. Chem. Phys.* **16**, 840.
Kuhn, H. (1949). *J. Chem. Phys.* **17**, 1198.
Linnett, J. W. (1960). *In* "Wave Mechanics and Valency". Methuen, London.
McConnell, H., Ham, J. S., and Platt, J. R. (1953). *J. Chem. Phys.* **21**, 66.
Merrifield, R. E., and Phillips, W. D. (1958). *J. Am. Chem. Soc.* **80**, 2778.
Scott, A. I. (1961). *Experientia*, **17**, 68.
Scott, A. I. (1964). *In* "Interpretation of the Ultraviolet Spectra of Natural Products".
    Pergamon, Oxford.

Sutton, D. (1968). *In* "Electronic Spectra of Transition Metal Complexes". McGraw-Hill, New York.
Woodward, R. B. (1941). *J. Am. Chem. Soc.* **63**, 1123.
Woodward, R. B. (1942). *J. Am. Chem. Soc.* **64**, 72.

Vibrational Spectroscopy

This chapter is concerned with spectra arising from vibrational energy changes in molecules. The infrared absorption spectrum of a diatomic molecule was largely explained (p. 5) in terms of a molecular model possessing classical mechanical and classical electrical properties. However, it was noted that the simple model does not explain the fact that the vibrational band for a molecule in the gas phase possesses fine structure, and that an overtone band occurs at a wave number value which is less than twice the wave number value of the fundamental vibration. In order to explain these observations it is necessary to consider the quantum theory of the interaction of radiation with matter.

Although a study of the absorption of infrared radiation is the most commonly used method for obtaining information on vibrational energy changes, an alternative method is to utilize the Raman effect. The techniques for obtaining infrared and Raman spectra will be described in this chapter. The principles of the two methods indicate that factors governing which transitions are allowed are not the same for infrared and Raman spectra. It has been indicated in Chapter 1 that molecular symmetry has an important role in determining whether transitions are active in infrared and/or Raman spectra. Molecules with a high degree of symmetry tend to have certain vibrations which are *either* Raman *or* infrared active. It will be seen (Expts V6–V8) that bands due to those vibrations which preserve the symmetry of the molecule tend to appear in the Raman spectrum and those which distort the symmetry tend to appear in the infrared spectrum. Molecules which possess a centre of symmetry obey the rule of mutual exclusion (p. 138). Molecules with little or no symmetry tend to have all vibrations active in *both* infrared and Raman, although many bands or lines arising from these vibrations will be too weak to be identified.

Although quantum methods are required to interpret vibrational spectra rigorously, a very large amount of experimental data can be rationalized using classical models. This is because it is easier to relate infrared and Raman spectra to various stretching and bending modes of molecular groupings than to vibrational energy levels. Physical models based on classical systems help in the understanding of both the wave number values and intensities of bands and lines in spectra.

I. PRINCIPLES OF VIBRATIONAL SPECTROSCOPY

It is convenient to discuss firstly the principles of vibrational spectroscopy in relation to diatomic molecules and then to extend the concepts developed to polyatomic molecules.

A. ABSORPTION SPECTRA OF DIATOMIC MOLECULES

I. Wave Number Values of Infrared Bands

The energy of a particular vibrational level is determined by solving the Schrödinger wave equation (p. 12) for the appropriate system. A potential-energy function E_x represented in terms of specified coordinates x may be substituted in equation 1.31 and the equation solved to give a set of vibrational wave functions ψ_V and energy levels E_V both of which are a function of an integer V. This integer is the vibrational quantum number. The choice of potential energy function is made on the basis of a consideration of a plausible classical model in terms of the experimental information on the probable way potential energy will vary with displacement of one atom relative to another. These considerations suggest two possible potential-energy functions which are based on either (a) a simple harmonic oscillator model or (b) a model which allows deviation from harmonic behaviour (p. 9).

The form of the potential-energy functions for these two models are given in Table 3.1 along with the expressions for the energy, E_V, of the system derived from the Schrödinger equation. The potential-energy function of an anharmonic oscillator can be represented as a power series in x (equation 3.1b) in which f_1, f_2, etc. are coefficients in the first, second and higher terms. This equation approximates to the Morse function or to the Lippincott function (p. 10).

Table 3.1. The frequency of infrared absorption bands

	(a) Harmonic oscillator	(b) Anharmonic oscillator	
Potential-energy function	$E_x = \frac{1}{2}fx^2$	$E_x = \frac{1}{2}f_1 x^2 - \frac{1}{3}f_2 x^3 + \ldots$	(3.1)
Quantum energy levels	$E_V = h\nu_e(V + \frac{1}{2})$	$E_V = h\nu_e(V + \frac{1}{2}) - h\nu_e x_e(V + \frac{1}{2})^2 + \ldots$	(3.2)
Zero point energy E_0	$E_0 = \frac{1}{2}h\nu_e$	$E_0 = \frac{1}{2}h\nu_e - \frac{1}{4}h\nu_e x_e$	(3.3)
Fundamental frequency $\nu_{0 \to 1} = \dfrac{1}{h}(E_1 - E_0)$	$\nu_{0 \to 1} = \nu_e$	$\nu_{0 \to 1} = \nu_e(1 - 2x_e)$	(3.4)
First overtone $\nu_{0 \to 2} = \dfrac{1}{h}(E_2 - E_0)$	$\nu_{0 \to 2} = 2\nu_e$	$\nu_{0 \to 2} = 2\nu_e(1 - 3x_e)$	(3.5)
First "hot" band $\nu_{1 \to 2} = \dfrac{1}{h}(E_2 - E_1)$	$\nu_{1 \to 2} = \nu_e$	$\nu_{1 \to 2} = \nu_e(1 - 4x_e)$	(3.6)

f = force constant of the bond
x = displacement from the equilibrium separation of the atoms
h = Planck's constant
V = vibrational quantum number
ν_e = equilibrium vibrational frequency of the molecule
x_e = anharmonicity constant for the vibration

The potential-energy functions, E_x, are represented in Fig. 3.1 by solid curves and the quantized energy, E_V is represented by horizontal lines.

A difference between the classical and quantum description of the vibrational energy of a diatomic molecule is that in the former, energy varies continuously with displacement of coordinate x (equation 3.1) whereas in the latter, energy varies discontinuously (equation 3.2). Quantum theory predicts zero-point energy (equation 3.3) which is of considerable significance in thermodynamic and kinetic studies.

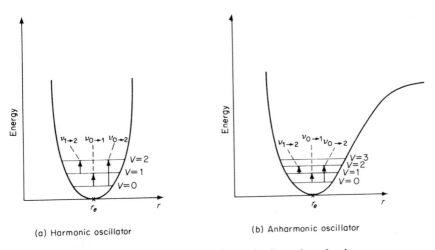

(a) Harmonic oscillator

(b) Anharmonic oscillator

Fig. 3.1. Potential energy functions of a diatomic molecule

In the simple harmonic oscillator approximation (a) the energy levels are given by equation 3.2a. In this equation ν_e is equivalent to a classical frequency and the equivalence emphasizes the dependence of the quantum treatment on a classical model. The frequency at which the fundamental band occurs is equal to the frequency at which the first "hot" band occurs, and is related to the force constant and reduced mass of the molecule (equation 1.5). The Maxwell–Boltzmann distribution law indicates that the population of the $V = 1$ and higher levels is negligible compared with population of the ground level ($V = 0$) except in cases where the energy differences between the $V = 0$ and $V = 1$ level is small (of the order 100 cm^{-1}). Therefore the number of "hot" band transitions ($\nu_{1 \to 2}$) will be very small compared with the number of fundamental transitions ($\nu_{0 \to 1}$) except for small energy differences between $V = 0$ and $V = 1$ levels. The first overtone band ($\nu_{0 \to 2}$) would be expected at twice the frequency of the fundamental band; the factors determining the intensities of overtones and fundamental bands will be considered in the next section.

5

Table 3.2. The intensity of infrared absorption bands

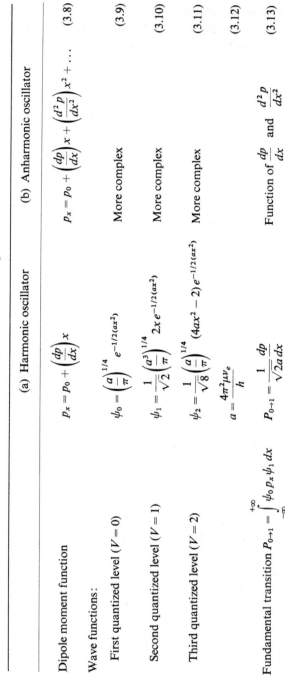

	(a) Harmonic oscillator	(b) Anharmonic oscillator	
Dipole moment function	$p_x = p_0 + \left(\dfrac{dp}{dx}\right)x$	$p_x = p_0 + \left(\dfrac{dp}{dx}\right)x + \left(\dfrac{d^2p}{dx^2}\right)x^2 + \cdots$	(3.8)
Wave functions:			
First quantized level ($V=0$)	$\psi_0 = \left(\dfrac{a}{\pi}\right)^{1/4} e^{-1/2(ax^2)}$	More complex	(3.9)
Second quantized level ($V=1$)	$\psi_1 = \dfrac{1}{\sqrt{2}}\left(\dfrac{a^3}{\pi}\right)^{1/4} 2x\, e^{-1/2(ax^2)}$	More complex	(3.10)
Third quantized level ($V=2$)	$\psi_2 = \dfrac{1}{\sqrt{8}}\left(\dfrac{a}{\pi}\right)^{1/4}(4ax^2 - 2)\,e^{-1/2(ax^2)}$	More complex	(3.11)
	$a = \dfrac{4\pi^2\mu\nu_e}{h}$		(3.12)
Fundamental transition $P_{0\to1} = \displaystyle\int_{-\infty}^{+\infty} \psi_0 p_x \psi_1\, dx$	$P_{0\to1} = \dfrac{1}{\sqrt{2a}}\dfrac{dp}{dx}$	Function of $\dfrac{dp}{dx}$ and $\dfrac{d^2p}{dx^2}$	(3.13)

First overtone transition $P_{0\to2} = \int_{-\infty}^{+\infty} \psi_0\, p_x\, \psi_2\, dx$ $\qquad P_{0\to2} = 0$ $\qquad$ Function of $\dfrac{dp}{dx}$ and $\dfrac{d^2p}{dx^2}$ $\qquad$ (3.14)

Selection rules $\qquad \Delta V = \pm1 \qquad \Delta V = \pm1, \pm2, \pm3$, etc. $\qquad$ (3.15)

Intensity of infrared band:

$$\left(\frac{dp}{dx}\right)^2 \quad \text{large—intense absorption}$$

$$\left(\frac{dp}{dx}\right)^2 \quad \text{small—weak absorption}$$

$$\left(\frac{dp}{dx}\right)^2 \quad \text{zero—no absorption}$$

p_0 = dipole moment at the equilibrium separation of the atoms
x = displacement from the equilibrium separation of the atoms
a = constant
μ = reduced mass of the molecule
ν_e = equilibrium vibrational frequency of the molecule
h = Planck's constant
V = vibrational quantum number

The values for the vibrational energy levels in the anharmonic oscillator are given by equation 3.2b and obviously differ from those in the harmonic oscillator approximation. These differences are particularly marked at higher values of the vibrational quantum number V. The classical analogy is that for large displacements, x, the potential energy function departs from the calculated value based on simple harmonic motion (equation 3.1a). The magnitude of the difference between the energy values for the two approximations depends on the magnitude of the coefficient of the cubic term in the potential-energy function (equation 3.2b) which involves x_e, the anharmonicity constant. If this constant has an appreciable positive value the fundamental frequency is less than ν_e (equation 3.4b) and the first overtone is less than $2\nu_e$ (equation 3.5b).

2. Intensity of Infrared Bands

The intensity of an infrared band associated with a transition from a lower level i to an upper level j is proportional to the square of the transition moment $P_{i \to j}$ (p. 23). The transition moment is given by the expression

$$P_{i \to j} = \int \psi_i p_x \psi_j \, d\tau \qquad (3.7)$$

where ψ_i and ψ_j are the wave functions of the i and j states and p_x is the dipole moment in the x direction. The dipole moment may be expressed as a linear function of x for small displacements (equation 3.8a) and a quadratic function of x for larger displacements (equation 3.8b). The wave functions of the three lowest vibrational energy levels are given in Table 3.2 for the harmonic oscillator model.

The transition moment for the transition from the $V = 0$ to the $V = 1$ levels and from the $V = 0$ and $V = 2$ levels may be calculated by substitution of the appropriate wave functions and dipole moment term in equation 3.7 and integrating this equation with respect to displacement along the molecular axis. The results of performing the operation using the wave functions for the harmonic oscillator are given in Table 3.2. It can be seen from the table that the transition moment for the fundamental transition $\nu_{0 \to 1}$ has a non-zero value (the value being proportional to the change in dipole moment with displacement in accord with the result from a classical description) while that for the first overtone transition ($\nu_{0 \to 2}$) is zero. These results can be generalized in the form of a selection rule $\Delta V = \pm 1$. This rule means that V must change by ± 1 for the transition to be active (allowed or permitted) in the infrared. Transitions between levels which are not adjacent are inactive (non-allowed or forbidden) according to the simple harmonic oscillator model.

For the anharmonic oscillator model the calculation of wave functions and transition moments is more complex because higher terms in the potential energy and dipole moment expressions must be taken into account. The effect of these higher terms is to relax the selection rules so that transitions between energy levels differing by one, two or three vibrational quantum number units may be observed ($\Delta V = \pm 1$, ± 2, ± 3). However, the bands are usually weak.

If the infrared spectrum reveals fundamental and overtone bands, the anharmonicity constant, x_e, may be calculated (Expt. V5). Departure from harmonic behaviour may be expressed by two effects.

(a) Mechanical Anharmonicity ($x_e \neq 0$). This arises from the effect of cubic and higher terms in the potential-energy expression (equation 3.2b).

(b) Electrical Anharmonicity ($P_{0 \to 2} \neq 0$). This arises from the effect of square and higher terms in the dipole-moment expression (equation 3.8b).

B. ROTATIONAL STRUCTURE OF INFRARED BANDS

I. Rotational Structure in Diatomic Molecules

The band in the infrared spectrum of carbon monoxide (Fig. 1.6) assigned to the fundamental stretching mode shows fine structure. This structure is due to simultaneous changes in vibrational and rotational energy levels. For a heteropolar diatomic molecule A—B in the gaseous state, transitions which lead to these changes are shown in Fig. 3.2 for the first four bands on each side of the centre of the band system.

The values of the energy levels in Fig. 3.2 may be calculated on the basis of the Born–Oppenheimer approximation which states that the molecules execute rotations and vibrations independently. The vibrational energy is given by equation 3.2b and is obtained by solving the Schrödinger equation for an anharmonic oscillator. The rotational energy may be obtained by solving the Schrödinger equation for a rotator in which the effect of centrifugal distortion at high rotational energies is taken into consideration. The energy E_J associated with the rotational quantum number J is given by the expression

$$E_J = hBJ(J + 1) - hDJ^2(J + 1)^2 \qquad (3.16)$$

where B is the rotational constant, h is Planck's constant and D is the centrifugal-distortion constant. The rotational constant B and the moment of inertia I are given by the following expressions

$$B = \frac{h}{8\pi^2 I} \qquad \tilde{B} = \frac{h}{8\pi^2 I c} \qquad (3.17)$$

$$I = \mu r^2 = \frac{m_A m_B r^2}{(m_A + m_B)} \qquad (3.18)$$

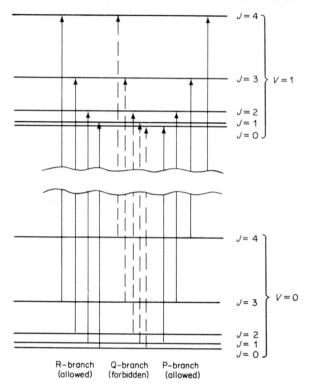

Fig. 3.2. Energy levels and associated vibration–rotation transitions for a molecule A—B

where c is the velocity of light, μ is the reduced mass, r is the internuclear separation and m_A and m_B are the masses of atoms A and B. If the constants in these equations are given values in appropriate units the numerical value of B will be in s^{-1} units and the numerical value of $\tilde{B}$ will be in cm^{-1} units. Similarly the numerical value of the centrifugal-distortion constant D may be in s^{-1} units or $\tilde{D}$ may be in cm^{-1} units.

The vibration–rotation energy $E_{V,J}$ of a diatomic molecule in a particular state V and J is given (according to the Born–Oppenheimer approximation) by the expression

$$E_{V,J} = E_V + E_J \tag{3.19}$$

Substitution of the values for E_V and E_J given by equations 3.2b and 3.16 in equation 3.19 gives

$$E_{V,J} = h\nu_e(V + \tfrac{1}{2}) - h\nu_e x_e(V + \tfrac{1}{2})^2 + hBJ(J + 1) - hDJ^2(J + 1)^2 \tag{3.20}$$

The values for the energy $E_{V,J}$ obtained from equation 3.20 have units of joules if the values of the other terms are in recommended SI units. It is often convenient to express the values of energy in units of cm^{-1} whence equation 3.20 would take the form

$$\tilde{E}_{V,J} = \tilde{\nu}_e(V + \tfrac{1}{2}) - \tilde{\nu}_e x_e(V + \tfrac{1}{2})^2 + \tilde{B}J(J + 1) - \tilde{D}J^2(J + 1)^2 \qquad (3.21)$$

The advantage of using equation 3.21 is that all the quantities are either in cm^{-1} units ($\sim$) or are dimensionless (V, x and J). It will be seen that subsequent manipulations may be more readily performed using expressions in this form. "Energy" expressions in which the values are in cm^{-1} units are sometimes described as *Term Values*. It may be seen that the following relationship exists

$$\tilde{E} = \frac{E}{hc} \qquad (3.22)$$

The wave number value, $\tilde{\nu}$, of any band arising from a transition between any two energy levels is given by

$$\tilde{\nu} = \tilde{E}_{V\text{upper, }J\text{upper}} - \tilde{E}_{V\text{lower, }J\text{lower}} \qquad (3.23)$$

The activity and selection rules for rotational energy changes are very similar to those for vibrational energy changes since both changes involve interaction between electromagnetic radiation and the dipole of a molecule. Similar methods to those leading to the results summarized in Table 3.2 for vibrational energy changes show that there must be a change in dipole moment during a rotation and the change in rotational quantum number must be ± 1. The selection rules for vibration–rotation transitions for an anharmonic oscillator are consequently $\Delta V = +1, +2, +3$, etc. and $\Delta J = \pm 1$.

If it is assumed that the rotational constant, $\tilde{B}$, has the same value in the vibrational states $V = 0$ and $V = 1$ and that the centrifugal-distortion constant, $\tilde{D}$, is negligibly small, then the wave number values of the absorption peaks corresponding to rotational structure of the fundamental vibrational transition may be calculated. The absorption peaks corresponding to the transitions from any rotational state $J = J^1$ in the vibrational state $V = 0$ to the rotational state $J = J^1 - 1$ in the vibrational state $V = 1$ (selection rule $\Delta V = +1$, $\Delta J = -1$) are described as the P branch. The wave number values $\tilde{\nu}_P$ of the peaks in the P branch are given by the expression

$$\tilde{\nu}_P = \tilde{E}_{V=1, J=J^1-1} - \tilde{E}_{V=0, J=J^1} \qquad (3.24)$$

Substitution of the appropriate expressions for $\tilde{E}$ into equation 3.24 gives

$$\tilde{\nu}_P = [\tfrac{3}{2}\tilde{\nu}_e - \tfrac{9}{4}\tilde{\nu}_e x_e + \tilde{B}(J^1 - 1)J^1] - [\tfrac{1}{2}\tilde{\nu}_e - \tfrac{1}{4}\tilde{\nu}_e x_e + \tilde{B}J^1(J^1 + 1)]$$
$$= \tilde{\nu}_e(1 - 2x_e) - 2\tilde{B}J^1 \qquad (3.25)$$

Similarly the absorption peaks corresponding to the transitions from any particular rotational state $J = J^1$ to the rotational state $J = J^1 + 1$ accompanying the fundamental vibrational transition (selection rule $\Delta V = +1$, $\Delta J = +1$) are described as the R branch for which the wave number values $\tilde{\nu}_R$ are given by

$$\tilde{\nu}_R = \tilde{E}_{V=1, J=J^1+1} - \tilde{E}_{V=0, J=J^1} \qquad (3.26)$$

Substitution for $\tilde{E}$ into equation 3.26 gives

$$\tilde{\nu}_R = [\tfrac{3}{2}\tilde{\nu}_e - \tfrac{9}{4}\tilde{\nu}_e x_e + \tilde{B}(J^1 + 2)(J^1 + 1)] - [\tfrac{1}{2}\tilde{\nu}_e - \tfrac{1}{4}\tilde{\nu}_e x_e + \tilde{B}J^1(J^1 + 1)]$$
$$= \tilde{\nu}_e(1 - 2x_e) + 2\tilde{B}(J^1 + 1) \qquad (3.27)$$

Equations 3.25 and 3.27 predict that the separation of adjacent peaks is $2\tilde{B}$ with a central gap of $4\tilde{B}$. It is observed in the vibration–rotation spectrum of carbon monoxide and other diatomic molecules that the spacing between adjacent peaks decreases as the wave number values of the peaks increases. This departure from the results predicted by equations 3.25 and 3.27 is a consequence of the limitations of the three approximations made. Firstly, the Born–Oppenheimer approximation is not strictly valid and an additional term should be included in equation 3.19 to take account of interaction between vibration and rotation energy levels. Secondly, the assumption that the internuclear separation does not increase due to centrifugal distortion at higher rotational energies is not strictly valid. Thirdly, it is clear from the form of the potential-energy curve of an anharmonic oscillator (Fig. 3.1b) that the mean internuclear separation, r_1, in the vibrational state $V = 1$ will be greater than the mean internuclear separation, r_0, in the vibrational state $V = 0$.

If the consequences of the third approximation are examined it follows that expressions for $\tilde{E}_{V,J}$ which are substituted in equation 3.23 should contain different rotational constants for the different vibrational levels. The rotational constant, $\tilde{B}_1$, in the vibrational state $V = 1$ will clearly be smaller than the rotational constant, $\tilde{B}_o$, in the vibrational state $V = 0$. The expressions for the wave number values of the peaks in the P and R branches may be obtained as

$$\tilde{\nu}_P = \tilde{\nu}_e(1 - 2x_e) - (\tilde{B}_1 + \tilde{B}_0)J^1 + (\tilde{B}_1 - \tilde{B}_0)J^{1^2} \qquad (3.28)$$

$$\tilde{\nu}_R = \tilde{\nu}_e(1 - 2x_e) + 2\tilde{B}_1 + (3\tilde{B}_1 - \tilde{B}_0)J^1 + (\tilde{B}_1 - \tilde{B}_0)J^{1^2} \qquad (3.29)$$

Equations 3.28 and 3.29 become equal to equations 3.25 and 3.27 when the rotational constants in the $V = 0$ and $V = 1$ levels are assumed to be equal.

The values of $\tilde{B}_1$ and $\tilde{B}_0$ for a diatomic molecule A—B may be obtained by fitting observed wave number values for the P and R branches to equations 3.28 and 3.29.

In Expt. V5 the vibration–rotation spectrum of HCl vapour is recorded. From the measurements of the wave number values of the peaks in the P and R branches, the parameters $\tilde{\nu}_e$, x_e, f, $\tilde{B}$ (assuming $\tilde{B} = \tilde{B}_0 = \tilde{B}_1$), I and r (assuming $r = r_0 = r_1$) are calculated. In addition a method is outlined for the calculation of $\tilde{B}_0$ and $\tilde{B}_1$ and hence r_0 and r_1.

2. Rotational Structure in Polyatomic Molecules

Examination of the two fundamental infrared bands of CO_2 reveals that the band at higher wave number value has a PQR structure very similar to that of CO but with a smaller spacing between the sub-bands (see Expt. V6). The

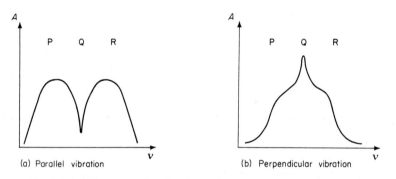

(a) Parallel vibration   (b) Perpendicular vibration

Fig. 3.3. Band contours for vibrations of a linear polyatomic molecule

fundamental band at lower wave number value has a prominent central maximum corresponding to a Q branch. The assignments of these bands indicate that the vibration *parallel* to the molecular axis has a missing Q branch and the vibration *perpendicular* to the molecular axis has a prominent Q branch. These results may be generalized for all linear molecules and are consistent with the selection rules $\Delta J = \pm 1$ for parallel vibrations and $\Delta J = 0, \pm 1$ for perpendicular vibrations.

The structures of infrared bands of linear molecules which have large moments of inertia are difficult to resolve. However, the contour of a vibration–rotation band for a linear molecule indicates whether the vibration is of the parallel or perpendicular type (Fig. 3.3).

The vibration–rotation spectra of non-linear molecules are more difficult to analyse. This is partly because it is necessary to consider the moments of inertia about each of the three principal axes of the molecule. Methods have been suggested for calculating the contours of bands arising from vibrations parallel to each axis (Badger and Zumwalt, 1938). The contours predicted by these methods may be compared with those which are observed experimentally. This is a valuable aid to assigning bands in the spectra of vapours.

C. THE RAMAN EFFECT

I. Raman Frequency Shifts

When monochromatic radiation from a suitable source of visible or ultraviolet light is incident upon a sample the radiation may be transmitted, absorbed or scattered by the sample. Scattered radiation may be caused by small particles of matter (Tyndall scattering) or by molecules (Rayleigh scattering); in both cases it has long been demonstrated that the frequency of the scattered radiation is the same as the frequency of the incident radiation.

In 1928 Raman discovered that molecular scattering of monochromatic radiation produces not only scattered radiation of unchanged frequency (Rayleigh scattering), but also a small proportion of scattered radiation at

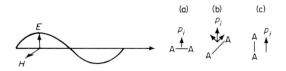

Fig. 3.4. Dipoles induced in a diatomic molecule at various possible orientations to incident radiation

different frequencies (Raman scattering). The difference in frequency between that of incident radiation and that of Raman-scattered radiation is known as the "Raman frequency shift". Raman shifts may only be observed from transparent, dust-free samples (liquids, solutions or—with more difficulty—gases and solids) in which the frequency of any absorption band is well-removed from that of the incident radiation. The magnitudes of the various Raman shifts observed for a particular sample are a property of the sample and are independent of the frequency of the incident radiation. These shifts in frequency correspond to frequencies of rotational or vibrational transitions of the scattering molecules which constitute the sample. In the case of a diatomic molecule only one vibrational Raman line is observed—that from the single fundamental stretching vibration. A number of rotational Raman lines may be observed from diatomic molecules in the gaseous state.

The Raman effect may, to some extent, be explained in terms of classical theory. Consider electromagnetic radiation incident upon a homonuclear diatomic molecule. Three of the many possible orientations of the molecules with respect to the incident radiation are shown in Fig. 3.4. If the frequency of the incident radiation is markedly different from the frequency of any absorption band in the spectrum of the molecule, then the oscillating electric field E will induce the electrons in the A—A bond to oscillate in phase. This leads to an oscillating dipole moment in the molecule A—A. The various induced dipole moments, p_i, are shown vectorially in Fig. 3.4 for the three chosen orientations. The magnitude and direction of the induced dipole moment is given by

$$p_i = \alpha E \qquad (3.30)$$

where α is the polarizability of the molecule and E is the amplitude of the electric vector of the incident radiation. Polarizability is a measure of the ease with which an electron can follow an electric field. The induced dipole moment is greater in orientation (c) than in (a) (Fig. 3.4) because electrons can be displaced along a bond more easily than across a bond. Polarizability is clearly an anisotropic property and will be discussed more fully in relation to polyatomic molecules.

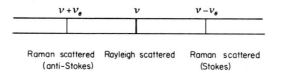

Raman scattered   Rayleigh scattered   Raman scattered
(anti-Stokes)   (Stokes)

Fig. 3.5. Vibrational Raman spectrum of a diatomic molecule

If a diatomic molecule is vibrating with a frequency ν_e it may be assumed that the polarizability will vary in a harmonic manner and will, therefore, vary with time according to the expression

$$\alpha = \alpha_0 + b \sin 2\pi\nu_e t \qquad (3.31)$$

where α_0 and b are constants and t is the time. The oscillating field, E, is chosen to occur at a much larger frequency ν than ν_e. The magnitude of the field as a function of time is given by

$$E = E_0 \sin 2\pi\nu t \qquad (3.32)$$

where E_0 is the peak amplitude of the oscillating field. Substitution of equations 3.31 and 3.32 into equation 3.30 gives the following expression for the variation with time of the magnitude of the dipole moment induced in the molecule

$$p_i = \alpha_0 E_0 \sin 2\pi\nu t + b E_0 \sin 2\pi\nu_e t \sin 2\pi\nu t \qquad (3.33)$$

Using the trigonometrical expression for the product of the sine of two angles equation 3.33 may be reduced to

$$p_i = \alpha_0 E_0 \sin 2\pi\nu t + \tfrac{1}{2}b E_0 \{\cos 2\pi(\nu - \nu_e)t - \cos 2\pi(\nu + \nu_e)t\} \qquad (3.34)$$

The oscillating dipole will emit radiation and equation 3.34 indicates that the emitted radiation will have components with frequencies ν, $\nu - \nu_e$ and $\nu + \nu_e$. The spectrum of the emitted radiation would contain three bands or lines at these frequencies and are referred to as Rayleigh, Raman (Stokes) and Raman (anti-Stokes) lines respectively. These lines are shown diagrammatically in Fig. 3.5.

The classical Rayleigh theory of light scattering shows the intensity of scattered radiation is proportional to the fourth power of the frequency (hence blue radiation of sunlight is scattered preferentially to other visible radiation by the Earth's atmosphere to impart a blue colour to a clear sky). This suggests that anti-Stokes Raman lines should be more intense than Stokes Raman lines.

Experimental results indicate that, contrary to classical predictions, the anti-Stokes Raman lines are very much weaker than the Stokes Raman lines particularly when the shift is large (> 1000 cm^{-1}). When the shift is small (≈ 10–100 cm^{-1}) the anti-Stokes and Stokes lines are of comparable intensity. Low-frequency shifts may be observed as a result of rotational changes in the scattering molecule. The existence of rotational Raman lines cannot be adequately explained by classical theories.

As in other contexts, quantum theory is required to provide an adequate description of the Raman phenomena. In fact the effect was predicted on this basis by Smekal in 1923 prior to its discovery by Raman.

Consider a diatomic molecule in one or other of the two lowest vibrational energy levels E_0 or E_1 (Fig. 3.6). In quantum terms an incident photon possesses energy ϵ_i given by the expression

$$\epsilon_i = h\nu_i \tag{3.35}$$

where h is Planck's constant and ν_i is the frequency of the incident radiation. On collision, the photon may perturb the molecule and the total combined energy of the molecule and photon, E_T, at the moment of collision is given by

$$E_T = \epsilon_i + E_V \tag{3.36}$$

where E_V is either E_0 or E_1. The energy of the photon scattered as a result of the collision with the sample molecule is given by

$$\epsilon_s = h\nu_s \tag{3.37}$$

where ν_s is the frequency of scattered radiation. This process can occur in four possible ways two of which correspond to elastic and the other two to non-elastic collisions (Fig. 3.6). Consideration of Fig. 3.6 shows that the magnitude of the Raman shift is related to the difference in energy between the levels E_0 and E_1. The values of the energy levels of a simple diatomic vibrator are given in Table 3.1. If the molecule is considered to be an anharmonic oscillator the Raman shifts in units of cm^{-1} will be given as

$$\text{Stokes} \quad \Delta\tilde{\nu} = \tilde{\nu}_i - \tilde{\nu}_s = \tilde{E}_1 - \tilde{E}_0 = \tilde{\nu}_e(1 - 2x_e) \tag{3.38}$$

$$\text{anti-Stokes} \quad \Delta\tilde{\nu} = \tilde{\nu}_s - \tilde{\nu}_i = \tilde{E}_1 - \tilde{E}_0 = \tilde{\nu}_e(1 - 2x_e) \tag{3.39}$$

where the terms in equations 3.38 and 3.39 have the same significance as in Table 3.1; the symbol ($\sim$) indicates the values are expressed in cm^{-1} units. It follows that the magnitude of the Raman shift corresponds to the wave number of the absorption band which would be observed if the particular transition was active in the infrared spectrum. It is normal practice to use the term "Raman line" or "Raman band", $\tilde{\nu}$, rather than "Raman shift", $\Delta\tilde{\nu}$.

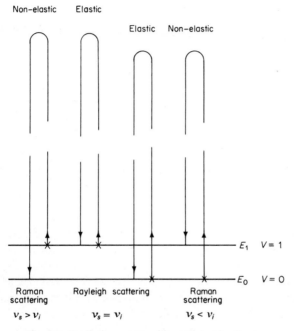

Non-elastic   Elastic

Elastic   Non-elastic

E_1 $V = 1$

E_0 $V = 0$

Raman   Rayleigh scattering   Raman
scattering   scattering

$\nu_s > \nu_i$ $\nu_s = \nu_i$ $\nu_s < \nu_i$

Fig. 3.6. Radiation scattered by a diatomic vibrator

The relative intensities of the Stokes and anti-Stokes Raman lines is determined by the relative population of the E_0 and E_1 levels. The relative population is determined by the Maxwell–Boltzmann distribution law (p. 8). When the difference in energy level is large ($E_1 - E_0 > 1000$ cm^{-1}) the population of the E_1 level is very much smaller than the population of E_0; when the difference in energy levels is small ($E_1 - E_0 \approx 10$–100 cm^{-1}) the populations of the two levels are comparable. The quantitative predictions of the Maxwell–Boltzmann law are in agreement with the experimental measurements. A quantum description of the Raman effect is, therefore, better than a classical description.

Factors which influence the choice of wave number, $\tilde{\nu}_i$, of exciting radiation will be considered in the experimental section of this chapter. Raman data is less easily obtained than infrared data because of the weak second-order nature of Raman scattering. There is, therefore, no apparent advantage in studying the vibrational spectrum of CO by this method since the information that the CO stretching vibration occurs at 2143 cm^{-1} can be obtained more readily by infrared absorption (Fig. 1.6). However, the Raman spectrum of O_2 and any other homopolar diatomic does provide information which cannot be obtained by infrared absorption studies because there is no change in dipole moment during the vibration, which is consequently infrared inactive. Furthermore, rotational lines are obtained in the Raman spectrum but there are no rotational absorption bands in the microwave spectrum of homopolar diatomic molecules because the transitions are also inactive. The selection rule for rotational Raman spectra of diatomic molecules is

$$\Delta J = \pm 2 \qquad (3.40)$$

Rotational transitions in diatomics are all Raman active because the polarizability and hence the *induced* dipole moment changes during a rotation. Rotational Raman spectra are not considered in any further detail in this book.

2. Polarizability

The idea of the polarizability of any molecule is a more difficult concept than that of some simple vector quantities which have been considered so far. Examples of the latter are an electromagnetic field and the displacement of atoms and electrons in a molecule. It is necessary to describe what we mean by polarizability before certain important aspects of the Raman effect can be understood, in particular the intensity and state of polarization of Raman lines.

An oscillating field, E, will induce an oscillating dipole moment p_i in a molecule. The magnitude of the induced dipole moment is proportional to the magnitude of the field (equation 3.30).

If a molecule is at any general orientation with respect to a set of space fixed cartesian coordinates, then the dipole moment, p_i, of the molecule and the amplitude of the oscillating electric field, E, can be represented as vectors which have three components in the direction of the axes. It is important to note that the vectors representing p_i and E generally have different directions. In other words there will be certain directions within a molecule in which electrons may be preferentially displaced by an electric field and these directions will not necessarily be in the direction of the electric field.

It follows that each of the three components of E contribute to each component of p_i by an amount which is determined by the components of polarizability for the molecule. Thus the induced dipole moment in the x direction, $p_{i(x)}$, is the sum of contributions of the dipole moment induced in the x

direction by each component of the field (equation 3.30a). $p_{i(y)}$ and $p_{i(z)}$ are expressed similarly and equation 3.30 represents a set of linear equations

$$p_{i(x)} = \alpha_{xx} E_x + \alpha_{xy} E_y + \alpha_{xz} E_z \qquad (3.30a)$$

$$p_{i(y)} = \alpha_{yx} E_x + \alpha_{yy} E_y + \alpha_{yz} E_z \qquad (3.30b)$$

$$p_{i(z)} = \alpha_{zx} E_x + \alpha_{zy} E_y + \alpha_{zz} E_z \qquad (3.30c)$$

It is clear that polarizability cannot be represented in terms of three components in the way the two *associated vectors* to which it relates may be described. This distinction classifies polarizability as a *tensor* property which must be expressed in terms of nine components. Equation 3.30 is strictly a matrix relationship which can be expanded into the set of linear equations 3.30a, b and c. The polarizability tensor, α, is a square matrix

$$\alpha = \begin{vmatrix} \alpha_{xx} & \alpha_{xy} & \alpha_{xz} \\ \alpha_{yx} & \alpha_{yy} & \alpha_{yz} \\ \alpha_{zx} & \alpha_{zy} & \alpha_{zz} \end{vmatrix} \qquad (3.41)$$

It can be shown that the effect of the component of the field in the y direction on the component of the induced dipole moment in the x direction is the same as the effect of the component of the field in the x direction on the component of the induced dipole moment in the y direction. Thus $\alpha_{yx} = \alpha_{xy}$, and similarly $\alpha_{zx} = \alpha_{xz}$ and $\alpha_{zy} = \alpha_{yz}$. The polarizability matrix is, therefore, symmetrical about its diagonal and the number of different entries is six.

3. Intensity of Raman Lines

The intensity of any Raman line for any molecule is determined by the transition moment (p. 23). For a diatomic molecule aligned along the x axis the transition moment becomes

$$P_{0 \to 1} = \int \psi_0 \, \alpha_{xx} \, \psi_1 \, dx \qquad (3.42)$$

This expression is similar to that for an infrared transition and the polarizability can be expanded as a function of displacement, x, in a manner similar to that for the dipole moment (equation 3.8) to give the expression

$$\alpha_x = \alpha_0 + \left(\frac{d\alpha}{dx}\right) x \qquad (3.43)$$

where α_0 is the value of the polarizability in the equilibrium configuration, and $d\alpha/dx$ is the change in polarizability with respect to displacement for small displacements x.

Substituting values of ψ_0 (equation 3.9) and ψ_1 (equation 3.10) and α_x (equation 3.43) into equation 3.42 provides an expression which can be integrated to give the transition moment

$$P_{0\to1} = \frac{1}{\sqrt{2a}}\frac{d\alpha}{dx} \tag{3.44}$$

where a is defined in Table 3.2.

In any diatomic molecule the stretching of the bond will be associated with a decrease in the charge density of bonding electrons. This means that the polarizability of the molecule will change continuously as the molecule vibrates. Hence diatomic molecules which are either heteropolar (A—B) or homopolar (A—A) have a finite value for $d\alpha/dx$ and hence a non-zero value for the transition moment $P_{0\to1}$. Consequently the Raman spectrum would contain a line associated with a fundamental vibrational mode.

If an overtone is observed in the Raman spectrum of a diatomic molecule, the anharmonicity constant, x_e, may be determined. The mechanical contribution to anharmonicity arises in the same way as in infrared absorption but the electrical contribution depends on terms in x^2 and higher powers of x which should strictly be included in equation 3.43. Overtone lines or bands are rarely observed in the Raman effect. It is considered that this is partly due to the weak second-order nature of the effect and partly because higher terms in the polarizability expression are less significant than higher terms in the dipole moment expression.

4. Polarization of Raman Lines

If, in a Raman experiment, the radiation incident on the sample is unpolarized and travelling in the z direction, the electric vectors of many different waves are randomly orientated in the xy plane (Fig. 3.7a). It is possible to polarize the radiation in any plane perpendicular to the xy plane and the electric vectors of the incident radiation polarized in the xz plane and in the yz plane are shown in Fig. 3.7b and c.

The degree of depolarization, ρ, may be defined as

$$\rho = \frac{I_\parallel}{I_\perp} \tag{3.45}$$

where $I_\parallel$ is the intensity of a Raman line produced using incident radiation polarized parallel to the cell (Fig. 3.7c) and $I_\perp$ is the intensity of the same Raman line produced using incident radiation polarized perpendicular to the cell (Fig. 3.7b). The values obtained for this parameter provide information

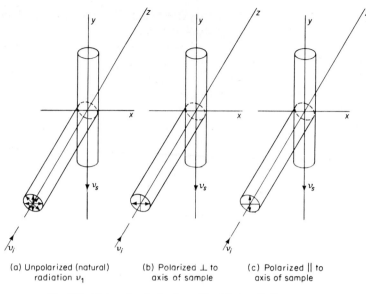

(a) Unpolarized (natural)   (b) Polarized ⊥ to   (c) Polarized ‖ to
    radiation ν_1 axis of sample   axis of sample

Fig. 3.7. States of polarization of incident radiation

on molecular vibrations which cannot be obtained in any other way. This is
summarized in Table 3.3.

The reasons for these results may be considered briefly as follows. Consider
firstly a totally symmetric vibration of an isotropic (spherically symmetrical)
molecule. Whatever the orientation of the molecule with respect to the direction
of the electric vector of the incident radiation, the *value* of the induced dipole
moment will be the same. However, the *direction* of the induced dipole moment
will be the same as the direction of the electric vector of the incident radiation.
The scattered radiation is perpendicular to the dipole moment induced in the
molecule. The intensity of the scattered radiation is, therefore, dependent on

Table 3.3

Degree of depolarization (ρ)	Type of vibration assigned to Raman line
0	Totally symmetric vibration of an isotropic molecule (e.g. CCl_4)
0–3/4	Totally symmetric vibration $\rho \rightarrow 0$ for molecules which are nearly isotropic $\rho \rightarrow 3/4$ for molecules which are highly anisotropic
3/4	Non-totally symmetric vibrations

the plane of polarization of the incident radiation. Figure 3.8 shows the dipole moment induced by incident radiation which is polarized firstly in a direction perpendicular to the cell and secondly in a direction parallel to the cell. Incident radiation polarized in the zx plane will be scattered in the zy plane. Incident radiation polarized in the zy plane will be scattered in the zx plane. If the scattered radiation is examined in the y direction, the intensity of the Raman line due to the totally symmetric vibration will have some finite value in the former case but will be zero in the latter case. The degree of depolariza-

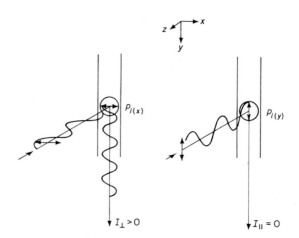

Fig. 3.8. Radiation scattered by a totally symmetric vibration of a spherically symmetrical molecule

tion $(\rho = I_{\parallel}/I_{\perp})$ will, therefore, be zero and the scattered radiation will be completely polarized.

Consider secondly a non-totally symmetric vibration in any molecule. The direction of oscillation of the induced dipole moment will depend on the orientation of the molecule which will be continually changing. Polarized incident radiation will be depolarized on being scattered by the sample. By averaging over all possible orientations it may be shown (Placzek, 1934) that the limiting value for the degree of depolarization is 3/4; hence the radiation is completely depolarized.

Consider thirdly a totally symmetric vibration of any molecule other than a spherically symmetric molecule. The degree of depolarization can have any value between the two possibilities described, and the scattered radiation will be partially polarized. If the molecule is of low symmetry, i.e. it tends to be highly anisotropic, then the scattered radiation tends to be only weakly polarized and ρ tends to 3/4. If the molecule is of high symmetry, i.e. it tends to be isotropic, then the scattered radiation tends to be strongly polarized and

ρ tends to zero. It follows that measurement of the degree of depolarization of vibrational Raman lines for any molecule provides information on the symmetry of the vibration associated with the line.

D. VIBRATIONAL SPECTRA OF POLYATOMIC MOLECULES

I. Normal Modes of Vibration

A polyatomic molecule can be regarded as a system of masses joined by bonds with spring-like properties. Each mass can be displaced in any direction and this displacement can be specified by three Cartesian coordinates. This means that each mass has three degrees of freedom and if the total number of masses is N then the total number of degrees of freedom of the system is $3N$. The $3N$ degrees of freedom are made up as in Table 3.4. A distinction between a

Table 3.4

Degrees of freedom	Non-linear molecule	Linear molecule
Translational	3	3
Rotational	3	2
Vibrational	$3N - 6$	$3N - 5$
Total	$3N$	$3N$

Table 3.5

Degrees of freedom	In-plane	Out-of-plane	
Translational	2	1	
Rotational	1	2	
Vibrational	$2N - 3$	$N - 3$	
	$2N$	N	Total $3N$

non-linear and a linear molecule arises because the rotational motion about the axis along the bonds of a linear molecule does not involve any detectable energy change.

The degrees of freedom of molecules which have only one plane of symmetry may be classified in terms of motion within and without the plane, as in Table 3.5.

It may be shown by constructing mechanical models or by a mathematical analysis that a system of masses joined by springs can only vibrate in certain modes, known as "normal" or "fundamental" modes of vibration. Each mode corresponds to a degree of vibrational freedom and has a characteristic frequency and form of vibration. By coupling a mechanical system to an oscillator of variable frequency it is possible to observe the successive excitations of each mode by scanning through the full range of frequencies. In general, if any mechanical or electrical system is coupled to an oscillator, the oscillator will only induce sympathetic vibrations in the system at certain natural frequencies. Typical examples of this phenomena occur when a tuning fork is induced to vibrate by sound waves of the correct frequency; when a bridge is induced to vibrate by a column of soldiers marching in step; or when a tuned radio circuit receives signals of the correct frequency. When an oscillator induces a sympathetic vibration in any system, the oscillator and the vibrator are said to be in resonance.

Normal modes of vibration (or fundamental modes of vibration) of any molecule are internal atomic motions in which all the atoms move in phase with the same frequency but with different amplitudes. The amplitude and direction of each atom may be represented by a displacement vector. The various displacements of the atoms in a given normal mode of vibration may be represented by a linear combination of the displacements of all atoms. For a particular normal mode this combination of atomic displacements is known as the normal coordinates, Q. There are, of course, $3N - 6$ ($3N - 5$ for a linear molecule) normal coordinates. In the case of a diatomic molecule the normal coordinate takes the form of a displacement between the atoms.

The intensity of the infrared band due to the nth normal mode depends on the square of the rate of change of dipole moment, p, of the molecule with respect to change in Q. The intensity of the Raman line due to the nth normal mode depends on the square of the rate of change of polarizability, α, of the molecule with respect to change in Q. A summary of these results for the nth normal mode of vibration is as follows:

Frequency	Normal Coordinate	Infrared Intensity	Raman Intensity
ν_n	Q_n	$\propto \left(\dfrac{dp}{dQ}\right)^2$	$\propto \left(\dfrac{d\alpha}{dQ}\right)^2$

An alternative statement of these results is that a vibration is *infrared* active if the *dipole moment* changes during the vibration and is Raman active if the *polarizability* changes during the vibration.

The approximate magnitudes of these changes can only be predicted for small molecules in which the approximate forms of the normal modes are known. In general the activity of normal modes of vibration in infrared and

Table 3.6. Distribution of normal modes of vibration for some selected molecules with C_{2v} symmetry

Number of modes belonging to each species

Species	bent triatomics (e.g. H_2O, H_2Se, OF_2, OCl_2, NO_2)						Activity
A_1	2	3	13	5	5	11	IR + R(pol.)
A_2		0	4	2	1	3	R(depol.)
B_1		1	7	1	2	6	IR + R(depol.)
B_2	1	2	12	4	4	10	IR + R(depol.)
Total ($3N - 6$)	3	6	36	12	12	30	

Raman spectra are predicted using symmetry concepts. The application of symmetry concepts to this problem was illustrated on page 24 for water, which belongs to the C_{2v} point group. It is instructive to consider a number of other molecules which belong to the C_{2v} point group (Table 3.6). The number of normal modes which occur in each symmetry species may be calculated (Wilson *et al.* 1955; Herzberg, 1945). These are listed in Table 3.6 for each molecule considered. The form of the normal coordinate for each normal mode may be calculated on the basis of the mechanical properties of a suitable model in which atoms are represented by point masses and bonds are represented by weightless springs. The approximate forms of the normal coordinates of water, or any other bent molecule of the type AB_2 are shown in Fig. 3.9. The approximate forms of the normal coordinates of the formate ion (which also has C_{2v} symmetry if the C—O bonds are assumed to be equivalent) may be predicted from consideration of the properties of C—H bonds (p. 145) and coupled vibrations of equivalent C—O bonds (p. 153); these are included in Fig. 3.9 for a general molecule ABC_2 possessing C_{2v} symmetry. In addition to molecules belonging to the C_{2v} point group normal modes of vibration of other types of molecules belonging to other point groups are included in Fig. 3.9. The approximate forms of the normal coordinates of each mode is also shown in Fig. 3.9 together with the symmetry species of that mode and its activity with respect of infrared or Raman. The modes of vibration which occur in the same symmetry species as the symmetric components of the polarizability tensor (the totally symmetric species) lead to polarized Raman lines (pol.); modes of vibration which occur in the same symmetry species as one or more of the antisymmetric components of the polarizability tensor lead to depolarized Raman lines (depol.). In Expts V6, V7 and V8 the vibrational spectra of some selected molecules are interpreted in terms of the symmetry properties of those molecules. The essential symmetry properties of the molecules which are considered are summarized in Fig. 3.9.

One useful generalization is the Rule of Mutual Exclusion. This rule states that if a molecule has a centre of symmetry those vibrations which are infrared active are Raman inactive and those vibrations which are Raman active are infrared inactive. In other words no vibration should appear in both the infrared spectrum and the Raman spectrum of a molecule possessing a centre of symmetry. Certain molecules possessing a centre of symmetry have modes of vibration which are inactive in both the infrared and Raman spectra. For example benzene contains two vibrations in the B_{2g} species, two in the B_{2u} species and two pairs in the E_{2u} species. It may be seen (Table 1.9) that these species contain neither a component of the dipole moment nor a component of the polarizability. These eight vibrations should not, according to symmetry considerations, be observed in the vibrational spectrum of benzene.

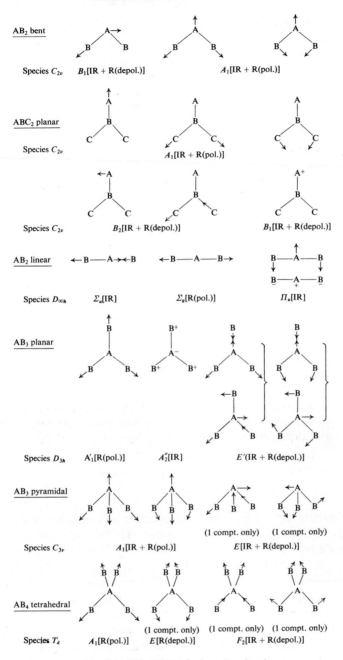

Fig. 3.9. Vibrations of selected molecules

2. Degenerate Vibrations

The bending mode of a linear triatomic molecule is shown in Fig. 3.9 as a doubly degenerate vibration. A doubly degenerate vibration is one for which there are two normal coordinates corresponding to a single frequency of vibration. It is instructive to compare the bending and rotational motions (about the axis along the bonds) of a linear triatomic molecule with the corresponding motions of a non-linear triatomic molecule.

It was noted that a linear molecule has one more degree of vibrational freedom and one less degree of rotational freedom than a non-linear molecule.

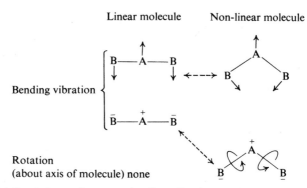

Fig. 3.10. Relation between degenerate bending vibrations and rotational motions in a triatomic molecule

For a non-linear triatomic it is readily apparent (Fig. 3.10) that a rotational degree of freedom in which the terminal atoms are moving in a negative direction (at right angles to the plane of the molecule) and the central atom is moving in a positive direction is equivalent to a bending vibration in a linear molecule. This bending vibration corresponds to another bending vibration at right angles to the former vibration. The two vibrations constitute a degenerate pair.

An important difference between degenerate and non-degenerate species exists when the symmetry operations comprising the particular point group to which the molecule belongs are applied to any directional property which belongs to that species. A degenerate directional property is transformed into a linear combination of the components of that directional property and the complete transformation produced by the application of the symmetry operation is represented by a matrix. The *character* of a particular transformation known as the *irreducible representation* is obtained by summing the diagonal entries of the matrix. In the case of a species which is non-degenerate any directional property within that species is transformed into the original directional property or into the inverse of the original directional property. This transformation reduces to a one-dimensional matrix for which the character is +1 or −1 for directional properties in which the direction is retained or inverted respectively.

The symbol used to describe a species of any point group which is degenerate is Π for a linear system, E for a doubly degenerate species and F for a triply degenerate species. Whenever energy states are degenerate their multiplicities must be taken into account in calculations involving the numbers of such states. For example, the number of vibrational modes in the AB_3 and AB_4 systems (Fig. 3.9) total $3N - 6$ when degeneracies are taken into account.

The planar and pyramidal AB_3 systems both possess a three-fold rotational axis of symmetry. In such cases molecules always possess doubly degenerate vibrations. Both components of the degenerate vibration are shown in Fig. 3.9 for the linear AB_2 system and the planar AB_3 system but only one component of the degenerate vibrations of the other systems are shown.

Frequently a band which is due to a degenerate vibration is split into its components by interaction between the molecules and some external field. The solid-state spectra of the SO_4^{2-} ion shows differences when compared with solution spectra (see Expt. V8) which are attributable to the splitting of degenerate energy levels by the crystal field within the solid lattice.

3. Combination Bands and Fermi Resonance

So far only fundamental modes of vibration of polyatomic molecules have been considered, although the presence of overtone bands in the spectra of diatomic molecules has been noted. In a polyatomic molecule, overtones may be observed with wave number values which are approximately twice those of the corresponding fundamentals. Also sum (or difference) combinations may appear at wave number values which are approximately the sum (or difference) of wave number values of the component fundamentals. These are all examples of binary combinations. Ternary combinations may be formed in similar ways with an increase in the number of possible combinations. The symmetry of any combination band may be determined by combining the symmetry properties of its components; this determines the activity of the combinations in the infrared and Raman spectrum. When permitted, these binary combinations frequently appear in the infrared spectrum with intensities of up to about five per cent of the fundamental components, but are usually too weak to be observed in the Raman spectrum. Ternary and higher combination bands are much weaker than binary combination bands, but can frequently be obtained in the infrared spectra of compounds if a sufficiently long path length of sample is used.

If a binary combination and a fundamental mode belong to the same symmetry species and have similar wave number values, interaction can occur leading to a pair of bands with comparable intensity values. This interaction is termed "Fermi resonance" and when this occurs the resultant bands each contain a contribution from each component of the interaction. The phenomena may be illustrated by the vibrational energy levels of carbon dioxide (Fig. 3.11). There are of course a different stack of vibrational energy levels corresponding to each fundamental mode of vibration.

The spectrum of CO_2 is considered in Expt. V6. Various overtone and combination bands in CCl_4 are examined in Expt. V9 in which Fermi resonance effects are also considered. The possibility of Fermi resonance should always be taken into account in molecules which show spectra with doublet bands in regions where only one band may otherwise be expected. The resonance may be considered in terms of intensity borrowing by a weak combination band

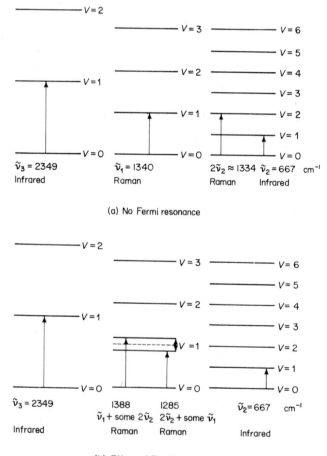

Fig. 3.11. Vibrational energy levels of carbon dioxide

from a near coincident fundamental followed by an increase in the separation of the energy levels.

4. Molecules Possessing Low Symmetry

Large molecules are likely to have fewer elements of symmetry than small molecules. Certain substituted benzene and ethylene compounds may have a plane of symmetry because the benzene and ethylene groups are themselves planar. Molecules with only a plane of symmetry belong to the C_s point group, The vibrations of such molecules fall into two species A' [in-plane vibrations. infrared and Raman active (polarized)] and A'' [out-of-plane vibrations, infrared and Raman active (depolarized)]. Out-of-plane vibrations of hydrogen and other groups will lead to large changes in dipole moments and will, therefore, give rise to intense bands in the infrared spectrum (see Expts V9, V12 and V13). Furthermore, these vibrations will not interact with the numerous in-plane vibrations because of the different symmetries of the vibrations. The wave number values of the out-of-plane vibrations will be characteristic of the arrangement of substituents in the molecule.

Molecules with no symmetry element other than the identity E should have all vibrations active in both the infrared and Raman spectra. Usually certain of the vibrations will lead to very weak bands or lines, others will overlap and some will occur at low wave number values which may be difficult to measure. It is, therefore, difficult to assign an observed spectrum to all the fundamental modes of vibration of a molecule.

An infrared spectrum of a polyatomic molecule is particularly irregular because the spectrum contains combination bands which may overlap with the bands due to fundamental modes of vibration, and the corresponding vibrations may interact with one another and distort the shape of the bands. Raman spectra show less irregularities because combination bands are less prominent. This is probably because electrical anharmonicity is more significant in the infrared absorption phenomena than in the Raman effect.

Although large molecules produce complex vibrational spectra, considerable information can still be obtained from these spectra. This is because certain vibrations are effectively localized in particular chemical groups and occur at wave number values and with band intensities characteristic of particular chemical groups.

E. CHARACTERISTIC GROUP VIBRATIONS

If the infrared spectra of a large number of different types of carbonyl compounds are examined then a strong band is observed in the region 1650–1850 cm^{-1} in each case. If the class of carbonyl compounds is restricted to aliphatic mono ketones the strong band occurs near 1719 cm^{-1}. These observations imply that a particular band in the spectrum of a carbonyl compound is associated with the carbonyl group. The band at 1719 cm^{-1} in the spectrum of acetone is referred to as the *carbonyl band* by virtue of the observation that all carbonyl compounds with a similar structure show a similar band.

Many other empirical correlations lead to the conclusion that certain infrared bands (and Raman lines) are characteristic of particular groups in molecules and represent *characteristic group vibrations.* Although it is not possible to deduce from these correlations the form of the particular molecular vibration associated with a characteristic band, it is possible to infer from other considerations that a particular stretching or bending motion may be the dominant motion of a particular fundamental mode of vibration. For example, it may be shown that the fundamental mode of vibration in acetone giving rise to the band at 1719 cm^{-1}, mainly involves the stretching of the C$=$O bond with a minor contribution from other groups. The band may, therefore, be assigned to the carbonyl *stretching vibration.* The extension of the concept of a carbonyl *band* to a carbonyl *stretching vibration* represents the development from an empirical correlation of data to the application of principles of molecular vibrations to that data.

Characteristic group vibrations may be explained in terms of two effects influencing normal modes of vibrations in polyatomic molecules. The first arises because of the presence of atoms with low masses in a molecule. The second effect arises because of the presence of bonds in the molecule with relatively large force constants. Consider a general model for a large molecule where A, B, C, D, etc. represent atoms of masses m_A, m_B, m_C, m_D, etc. which are joined by bonds with force constant $f_{A-B}, f_{B-C}, f_{C-D}$, etc.

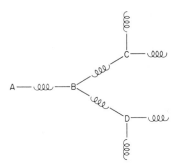

In the case of molecules in which all the masses and force constants are of similar magnitude

$$m_A \approx m_B \approx m_C \approx m_D \approx \text{etc.}$$

$$f_{A-B} \approx f_{B-C} \approx f_{C-D} \approx \text{etc.}$$

In this situation each of the $3N - 6$ normal modes of vibration has a complex form which cannot be described solely in terms of any particular group. In such molecules all atoms make significant contributions to each mode of vibration and the various groups in the molecule are said to be coupled with one another. It is not possible to associate a particular group with a particular band in the spectrum, but rather all groups are associated with all bands in the spectrum. It follows that if one atom is replaced by another of different mass and different associated force constant then every mode will be affected.

The two effects which lead to characteristic group vibrations will now be considered. The effect arising from differences in masses of atoms in a molecule, which leads to characteristic vibrations for groups containing those atoms, will be described as Type I group vibrations. The effect arising from differences in values of force constants for bonds in molecules, which leads to characteristic vibrations in molecules for groups containing those bonds, will be described as Type II group vibrations.

I. Type I Group Vibrations

Consider the situation where the atom A in the molecule represented on page 144 has a much smaller mass than the other atoms in the molecule.

$$m_A \ll m_B, m_C, m_D, \text{etc.} \qquad (3.46)$$

Owing to the relative inertia of the molecule as a whole compared to atom A, it is to be expected that one of the various modes of vibration would be the movement of atom A relative to the rest of the molecule, i.e. one mode will be dominated by the stretching of a bond between atom A and the adjacent atom in the molecule. The system approximates to a diatomic molecule

$$f_{A-X}$$

$$A \text{———} \underline{\text{QQQQQQQQQQQ}} \text{———} X$$

where X represents the combination of all other atoms and f_{A-X} is the force constant of the bond between atom A and X. The wave number value of the A—X stretching vibration is given by the Hooke's law expression

$$\tilde{\nu}_{A-X} = \frac{1}{2\pi c} \left(\frac{f_{A-X}}{\mu_{A-X}} \right)^{1/2} \qquad (3.47)$$

where

$$\mu_{A-X} = \frac{m_A m_X}{m_A + m_X} \approx m_A \qquad (3.48)$$

Thus to a good approximation

$$\tilde{\nu}_{A-X} = \frac{1}{2\pi c} \left(\frac{f_{A-X}}{m_A} \right)^{1/2} \qquad (3.49)$$

It follows from equation 3.49 that since m_A is the smallest mass in the molecule the A—X stretching vibration will be the fundamental mode with the highest wave number value. This assumes that all the force constants are approximately equal since a very large value for a particular force constant can lead to a large wave number value.

The most commonly observed example of the stretching vibration of the bond between a relatively light atom and a heavier atom occurs when a hydrogen atom is linked to carbon, oxygen, nitrogen or any other relatively heavy atom. The spectrum of $CHCl_3$ (see Expt. V9, No. 2) shows a band near 3000 cm^{-1} which is assigned to the C—H stretching mode.

Bending modes occur at lower frequencies than stretching modes since less energy is involved in bending than in stretching motions. Comparison between the spectra of $CHCl_3$ and CCl_4 (see Expt. V9, Nos. 1 and 2) allows the C—H bending mode to be assigned readily.

One method of establishing whether bands in the spectra of compounds containing an X—H group involve X—H vibrations is to prepare the corresponding X—D compound and compare the spectra. In Expt. V11 the spectra of compounds containing a single C—H, N—H and O—H group is compared with the spectra of the corresponding deuterated compound. It follows from equation 3.49 that the ratio of the wave number of the stretching vibration of the X—D group to that of the X—H group is given by

$$\frac{\tilde{\nu}_{X-D}}{\tilde{\nu}_{X-H}} = \left(\frac{m_H}{m_D}\right)^{1/2} = 2^{-1/2} \tag{3.50}$$

It can be seen from equation 3.50 that substitution of D for H in a C—H, N—H or O—H group will result in a shift of the corresponding stretching vibration to a lower wave number value by a factor of $2^{-1/2}$.

X—H bending motions may contribute to lower wave number bands in the spectra and these bands will move to still lower wave number values on substituting H by D. The magnitudes of these changes are not given by equation 3.50 since this is only applicable to stretching vibrations. In Expt. V11 examples are given (phthalimide, sodium acetylacetonate) where the X—H group is in a planar environment. The effect of the plane of symmetry is to produce three modes which have very little contribution from groups other than the X—H group and are of the form:

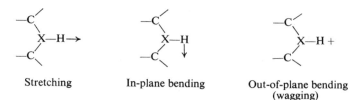

Stretching   In-plane bending   Out-of-plane bending
                                              (wagging)

These modes are shown as pure in-plane stretching and bending and out-of-plane bending (or wagging) and the wave number values of bands due to these modes follows this sequence. Bands may be assigned to these three modes by observing which bands move to lower wave number values when an X—H group is converted to an X—D group (provided the modes have a simple form).

When molecules have little symmetry there is a tendency for X—H bending motions to couple with other vibrations. Spectra of compounds containing X—H and X—D groups may then show considerable differences. Certain bands in compounds containing X—D groups may be at higher wave number values than their apparent counterparts in spectra of the undeuterated compounds. This is because an X—D bending motion may couple with other motions in a different way to an X—H bending motion and some of the resultant modes may have relatively high wave number values.

2. The Effect of Hydrogen Bonding on X—H Stretching Vibrations

The spectra of solutions of compounds containing O—H, N—H and other polar X—H groups normally show changes as the solute concentration

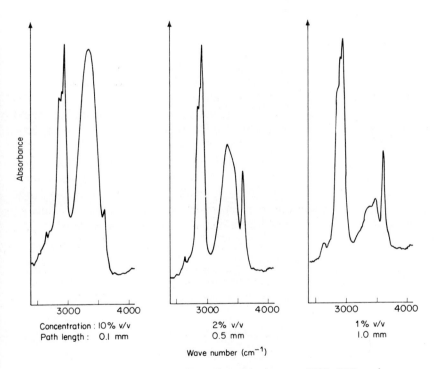

Fig. 3.12. Infrared spectrum of isopropanol in the range 2500–4000 cm^{-1}

changes. For example, Fig. 3.12 shows the spectrum of a solution of iso-propanol in CCl_4 at three different concentrations and path-lengths.

The values of concentrations and path-lengths chosen provide the same number of molecules in the path of the radiation for all three spectra. The absorbance of the C—H stretching vibrations in the region 2800–3000 cm^{-1} is approximately the same for the three solutions but the absorbance of the O—H stretching frequencies is markedly different for the three solutions. These differences are due to concentration dependent equilibria of the type:

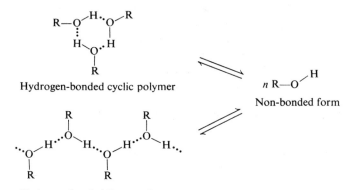

Hydrogen-bonded cyclic polymer

Non-bonded form

Hydrogen-bonded linear polymers

The effect of hydrogen bonding of the hydrogen atom of one molecule to the oxygen atom of another molecule leads to a reduction in wave number value and an increase in width and intensity of the band due to the O—H stretching vibration. Thus the sharp band near 3600 cm^{-1} which is evident in dilute solutions is due to a non-bonded O—H stretching mode and the broad band between 3000–3500 cm^{-1}, which is evident in concentrated solutions, is due to a hydrogen bonded O—H stretching mode. If the bonding is inter-molecular as shown for isopropanol the equilibrium is concentration depen-dent. In polyfunctional compounds intramolecular bonding may be possible. In this case the equilibrium is independent of concentration. In Expt. V10 inter- and intra-molecular hydrogen bonding in acids is investigated.

3. Type II Group Vibrations

Consider the situation where the bond A—B in the molecule represented on page 144 has a much higher force constant (i.e. is much stiffer) than the other bonds in the molecule

$$f_{A-B} \gg f_{B-C}, \quad f_{C-D}, \quad \text{etc.}$$

The motion of the masses separated by a stiff bond is determined mainly by that bond and is less dependent on other groups in the molecule. The group constituted by the system A—B approximates to a diatomic molecule, A—X, and the wave number of the vibration can be considered in terms of equation 3.47. The wave number of the band corresponding to the A—B stretching vibration will be higher than the wave number of other bands in the spectrum since this band has the highest force constant. This assumes the effects described as Type I and Type II are independent.

Examples of characteristic group vibrations of Type II occur in the spectra of molecules containing double and triple bonds. The force constant of a bond is related to the bond order. The wave number values of bands due to typical groups follow in sequence in accord with equation 3.47.

C—C	C=C	C≡C
C—N	C=N	C≡N
C—O	C=O	
400–1300 cm^{-1}	1550–1850 cm^{-1}	2100–2300 cm^{-1}

Single bonds do not fulfil the requirements that the force constant of the bond is large in relation to other bonds in the molecule. It is not, therefore, possible to observe bands which are particularly characteristic for these groups. However, compounds containing a very polar group such as C—O (see Expt. V9) show strong bands in the 950–1300 cm^{-1} region which arise from vibrations with considerable C—O stretching contribution. Bands at similar or lower wave number values are observed in compounds containing C—N groups but these are less intense.

The vibrations of double and triple bonds are less coupled with other molecular motions than vibrations of single bonds. This is because the wave number of the vibration has been raised out of the region of the spectrum where interactions can take place. The intensity of these vibrations depends upon the change of dipole moment with respect to the normal coordinate of the mode. Therefore the intensity is zero if a group is symmetrically substituted as in the following examples:

$$\underset{H}{\overset{H}{\diagdown}}C=C\underset{H}{\overset{H}{\diagup}} \qquad \underset{H}{\overset{Cl}{\diagdown}}C=C\underset{Cl}{\overset{H}{\diagup}} \qquad CH_3-C\equiv C-CH_3$$

The intensity is large if the multiple bond is asymmetrically substituted (see Expt. V9, Nos. 8–13).

The multiple bond most extensively investigated by infrared spectroscopy is the carbonyl group. This group is highly polar ($>C^{\delta+}=O^{\delta-}$) and, therefore, gives rise to an intense infrared absorption band. Furthermore, the precise wave number value is characteristic of the type of carbonyl compound being studied. This value depends on the physical state of the compound and on which solvent is used. For comparison purposes it is convenient to examine carbonyl compounds in dilute solution in CCl_4. In Expts V9 and V12a the infrared spectra of various types of carbonyl compounds are examined. The wave number value of the carbonyl bands may be related to structural features of the compounds under examination.

It is normally considered that substituents affect the wave number value of carbonyl stretching vibrations in two ways; by the inductive effect which involves electrons in σ bonds, and by the resonance (or mesomeric) effect which involves electrons in π orbitals and non-bonding orbitals. In the case of electronegative groups (—OR, —halogen, —NR_2) the inductive effect confers electron-acceptor properties on the group whilst the resonance effect may confer electron-donor properties on the group. In relation to the effect on the force constant of adjacent groups it is considered that the inductive effect is the dominant factor for oxygen-containing substituents and halogen substituents, whereas the resonance effect is the dominant factor for nitrogen-containing substituents. The effect of a large number of different types of substituents on the wave number values of carbonyl bands can be rationalized in terms of competing inductive and resonance effects (Bellamy, 1968). It is helpful to recognize four of the principal factors as an aid to the general prediction of the wave number value of a carbonyl vibration in any particular compound compared to the value in acetone.

(a) Adjacent Electronegative Group

$$\begin{matrix} X \\ \diagdown \\ \diagup \\ H_3C \end{matrix} C^{\delta+}\!\!=\!\!O^{\delta-} \qquad X = H,\ OH,\ OR\ \ (R = \text{alkyl}),\ \text{halogen}$$

The effect of electronegative groups substituted on a carbonyl group is to cause inductive withdrawal of the centre of negative charge associated with the bonding electrons away from the oxygen atom. Hence the polarity of the bond is reduced. This leads to a shortening of the bond because covalent-bond radii are smaller than ionic-bond radii. A shortening of a bond results in an increase in the force constant and hence of the wave number of the vibration. The occurrence of the carbonyl vibrations of the above compounds at *higher* wave numbers than in acetone can hence be rationalized. Furthermore, the occurrence of the carbonyl vibration of alkyl esters at lower wave number values than in the corresponding (unbonded) acids can be understood in terms of the electron-acceptor properties of the oxygen atom being *reduced* by the inductive supply of electrons from the alkyl group.

(b) Adjacent Groups Containing Lone-Pair Electrons

$$\begin{matrix} Y \\ \diagdown \\ \diagup \\ H_3C \end{matrix} C\!\!=\!\!O \quad \longleftrightarrow \quad \begin{matrix} Y^+ \\ \diagdown \\ \diagup \\ H_3C \end{matrix} C\!\!-\!\!O^- \qquad Y = NH_2,\ NHR_1,\ NR_2\ \ (R = \text{alkyl})$$

It is assumed that Y is a relatively weak electronegative group or atom; if Y is a relatively strong electronegative group or atom, then the inductive effect is dominant (effect of adjacent electronegative group). Lone-pair electrons in adjacent groups and electrons occupying π orbitals can interact by means of the resonance effect. These electrons are delocalized and the true structure can be represented in terms of contributions from canonical forms as shown above. This interaction leads to a reduction in the bond order and to a weakening of the C—O bond and hence to a lowering of the wave number value of the carbonyl vibration with respect to that in acetone. Hence the occurrence of the carbonyl band in amides at lower wave number values than in acetone may be rationalized on this basis.

(c) Adjacent Unsaturated or Aromatic Group

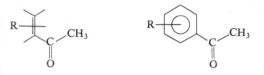

R = substituents on olefinic or aromatic group

The presence of adjacent groups containing electrons occupying π orbitals is a further example of a resonance effect. It is possible to show canonical forms which illustrate the reduction in bond order of the carbonyl bond as a consequence of resonance interaction. An alternative description of this effect is conjugation between the electrons occupying π orbitals in the carbonyl group and in the adjacent group; this leads to an increase in double-bond character of the C—C bond separating these two groups and a decrease in the double-bond character of the carbonyl group. Hence there is a reduction in the force constant of the carbonyl bond. The occurrence of carbonyl bands at lower wave number values when an α,β-unsaturated or aromatic group is present compared with the corresponding saturated molecules may, therefore, be explained.

(d) Ring Strain

 cyclic ketones, lactones, lactams

A carbonyl group normally has three bonds in a plane and at an angle of about 120° to one another. The energy of the carbonyl stretching vibration is partly associated with motion of the oxygen atom and partly with motion of the carbon atom. If the adjacent C—C bonds are part of a ring system the whole of the motion of the C atom must be taken up by distortion of C—C bonds. The smaller the ring the more energy is associated with this distortion and hence the higher the wave-number value of the carbonyl band. It is observed that the six- and higher-membered rings of cyclic ketones, lactones and lactams exhibit carbonyl bands at about the same wave number value as the corresponding alicyclic compounds. Five- and four-membered ring compounds exhibit progressively higher values for the wave number of the carbonyl band.

In Expt. V12a the wave number values of a number of carbonyl compounds are determined and the results are considered in terms of factors (a), (b), (c) and (d). When several factors operate they may be assumed to be additive. On this basis it is possible to attempt to predict the wave number values of the carbonyl stretching vibration of carbonyl compounds.

The rationalization of the wave number values of the carbonyl stretching vibration in terms of the *four* factors listed is an oversimplification. There is much evidence that interaction can occur through space as well as through bonds, for example by the appearance of carbonyl bands at different wave number values in different conformers. So far explanations have been provided largely in terms of an ideal carbonyl stretching vibration considered as a characteristic group vibration. In reality other motions within the molecule make a contribution to the particular mode largely dominated by stretching of the C$=$O bond. Some changes in the wave number value of the carbonyl band with change of substituent are due to a change in the other contributions to the mode. Thus the difference in wave number value of the carbonyl stretching vibration between acetone and acetaldehyde arises to some extent, because of a large difference in the mass effect contribution between a methyl group and a hydrogen. The differences between acetone and formamide are, to some extent, explicable in terms of contribution of N—H deformation and C—N stretching contributions to the mode assigned to the carbonyl stretching motion.

4. Vibrations of Equivalent Groups

The infrared spectrum of a compound containing a methylene group contains two bands near 3000 cm^{-1}. These bands arise from symmetric and anti-symmetric CH$_2$ stretching vibrations. These and a number of other modes of vibration of a CH$_2$ group are included in Fig. 3.13. The rocking mode is observed near 720 cm^{-1} in compounds containing the $+$CH$_2+_n$ group where n is greater than 3 (see Expt. V9, Nos. 5, 10, 11 and 27). When fewer CH$_2$ groups are coupled together the wave number value of the rocking mode is at a higher value.

The vibrations of a methyl group consist of symmetric CH$_3$ stretching, antisymmetric CH$_3$ stretching (degenerate), symmetric CH$_3$ bending, anti-symmetric CH$_3$ bending (degenerate) and CH$_3$ rocking (degenerate) vibrations.

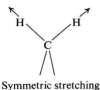

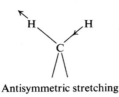

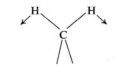

Symmetric stretching   Antisymmetric stretching   Symmetric bending (scissoring)

Antisymmetric bending   Symmetric out-of-plane   Antisymmetric out-of-plane
(rocking)   bending (wagging)   bending (twisting)

Fig. 3.13. Modes of vibration of a methylene group

Symmetric vibrations give relatively strong Raman lines compared with anti-symmetric vibrations, whereas antisymmetric vibrations give relatively strong infrared bands compared with symmetric vibrations.

The assignment of the principal modes of CH_2 and CH_3 groups are shown in Fig. 3.14 for a series of saturated hydrocarbons.

Bands due to modes involving CH_2 and CH_3 groups other than those which have been shown in Fig. 3.14 are generally not particularly prominent.

Vibrational coupling occurs whenever two or more identical atoms are linked to some common atom. The spectra of CCl_4 and $CHCl_3$ both show a strong band near 800 cm^{-1}. This band is assigned to an antisymmetric C—Cl stretching vibration involving four and three chlorine atoms respectively. When the vibration involves only one or two chlorine atoms, bands assigned to the C—Cl stretching vibration are weaker and occur at a lower wave number value than the corresponding band in more highly substituted groups. In the Raman spectrum of CCl_4 (Expt. V8) the line assigned to the symmetric C—Cl stretching vibration is chosen on the basis of its very low degree of depolarization.

Another example of vibrational coupling occurs for the carboxylate anion. This may be represented in two ways

Asymmetrical ion   Symmetrical ion

It would be expected that the asymmetrical ion would give rise to a single absorption band in the region 1700–1800 cm^{-1}. However, the fact that the spectrum of the carboxylate ion of acetic acid shows two strong bands at lower

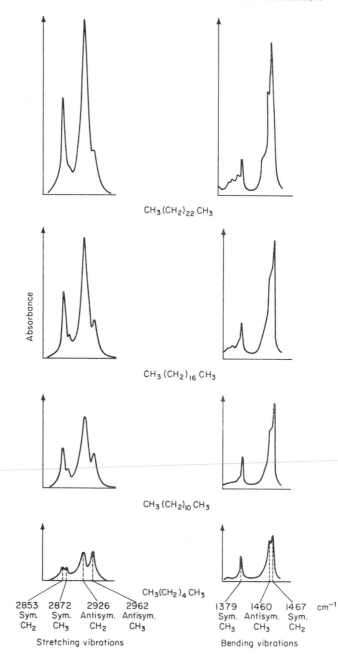

$CH_3(CH_2)_{22}CH_3$

$CH_3(CH_2)_{16}CH_3$

$CH_3(CH_2)_{10}CH_3$

$CH_3(CH_2)_4CH_3$

2853	2872	2926	2962		1379	1460	1467	cm^{-1}
Sym.	Sym.	Antisym.	Antisym.		Sym.	Antisym.	Sym.	
CH_2	CH_3	CH_2	CH_3		CH_3	CH_3	CH_2	

Stretching vibrations   Bending vibrations

Fig. 3.14. CH_2 and CH_3 stretching and bending modes in saturated normal hydro-carbons. Adapted with permission from John Wiley and Sons Inc.

wave number value suggests that the symmetrical ion structure is preferred. The lower wave number band is assigned to the symmetric stretching vibration and the higher wave number band is assigned to the antisymmetric stretching vibration. Corresponding bands in sodium benzoate are similar to those in the isoelectronic molecule nitrobenzene (see Expt. V9, Nos. 30 and 31).

The absorption bands due to the out-of-plane vibrations of H atoms in substituted ethylenic compounds are characteristic of the substitution patterns

Table 3.7

Olefin	Methylene wagging	Wagging of *trans* H	Wagging of *cis* H
(1) R—CH=CH$_2$	R_C=C$^\ominus$/H$^\oplus$ H/ \H$^\oplus$ ≈912 cm^{-1}	R_C=C/H$^\oplus$ $^\oplus$H/ \H ≈990 cm^{-1}	
(2) R—CH=CH—R (*trans*)		R_C=C/H$^\oplus$ $^\oplus$H/ \R ≈965 cm^{-1}	
(3) R—CH=CH—R (*cis*)			$^\oplus$H_C=C/H$^\oplus$ R/ \R 675–730 cm^{-1}
(4) R$_2$C=CH$_2$	R_C=C$^\ominus$/H$^\oplus$ R/ \H$^\oplus$ ≈890 cm^{-1}		

around the C=C group. In olefins (see Expt. V12, Nos. 10, 11 and 12) strong bands in the region 700–1000 cm^{-1} are assigned to out-of-plane bending vibrations (or wagging vibrations) as shown in Table 3.7.

The basis of these assignments is discussed by Potts and Nyquist (1959). Furthermore, normal coordinate calculations have been performed on the olefinic C—H out-of-plane bending vibrations of the vinyl group by Scherer and Potts (1959). It may be shown empirically that the effects of various substituents are approximately additive. Furthermore, the bands near 912 cm^{-1} and 890 cm^{-1} arising from structures 1 and 4 respectively are shifted in approximately the same way as each other by the same substituents and also the bands near 990 cm^{-1} and 965 cm^{-1} arising from 1 and 2 are shifted in approximately the same way as each other by the same substituents, but in a different manner to the previous pair. This evidence suggests each pair of bands corresponds to two similar normal modes of vibration. The normal coordinate calculations support this conclusion in that the bands near 912 cm^{-1} and 990 cm^{-1} in a vinyl group correspond respectively to an in-phase wagging motion of the terminal methylene group, and an in-phase wagging motion of the two hydrogens *trans* to one another on the double bond (this motion is sometimes described as a carbon–carbon double bond twisting motion, since the torsional angle between the two trigonal carbon atoms changes during the vibration). The mode of vibration to which the band near 912 cm^{-1} in 1 is assigned is clearly similar to the mode of vibration to which the band near 890 cm^{-1} is assigned in 4. Also the mode to which the band near 990 cm^{-1} in 1 is assigned is clearly similar to that to which the band near 965 cm^{-1} is assigned in 2. In Expt. V12b the effect of various polar substituents on the out-of-plane C—H bending vibrations are examined. In Expt. V16, the utilization of these bands in the analysis of various structural units of polychloroprene is investigated.

A number of characteristic coupled C—H vibrations are also observed in aromatic compounds. For example monosubstituted benzene derivatives show bands between 3000 and 3150 cm^{-1} which are assigned to coupled aromatic C—H stretching vibrations (see Expt. V10). The spectrum of polystyrene (Expt. V9, No. 7) shows five bands in this region when recorded under medium or high resolution conditions. These bands are due to coupled vibrations between the five hydrogens in the aromatic ring. In-plane C—H bending vibrations lead to bands between 1200 and 1500 cm^{-1} which are coupled with other in-plane ring vibrations. An out-of-plane C—H bending vibration and an out-of-plane ring vibration are responsible for two strong bands near 750 and 700 cm^{-1} respectively in nearly all monosubstituted aromatic compounds. Combinations of these vibrations produce a characteristic pattern of bands between 1600 and 2000 cm^{-1}. Other types of benzene substitution patterns lead to out-of-plane vibrations at different wave number values. The bands due to these vibrations and their combinations may, to some extent, be used to characterize an unknown substitution pattern in an aromatic compound (see Expt. V13).

5. Contributions from Other Groups to Characteristic Group Vibrations

Correlations between characteristic group vibrations and other molecular properties suggest that the vibrational mode associated with the characteristic group vibration is localized within a particular group in the molecule. The simple treatment already outlined has shown that this is a reasonable assumption under certain limiting conditions. More refined calculations indicate that contributions from other groups in the molecule are always made to any particular mode. For example in secondary amides, which are a class of compounds of particular interest, the principal amide modes have been shown to be of a complex form. Calculations of the wave number values and forms of vibrations in *N*-methylacetamide have been made by Miyazawa *et al.* (1958). The results of these calculations in relation to the spectrum of *N*-methylacetamide are shown in Fig. 3.15.

The calculated atomic displacements shown in Fig. 3.15 emphasize the importance of contributions from other molecular motions to a vibration

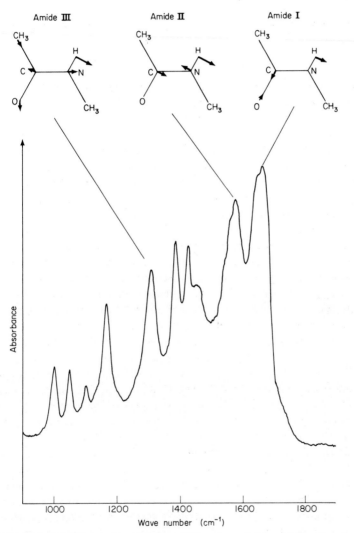

Fig. 3.15. Calculated forms of amide I, II and III modes and observed vibrational spectrum of *N*-methylacetamide

which is largely dominated by a particular group. In particular the band at 1650 cm^{-1} in N-methylacetamide is not exclusively a C=O stretching vibration, but contains an appreciable contribution from an N—H bending motion. It is, therefore, not strictly justifiable to compare this vibration with carbonyl vibrations in compounds of different types in which the degree of vibrational coupling will be different. The three modes shown in Fig. 3.15 are designated Amide I, II and III to avoid a commitment to the nature of the mode.

II. INSTRUMENTATION AND SAMPLE HANDLING PROCEDURES

A. DESCRIPTION OF INFRARED SPECTROMETERS

Infrared spectrometers may be divided into two categories. Firstly, instruments which are designed for routine use in which a standard low-resolution spectrum can be quickly obtained with a minimum of setting up procedures; secondly, instruments which are designed to obtain medium or high-resolution spectra.

The optics of typical instruments are shown in Fig. 3.16. These instruments have the following features in common. Radiation from a source is divided into two beams: a sample beam and a reference beam. These beams pass through the sample and reference paths of the cell compartment. Each beam then passes through an attenuator. The sample-beam attenuator is operated by a mechanical control which can be used to balance both beams and hence to zero the pen on the recorder. The reference-beam attenuator is operated by a self-balancing servo mechanism. After passing through the attenuators the two beams are combined in space, but separated in time by a rotating sector mirror acting as a beam switch which either reflects the reference beam or transmits the sample beam via a system of mirrors to a focus at the entrance slit of the monochromator. By rotation of a suitable component of the monochromator the spectrum is scanned across the exit slit, the scanning mechanism being linked to the chart drive mechanism and hence to the wave number scale. The width of the entrance and exit slits of the monochromator may be varied and are coupled together. The slit widths are programmed to compensate for variation of energy of source with wave number. The purpose is to maintain constant energy of radiation at the detector when there is no absorption in the sample beam. Absorption of radiation in the sample beam at any particular wave number leads to an a.c. signal at the detector. The frequency of this signal is controlled by the speed of rotation of the sector mirror. The signal is amplified by a phase-sensitive amplifier tuned to the a.c. frequency, and is then used to drive the attenuator in the reference beam into a position for which the a.c. signal is zero (this is the self-balancing servo principle). Hence the movement of the attenuator in the reference beam equates the transmission of the reference beam to that of the sample beam. This movement is registered

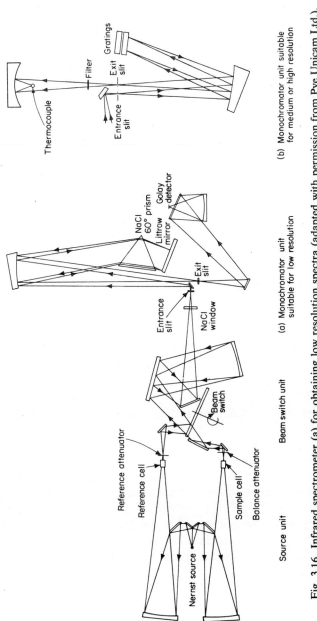

Fig. 3.16. Infrared spectrometer (a) for obtaining low resolution spectra (adapted with permission from Pye Unicam Ltd.), (b) for obtaining high resolution spectra (adapted with permission from Perkin–Elmer Ltd.)

at the recorder as the transmittance (or absorbance) of the sample as a function of wave number of radiation.

The principal difference between a low-resolution instrument and a high-resolution instrument is in the design of the monochromator. However, the sources and detectors of instruments designed for a particular standard of performance may also have special features.

I. Sources

The two commonly used sources of infrared radiation are the Nernst filament and the Globar. The Nernst filament is a mixture of oxides of rare earth metals (zirconium, yttrium and erbium). The filament conducts electricity only at elevated temperatures and has to be heated externally for starting up. Electrical heating then causes emission of radiation. It is important to avoid switching off the source since thermal shock can crack the filament. The power can usually be reduced when the instrument is not in use. A Nernst source may be air cooled or, if running at high temperature, it may be necessary for the source to be water cooled. For a power level of 150 J s^{-1} the temperature of the filament is about 1800°C. Peak emission of energy is at about 4000 cm^{-1} while at 650 cm^{-1} the emission of energy is only about 0·5 % of the peak value.

The globar is made of silicon carbide and conducts at room temperature. It can be operated by applying a voltage across the bar and can be switched on and off with little risk of damaging the filament. The globar source operates at a lower temperature and the intensity of radiation emitted in the far infrared is somewhat less than is the case for the Nernst source.

2. Monochromators

Fig. 3.16 shows typical prism and grating type monochromators respectively. A monochromator consists of an entrance slit, a collimator, a dispersion element and an exit slit. A typical scanning arrangement is shown in Fig. 3.16a using the Littrow system in which a mirror reflects the dispersed radiation back through the prism. The Littrow mirror is rotated by a drive system causing the spectrum to be scanned across the exit slit. Prisms constructed of sodium chloride or potassium bromide are commonly used. The dispersion of a prism increases near its cut-off point so that moderately good resolution spectra may be recorded at wave numbers as low as 650 cm^{-1} or 375 cm^{-1} using sodium chloride and potassium bromide prisms respectively. At high wave numbers (1500–4000 cm^{-1}) the dispersion of these materials is poor and other prism materials or, alternatively, diffraction gratings may be used as dispersing elements.

Fig. 3.16b shows a typical grating arrangement in a monochromator. Two gratings are available to cover the required spectral range. In any particular spectral region a particular grating and order of grating is selected together with a filter which removes radiation from other orders of the grating. Instead of a filter some instruments incorporate a double monochromator system.

In both types of monochromators the slit widths are usually programmed to provide sufficient energy at the detector to enable accurate absorbance or transmittance measurements to be made. If the slit is too wide, poor spectral resolution is obtained, and if the slit is too narrow, insufficient energy reaches the detector. In order to obtain the best results the slit widths must be selected by manual adjustment.

3. Detectors

Either a thermocouple or the Golay cell are commonly used as detectors. The thermocouple is robust and usually gives lengthy service. Those used as infrared detectors have a very small active surface and the radiation has to be carefully focused to maintain maximum thermal response. In the Golay cell radiation causes a thermal effect on a closed gas system to operate a flexible diaphragm which is mirrored on one side. The mirror relays a secondary signal to a photocell detector. The Golay cell can gather radiation over a large aperture and it is intrinsically more sensitive than a thermocouple which has a very small aperture for its sensitive region. However, the diaphragm deteriorates with time and Golay cells require more frequent replacement than thermocouples.

The detector in an infrared spectrometer is activated alternately by sample and reference beams at the frequency of beam switching and differences in energy of emission at the 10^{-10} J s^{-1} level can be detected. It is important to tune the amplifier to the frequency of the beam switch to avoid extraneous signals affecting the servo mechanism. The amplifier must also be sensitive to the phase of the radiation, so that the information concerning which beam has least energy is used to operate the servo mechanism in the correct direction to annul the energy difference.

4. Presentation of Spectra

There are a number of possible ways of presenting infrared spectra. The ordinate scale may be in transmittance units or in absorbance units and the abscissa scale may be in wave number units (cm^{-1}) or in wavelength units (μm). Either scale may be in either direction. A recommended standard presentation is shown in Fig. 3.17 where the infrared spectrum of indene is shown as recorded on a grating instrument. The wave number values of principal bands are provided as a recommended calibration standard (IUPAC, 1961).

If quantitative measurements are required it is convenient to use an ordinate scale in absorbance units. This allows the direct measurement of absorbance which is related to concentration of sample by the Beer–Lambert law equation

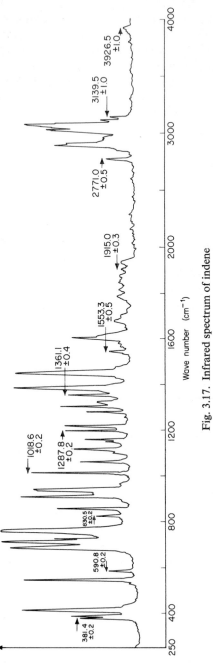

Fig. 3.17. Infrared spectrum of indene

(p. 68). In Expts V14, V15 and V16 the use of infrared spectroscopy for quantitative analysis is illustrated.

B. SAMPLE HANDLING PROCEDURES FOR INFRARED SPECTROSCOPY

An important feature of infrared spectroscopy is that spectra may be readily obtained from samples in all physical states. The amount of sample required is in the region 10^{-3}–10^{-6} g. A comprehensive account of many aspects of sample handling procedures has been provided by Miller (1965).

I. Samples and Sample Cells

(a) *Liquids*

The infrared spectra of liquids are normally recorded using cells with path-lengths in the region 0·01–1·0 mm. Alternatively a thin film of liquid may be held between two optical plates (or windows). This becomes necessary in the case of liquids which strongly absorb because cells of 0·01 mm path-length or shorter are difficult to construct, fill and maintain.

(b) *Solutions*

Solutions of solids, liquids or gases in suitably transparent solvents may be examined in the same way as pure liquids. Carbon tetrachloride, cyclohexane and carbon disulphide are frequently used as solvents for non-polar solutes and chloroform and bromoform are frequently used as solvents for relatively polar solutes. Because of absorption by the solvent, the path-length of solution that may be used is restricted to about 0·1–1·0 mm. However, solutions in carbon tetrachloride may be examined at up to 100 mm path-lengths for the region above 1600 cm^{-1} (see Expts V9 and V10). When samples are examined in solution the presence of solvent bands should be noted. These bands may be removed from the spectrum by placing a matched cell containing solvent in the reference beam (that is a cell containing an equivalent path-length of solvent to that in the sample beam). When using this technique, which is known as the difference or solvent cancellation method, it should be borne in mind that there may be regions of the spectrum where no energy is registering on the detector because of strong absorption in both beams. In these regions the apparent transmittance reading recorded by the pen is spurious. Caution must be exercised in examining the spectra of solutes (with or without a matching cell of solvent in the reference beam). To avoid mistakenly assigning solvent bands to the solute a spectrum of the solvent recorded under the same conditions as the sample should be available for comparison.

(c) Solids

The spectrum of a solid may be obtained in one of three ways: (i) as a thin film, (ii) as a dispersion in a suitable medium and (iii) as a dispersion in an alkali halide matrix (halide disc).

Thin films of certain solids (usually polymers) may be prepared by pressing or rolling the solid or by deposition of the solid from solution by evaporation of the solvent. When stable films of thicknesses in the region 0·01–0·1 mm are formed these can be mounted directly in the beam of a spectrometer.

Dispersions of solids in hydrocarbon (Nujol or liquid paraffin B.P.) or halogenated hydrocarbon ("Fluorolube" or fluorinated hydrocarbon, hexachlorobutadiene) may be prepared. The halogenated hydrocarbon is used when information concerning C—H vibrations in the solid is needed. The suspension or mull may be formed by grinding the sample and mulling agent either with a pestle and mortar or between glass plates until a smooth paste is formed. The mull may then be pressed as a thin film between optical plates and mounted in the beam of a spectrometer.

A dispersion of a solid in an alkali halide disc may be formed by grinding a mixture of the solid and an alkali halide (KBr or KCl) until an intimate mixture is obtained. This is then converted into a clear, glass-like disc by pressing under vacuum at high pressure. Several commercial sets of apparatus exist for preparing halide discs in this way.

(d) Gases

The spectrum of a gas may be obtained by using a cell of up to 100-mm pathlength. Multiple-pass cells which reflect the beam of radiation through the sample several times have been constructed. These provide greater sensitivity in the measurement of absorbance.

(e) Cells

The optical materials (NaCl, KBr) used for infrared cells are readily attacked by moisture or corrosive compounds. The windows of a cell used for a variety of samples usually become etched after a period of time. In this case the cell may be dismantled and the windows restored to good optical quality by polishing with jewellers rouge moistened with alcohol. After reassembling the cell it is desirable to remeasure the path-length. This procedure is outlined in Expt. V3.

2. Reflectance Measurements

So far only the measurement of absorption of infrared radiation by a sample has been considered. It is also possible to obtain spectra by reflection of infrared radiation. The application of the method of total attenuated reflectance to the

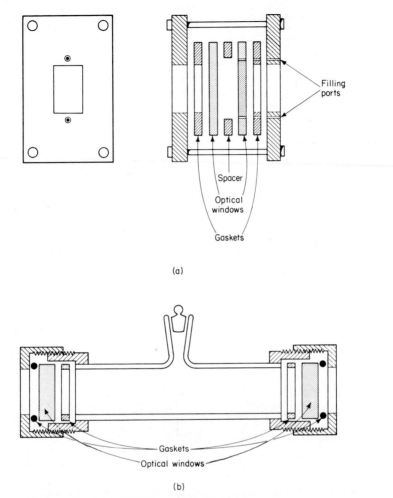

Filling ports

Spacer

Optical windows

Gaskets

(a)

Gaskets

Optical windows

(b)

Fig. 3.18. Cells for obtaining infrared spectra

determination of the spectra of various polymeric and paste-type samples and aqueous solutions is illustrated in Expt. V4.

C. RAMAN SPECTROSCOPY

The basic instrumentation for obtaining a Raman spectrum is illustrated in Fig. 3.19. Light from an intense line source (for example a mercury lamp) irradiates a sample which is protected from unwanted lamp radiations by suitable filters and which is kept close to room temperature by passage of cooling water through an annulus surrounding the sample cell.

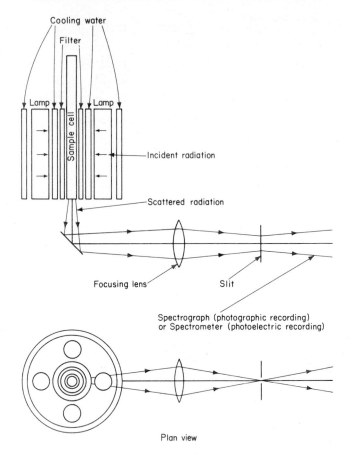

Fig. 3.19. Arrangement for obtaining a Raman spectrum

The spectrum of the scattered radiation may be examined using a conventional spectrograph designed to photographically or photoelectrically record spectra in the visible region. The Raman lines due to the sample may be distinguished from the other lines due to the source by comparison of the emission spectrum of the source (obtained by inserting a suitable reflector into the source unit) with a spectrum obtained by light scattered by the sample.

I. Sources

Conventional medium-pressure mercury lamps provide a line at 435·8 nm which is particularly suitable for the excitation of Raman spectra. This is because the line is intense and the source has few lines in the region where Raman lines are likely to occur. The Raman lines excited by other mercury lines will be considerably weaker than those excited by the line at 435·8 nm. Filters which reduce the intensity of these other mercury lines include potassium nitrite, rhodamine and *p*-nitrotoluene.

Improved mercury sources have been developed which provide a relatively sharper, more intense line at 435·8 nm with less source background in the Raman region than conventional mercury lamps. This has been achieved by cooling the envelope of the electrode and the arc. The arrangement is known as the Toronto arc. Sources have also been developed from emission spectra of elements other than mercury where it is desirable to excite a Raman spectrum in regions of the spectrum other than the blue region. This is useful in the case of compounds which have absorption bands in the blue region but are transparent in other regions. Recently laser excitation methods have been developed for Raman spectroscopy. Continuous gas laser beams are available for a number of different wavelengths and the properties of these laser beams make it possible to excite Raman spectra from small quantities of samples (10^{-3}–10^{-6} g) in various physical states.

2. Monochromators

The monochromator in a Raman spectrometer may consist of a slit, collimator, one or more prisms and a focusing lens. The prisms may be glass for the visible region and quartz for the ultraviolet region. Higher resolution can be obtained using a diffraction grating but this results in loss of energy because of the distribution of energy amongst the various orders of the grating.

3. Detectors

A conventional photographic method of detection is the simplest way of obtaining a Raman spectrum in the visible region of the spectrum. Very weak Raman lines can be observed by increasing the exposure time, provided that the intensity of the Raman line is appreciably greater than the intensity of the background radiation of the exciting radiation. The disadvantage of photographic detection is the time necessary to process the plate and measure the Raman wave number shifts.

Photoelectric recording methods have been incorporated in many commercial Raman spectrometers using a suitable scanning arrangement, exit slit, photomultiplier detector and amplifier. The method is intrinsically less sensitive than the photographic method since measurements are made over a very much shorter period of time. However, it is more rapid and convenient to present a spectrum as a trace on a calibrated chart than as a set of lines on a plate.

4. Presentation of Spectra

The Raman spectrum of carbon tetrachloride recorded on a dual-purpose photographic and photoelectric instrument is shown in Fig. 3.20. The spectrum of the exciting radiation is included for comparison.

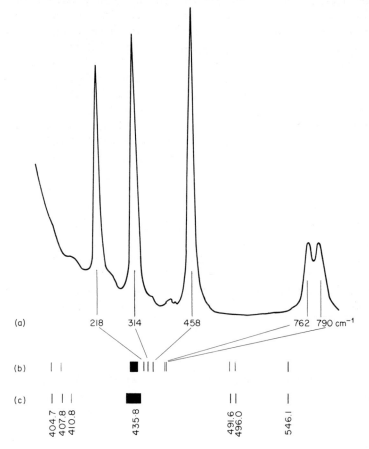

Fig. 3.20. Raman spectrum of carbon tetrachloride, (a) photoelectrically recorded, (b) photographically recorded, (c) mercury-emission spectrum. Adapted with permission from Rank Precision Industries Ltd.

5. Samples and Sample Cells

Because Raman scattering is a very weak effect, relatively large quantities of sample are normally required (1·0–10 g). However, very much smaller quantities of sample may be studied if the intensity of the exciting radiation can be increased accordingly, for example, using a high-powered laser beam. The sample must be dust-free to avoid the incident radiation being scattered by suspended material and detected together with the Raman-scattered radiation. The sample must not have an absorption band at a wavelength near that of the exciting line.

A glass cell suitable for liquids and solutions consists of a length of tube with an optical window adhered to one end. A cell with a re-entrant cone can be used for certain powdered solids. For gases multi-pass cells are normally required.

6. Measurement of Degree of Depolarization

A convenient method for measuring the state of polarization of Raman lines is known as the method of polarized incident radiation (Rank and Kagarise, 1950). In this method the intensities of the Raman lines are measured, firstly, when the incident radiation is polarized so that its electric vector is parallel to the Raman cell and secondly, when this vector is perpendicular to the Raman cell. Polarization of the incident radiation is achieved by inserting a cylinder of "polaroid" (a material which transmits light polarized in one particular direction) between the cell and the source of exciting radiation.

The degree of depolarization is the ratio of the intensity of the line excited by radiation polarized parallel to the cell to that excited by radiation polarized perpendicular to the cell (p. 132). Intensities of photographically recorded lines may be assessed semi-quantitatively by visual inspection or, more quantitatively, by the use of a microdensitometer to measure the degree of blackening of the plate for each line. For polarization measurements direct photoelectrical recording is much more convenient than photographic methods.

7. Calibration of Photographic Plates

The wavelength, λ, of a line on a photographic plate may be obtained in terms of the distance of the line from a chosen reference line, d, from the Hartmann equation (Sawyer, 1963)

$$\lambda = \lambda_0 + \frac{C}{d_0 - d} \tag{3.52}$$

where λ_0, C and d_0 are constants. These constants depend on the region of the spectrum under examination and on the spectrograph used. Hence the constants in the Hartmann equation have to be calculated for the conditions under which the lines of unknown wavelength are photographed. This may be achieved by superimposing the spectra of a mercury arc and a copper arc, and by measuring the distance on the plate of selected copper lines from a chosen mercury line. For the region 404·6–546·1 nm, the appearance of the plate is shown in Fig. 3.21.

The mercury line at 434·3 nm is a convenient reference line in this region and the distance of three copper lines (of known wavelength, spanning the particular region) from the mercury reference line may be measured using a travelling microscope. The constants, λ_0, C and d_0 can be calculated by substitution of values of λ and d in equation 3.52 to obtain three equations which may be solved for the three unknowns.

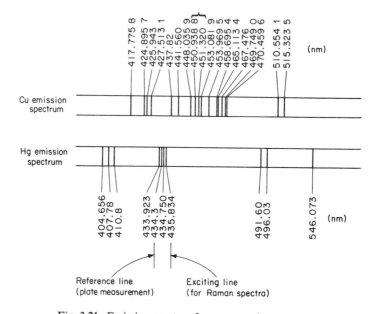

Fig. 3.21. Emission spectra of mercury and copper arcs

Since the values of the constants in the Hartmann equation may vary slightly from plate to plate the constants should, strictly, be derived for each plate used. However, the constants derived from the calibration plate can often be applied to other plates to a reasonable degree of accuracy. If accurate values of wavelength are required then a copper arc (or the arc of some other suitable element such as iron) may be superimposed on the spectrum, and the wavelength of selected lines calculated from the approximate Hartmann equation. The calculated values and the literature values may be compared and the differences used to construct a correction curve.

The wavelength of lines, λ, may be converted into wave number shifts, $\tilde{\nu}$, from the Raman excitation line (Hg 435·834 nm) by

$$\tilde{\nu} = \frac{10^7}{435 \cdot 834} - \frac{10^7}{\lambda} \ \text{cm}^{-1} \tag{3.53}$$

The wavelength and wave number values of lines or bands other than in Raman spectroscopy, which have been photographically recorded, may be calculated by this method by using an appropriate calibration arc spectrum and reference line. For example the visible spectrum of iodine may be photographically recorded (Expt. E3).

III. EXPERIMENTS

A. PERFORMANCE OF INSTRUMENTS AND EXPERIMENTAL TECHNIQUES

The performance of instruments for recording either routine spectra or spectra under medium or high resolution conditions is investigated in Expts V1 and V2 respectively. These experiments are designed to develop an understanding of the optical, mechanical and electrical features of particular instruments, and to gain experience in checking the specifications and making simple adjustments when necessary.

Experiments are also provided in the measurement of path-length of cells (V3) and in the use of attenuated total reflectance for obtaining spectra (V4).

VI. Performance of a Low Resolution Infrared Spectrometer

Object

To check the performance of an infrared spectrometer used for routine work.

Theory

Certain infrared spectrometers have been designed to produce a spectrum on a precalibrated chart within a scanning time of 1 to 12 minutes over the range 650 to 5000 cm^{-1}. These instruments frequently utilize a prism as a dispersion medium. Alternatively, higher resolution (or resolving power) may be obtained using a diffraction grating. The general layout of a typical instrument is shown in Fig. 3.16 and a brief account of the mode of operation of instruments of this type is given on page 158. Details of the optics, electronics and mechanics of a simple infrared spectrometer differ according to the model and manufacturer. The performance of a given instrument will depend on the components used and, in particular, on the dispersing system of the monochromator. The manufacturer's handbook on the instrument generally gives the standard of performance to be expected and suggests tests which can be carried out to confirm that the performance of a given instrument is within specification.

The following procedure outlines suggested tests for an instrument which utilizes a sodium chloride prism in the monochromator (this is a suitable arrangement for routine studies). Typical specifications are indicated, but the actual performance should be compared with the specified performance for the particular instrument in use. When an instrument fails to meet the performance in any respect, simple fault-finding procedures are usually provided.

Apparatus

Infrared spectrometer (low resolution, 650–5000 cm^{-1}); stray light filter (lithium fluoride); polystyrene film.

Procedure

PEN RESPONSE. Set the instrument to a selected wave number value (say 1000 cm^{-1}). Blank off the sample beam to give about 15% transmittance and quickly remove the blanking device. Estimate the time for the pen to return to balance. (Under damping conditions leading to less than 1% overshoot of pen the time of pen response should be less than 1 second.)

ENERGY PROGRAMME. Set the instrument to a series of wave number values in turn (650, 1000, 1900, 3000, 4000 cm^{-1}). Repeat the pen response test at each wave number value. (If the energy programme is correct the pen response should be the same at each wave number value.)

STRAY LIGHT. Select a sample which absorbs radiation completely below some suitable wave number value but has a high transmittance above this value (for example, a lithium fluoride plate cuts out all radiation below 1400 cm^{-1}). Measure the apparent percentage transmittance value recorded at a wave number value where the plate absorbs completely. Measure the percentage transmittance of a suitable screen which is opaque to all wave number values of radiation emitted by the source. Estimate the difference of these two values. This difference is the "stray light". (The "stray light" should be less than 5% transmittance at all wave number values.)

THE PERCENTAGE NOISE LEVEL. Offset the pen to about 97% transmittance and record a spectrum with the sample and reference compartments empty. Estimate the peak-to-peak distance within which the noise fluctuation is contained for 90% of the time. The percentage noise level is this distance as a percentage of the pen deflection from the 0% transmittance value. (This percentage noise level should be within 1%.)

BEAM CANCELLATION. Examine the variation of the mean pen position over the whole wave number range of the spectrum as recorded in the percentage noise level test. (This should be within 1·5% of the overall mean value.)

Superimpose three spectra of a *standard* polystyrene film on the same chart. A spectrum of a polystyrene film is shown in the figure.

WAVE NUMBER ACCURACY. Compare the wave number values of the appropriate bands in the spectrum of polystyrene with the standard values shown in the figure (the accuracy should be within the following limits: ± 2 cm^{-1} at

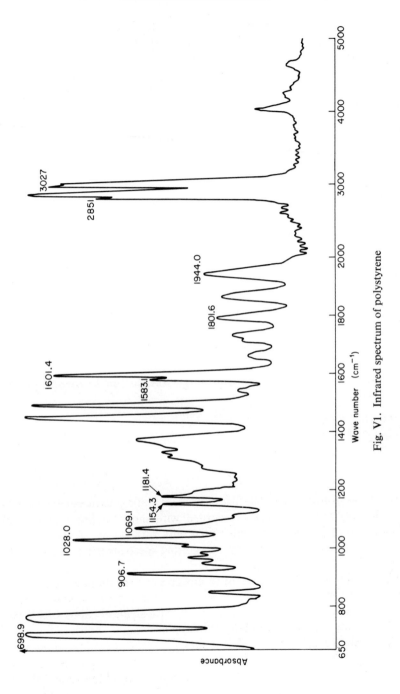

Fig. V1. Infrared spectrum of polystyrene

650 cm^{-1}, ± 4 cm^{-1} at 1000 cm^{-1}, ± 8 cm^{-1} at 1600 cm^{-1}, ± 12 cm^{-1} at 2000 cm^{-1}, ± 28 cm^{-1} at 3000 cm^{-1}, ± 50 cm^{-1} at 5000 cm^{-1}.)

WAVE NUMBER REPRODUCIBILITY. Estimate the range of values of wave numbers of the superimposed spectra of polystyrene for bands in different spectral regions (the reproducibility limits should be within half of the accuracy limits listed in the wave-number accuracy test).

TRANSMITTANCE REPRODUCIBILITY. Compare the range of values of transmittance of the superimposed spectra of polystyrene in different spectral regions. (The reproducibility limits should be better than $\pm 0 \cdot 5 \%$ between 50 and 100% transmittance readings.)

RESOLUTION. In the spectrum of polystyrene measure the difference between the trough of the pairs of bands at 1583 and 1601 cm^{-1}, at 2851 and 2924 cm^{-1} and at 3027 and 3035 cm^{-1} and the weaker, lower wave number band in each case. (This difference should be greater than 6% of transmittance for the first two pairs and should be greater than 1% of transmittance for the last pair.)

Results and Discussions

Compare the experimental data on the performance of the instrument with the typical specifications given for each test or, if possible, with the manufacturer's specification.

Comment on the performance of the instrument.

V2. Performance of a Medium or High Resolution Infrared Spectrometer

Object

To study the effect of operating conditions on infrared spectra recorded under medium or high resolution conditions.

Theory

Infrared spectrometers which utilize diffraction gratings as dispersion elements generally enable spectra to be recorded with much higher resolution (or resolving power) than instruments fitted only with prisms as dispersion elements. The general layout of a typical grating spectrometer is shown in Fig. 3.16.

The resolution of an instrument in any spectral region may be defined as the smallest wave number interval which is observed in the spectrum of a suitable compound (one in the spectrum of which a pair of bands are observed which are only just distinguishable one from another as shown in the figure).

The resolution of an instrument is limited by the slit-width of the monochromator unit; as these are narrowed the resolution improves. The width of the slit can only be reduced to a certain value, beyond this limit diffraction effects prevent any further improvements in resolution. The slit-width determines the energy of the radiation reaching the detector; at narrow slit-widths the energy may be insufficient to operate the servo-mechanism effectively. This difficulty may be overcome to some extent either by increasing the amplifier gain or by slowing the scan speed (thereby increasing the energy incident to the detector per unit time). The disadvantage of these adjustments are that the former leads to a reduction of the signal to noise ratio and hence to a reduction in the accuracy with which absorbance measurements may be made while the latter leads to an increase in the time necessary to record the spectrum.

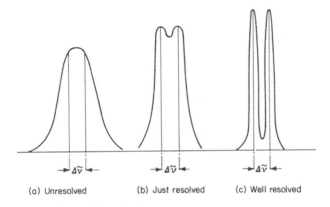

(a) Unresolved   (b) Just resolved   (c) Well resolved

Fig. V2. Pair of bands recorded under different resolution conditions

Thus the three most important controls on an infrared spectrometer suitable for medium or high performance work are the slit-widths, the gain control and the scan speed. These determine the resolution, signal to noise ratio, accuracy of absorbance measurements and time necessary to record a spectrum. If one or more of these four parameters is specified the best compromise of the three operational controls may be determined by experiment. It is found that in order to record the true absorbance and shape of any particular band, the instrument should be capable of resolving bands separated by about one fifth of the half-band width of the spectral band. Thus the recorded shape of a band which has a narrow half-band width is particularly dependent on the resolution of the instrument. The spectral slit-width at different wave number values for optimum conditions is usually specified by the instrument manufacturers.

The wave number accuracy of any instrument may be checked by recording the spectra of standard compounds with bands of known wave number values. In this experiment tests are suggested for the investigation of the performance of an instrument in terms of resolution and wave number accuracy and methods are proposed for investigation of the inter-relationship of the various operating parameters.

Apparatus

Infrared spectrometer (medium or high resolution 250 or 400 cm^{-1}–4000 cm^{-1}); 0·02 mm CsI or KBr cell, 100 mm NaCl cell; indene, carbon monoxide, ammonia, cyclohexane.

Procedure

Record the spectrum of indene at 0·02-mm path-length under normal operating conditions from 250 to 4000 cm^{-1} using a CsI cell or from 400 to 4000 cm^{-1} using a KBr cell.

Table V2

Gas	Pressure (N m^{-2})	Spectral range (cm^{-1})
NH$_3$	10 000	3400–3500
NH$_3$	6 700	800–1200
CO	20 000	2050–2250

Fill a 100-mm path-length gas cell with *one* of the gases at the pressure stated in Table V2 and record the spectrum of the gas in the stated region of the spectrum.

Select two bands in the spectrum of the gas with a small wave number separation and vary the operating parameters in the following order:

(i) Select as high an amplifier gain as possible consistent with the noise level of the background being reasonably acceptable (i.e. of the order 1–3% random-noise fluctuation compared with the peak signal).

(ii) Scan the spectrum over the two chosen bands at various decreasing slit-widths. Select the optimum slit-width.

(iii) Scan the spectrum over the two chosen bands at the optimum slit-width and at various decreasing scan speeds. Select the optimum scan speed (consistent with the spectrum being recorded in a reasonable period of time).

(iv) At the optimum scan speed, repeat (ii) and note if the optimum slit-width is different.

(v) At the optimum slit-width, repeat (iii) and note if the optimum scan speed is different.

Record the spectrum of cyclohexane between 800 cm^{-1} and 940 cm^{-1} at 0·02 mm path-length and under the following conditions:

(i) At a series of different slit-widths at constant scan speed.

(ii) At a series of different scan speeds and at constant slit-width.

Results

Tabulate the observed and literature wave number values (Fig. 3.17) of bands in the spectrum of indene.

Record the apparent resolution of the instrument at the conditions under which the spectrum of indene was recorded.

Record the apparent resolution of the instrument at the conditions under which the spectrum of the gas was recorded.

Record any differences in relative absorbance values and relative widths at half-band height of the pair of cyclohexane bands.

Discussion

1. Comment on the comparison between the measured wave number values and the literature values for the specified bands in indene.

2. Comment on the comparison between the estimated value for the resolution of the instrument and the manufacturers specified value at various operating conditions.

3. Relate differences in relative intensities of cyclohexane bands to differences in half-band width under the differing operating conditions. What conclusion can you draw concerning the use of infrared spectroscopy for quantitative analysis?

V3. Measurement of Cell Path-length

Object

To measure the path-length of an infrared cell by the method of interference fringes.

Theory

Rock-salt plates which are commonly used for infrared cell windows are easily damaged by exposure to water and other substances likely to etch the surface. It is frequently necessary to dismantle cells, polish the plates and reassemble the cell. The path-length of the cell may then be determined by the method of measuring interference fringes

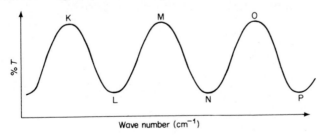

Fig. V3.1. Interference pattern

178   MOLECULAR SPECTROSCOPY: THEORY AND EXPERIMENT

If a cell with a path-length in the region 0·01–1·0 mm is placed in the sample beam of an infrared spectrometer, the recorded spectrum may have the form shown in the figure provided the windows are plane and parallel to one another. The amplitude of the wave form shown in the figure may vary from 15% to 2% according to the state of the cell windows. The relationship between the cell path-lengths and the difference in wave number $\tilde{\nu}$ between adjacent peak maxima or adjacent minima may be derived as follows.

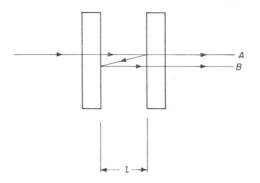

Fig. V3.2. Passage of radiation through a cell

Consider a beam of radiation of wavelength λ incident to one face of the cell. Most of the radiation will pass through the cell (beam A) but some will emerge having undergone a double reflection at the inner surface (beam B) as shown in the above figure. In this case beam B will have traversed an extra distance $2l$. If the distance $2l$ is equal to a whole number of wavelengths $n\lambda$ (where n is a positive integer) then A and B will be in phase and the intensity of the combined beams will be at a maximum (points K, M and O in the figure correspond to adjacent n values). Similarly the intensity of the combined beams will be at a minimum when the two beams A and B are 90° out of phase, i.e. when $2l = (n + \frac{1}{2})\lambda$ (points L, N and P in the figure).

It follows that the path-length of the cell, l, is given by

$$2l = n\lambda_1 = \frac{n}{\tilde{\nu}_1} \tag{V3.1}$$

$$2l = (n + \frac{1}{2})\lambda_2 = \frac{(n + \frac{1}{2})}{\tilde{\nu}_2} \tag{V3.2}$$

where λ_1 and $\tilde{\nu}_1$ are the wavelength and wave number values at M, λ_2 and $\tilde{\nu}_2$ are the wavelength and wave number at N. Eliminating n from equations

V3.1 and V3.2 gives

$$l = \frac{1}{4(\tilde{\nu}_2 - \tilde{\nu}_1)} \qquad \text{(V3.3)}$$

It follows that the path-length, l, can be expressed as

$$l = \frac{m}{2\Delta\tilde{\nu}} \qquad \text{(V3.4)}$$

where m is the number of complete peak-to-peak fringes between two maxima separated by an interval $\Delta\tilde{\nu}$. If the value of $\Delta\tilde{\nu}$ is in cm^{-1} units the calculated value of l will be in cm units

Apparatus

Infrared spectrometer (1000–4000 cm^{-1}) resolution about 1 cm^{-1}; one or more infrared cells of fixed or variable path-length.

Procedure

FIXED PATH-LENGTH CELL. Record the spectrum of a cell which has windows of good optical quality. Measure the wave number interval, $\Delta\tilde{\nu}$, between suitable pairs of peaks or troughs and count the number of fringes, m, between them. Repeat the procedure for four or more different pairs of peaks.

VARIABLE PATH-LENGTH CELL. Record the spectrum of an empty variable path-length cell using the drum settings 0·50, 1·0, 1·5 and 2·0 mm. Measure appropriate m and $\Delta\tilde{\nu}$ values for each drum setting.

Note: It is important not to screw up the cell to a sufficiently close value to the zero setting to cause the cell faces to rub together.

Results and Discussions

Calculate the various l values using equation V3.4 for the fixed path-length cell and/or the variable path-length cell.

In the case of the variable path-length cell plot a graph of l against drum readings and determine the drum reading corresponding to the zero path-length of the cell.

Comment on any differences between the calculated path-lengths for the cells and the marked path-length values. Discuss any differences in relation to the condition of the cell.

V4. Attenuated Total Reflectance

Object

To obtain the attenuated total reflectance spectra of certain polymeric samples and of some aqueous solutions.

Theory

In general it is easier to record transmission spectra in the infrared than to record reflectance spectra. However, certain samples may be advantageously studied by measuring the amount of reflected radiation as a function of wave number. A suitable arrangement for obtaining such spectra is shown in the figure.

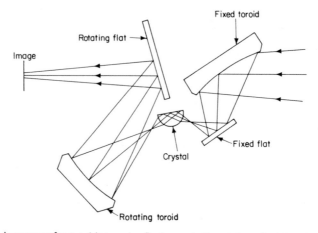

Fig. V4. Apparatus for total internal reflectance studies. Adapted with permission from Perkin–Elmer Ltd.

Since the energy of the radiation reflected from the sample is low, the energies of the sample and reference beams are balanced by having an attenuator mounted in the reference beam. Mechanical control of the attenuator will scale-expand the pen reading due to the limited energy reflected from the sampled mounted against the face of the crystal.

If radiation passing through a medium of high refractive index strikes an interface with a medium of low refractive index, TOTAL internal reflection occurs if the angle of incidence is greater than the critical angle (the value of the angle of incidence above which total internal reflection occurs). The ATTENUATION (that is the decrease in intensity) of the beam on reflectance is a function of the change in refractive index between the media at the interface. The refractive index changes markedly in the vicinity of an absorption band and consequently the ATTENUATED TOTAL REFLECTANCE (ATR) spectrum will be similar to an absorption spectrum.

The crystal normally consists of a prism or hemicylinder of KRS-5 (a mixed crystal containing thallium bromide and thallium iodide), or some other high refractive index material. It is important to achieve high optical contact between the sample and the crystal. Pastes, solid film, coatings, fibrous materials, etc. may be clamped firmly against the face of the crystal. Aqueous and other solutions may be mounted in position by a cell which is mounted behind the crystal and which allows direct contact between the solution and the crystal face.

The ATR spectrum obtained of a sample is independent of the thickness of sample but varies considerably with the angle of incidence of the beam. The smaller the angle of incidence, the greater the penetration into the sample. However, since this must exceed the critical angle, it is necessary to vary the angle of incidence through several degrees above the critical angle to find the best angle of incidence for optimum performance for any particular sample. The optics of the ATR unit are arranged so that the angle of incidence can be varied whilst the geometry of the beam leaving the unit is maintained constant.

Apparatus

Infrared spectrometer (650–4000 cm^{-1}); ATR attachment; reference beam attenuator, BaF_2 plates; samples of polymeric materials such as polyvinylchloride, polytetrafluoroethylene, polystyrene; miscellaneous commercial and unknown samples, polyester resins, epoxy resins, adhesive tape, tobacco flake, tile, toothpaste; 10% v/v solution of acetone in water.

Procedure

SOLIDS. Place the ATR unit and the reference beam attenuator in the sample and reference beam respectively. Adjust the mirror optics according to the manufacturer's instructions to give maximum energy.

Press a sample of polytetrafluoroethylene against the prism so that maximum contact with the reflecting face is made. Record the spectrum between 1000 and 1400 cm^{-1} at a series of different values for the angle of incidence and obtain the optimum angle of incidence. Record the reflectance spectra and where possible the transmission spectra of selected solid samples at the optimum angle of incidence.

SOLUTIONS. Using the attachment for solution samples record the reflectance spectrum of a 10% solution of acetone in H_2O between 800 and 2000 cm^{-1}. Using BaF_2 plates record the transmission spectrum between 800 and 2000 cm^{-1} of a thin film of the solution of acetone in water.

Results and Discussion

1. Where possible, comment on any differences between the reflectance and transmission spectra.

2. Is the limit of detection of a carbonyl group in aqueous solution likely to be higher in reflectance than in transmittance?

7

B. VIBRATIONAL SPECTRA OF SMALL MOLECULES

In this section the vibrational spectra of small molecules will be considered. For a diatomic molecule a number of fundamental parameters may be calculated from the infrared spectrum as illustrated in Expt. V5 for HCl. The structure and symmetry of triatomic molecules may be related to the infrared vapour spectra (V6), and to the Raman spectrum of the compound in solution (V7). The structure and symmetry of some tetra and penta atomic molecules and ions may be related to both the infrared and Raman spectra (V8).

V5. The Vibration–Rotation Spectrum of HCl

Object

To determine the following parameters for the HCl molecule.

(a) The wave number value, $\tilde{\nu}_e$, for the equilibrium vibration, the anharmonicity constant, x_e, the force constant, f, of the bond, the average rotational constant, $\tilde{B}$, the average moment of inertia, I, and the average internuclear separation, r, for the molecule.

(b) The values of the rotational parameters $\tilde{B}_0$, $\tilde{B}_1$, I_0, I_1, r_0 and r_1 in each of the vibrational states corresponding to $V = 0$ and $V = 1$.

Theory (a)

To a reasonable degree of approximation, the HCl molecule can be considered as an anharmonic oscillator and as a rigid rotor. The energies of the various vibration–rotation levels may be obtained using the Born–Oppenheimer approximation. The wave number values of the various vibration–rotation bands in the spectrum may be obtained from the Bohr relationship and the appropriate selection rules. It follows from these considerations (p. 121) that the vibration–rotation spectrum of HCl consists of a P branch with peaks having wave number values

$$\tilde{\nu}_P = \tilde{\nu}_e(1 - 2x_e) - 2\tilde{B}J^1 \tag{V5.1}$$

and an R branch with peaks

$$\tilde{\nu}_R = \tilde{\nu}_e(1 - 2x_e) + 2\tilde{B}(J^1 + 1) \tag{V5.2}$$

where J^1 is the rotational quantum number of the rotational level from which each transition originates and $\tilde{B}$ is the average rotational constant for the two vibrational levels $V = 0$ and $V = 1$ and is given by

$$\tilde{B} = \frac{h}{8\pi^2 Ic} \tag{V5.3}$$

where I is the moment of inertia about an axis perpendicular to the axis of the molecule and passing through the centre of gravity. Hence from equations V5.1 and V5.2 the separation between adjacent peaks in the P branch and in the R branch is $2\tilde{B}$. The central gap between the P and R branches is $4\tilde{B}$ and corresponds to the missing Q branch. The band centre for the fundamental absorption will be

$$\tilde{\nu}_{0\to1} = \tilde{\nu}_e(1 - 2x_e) \qquad (V5.4)$$

If the first overtone is observed it will have a similar rotational structure with a band centre at

$$\tilde{\nu}_{0\to2} = 2\tilde{\nu}_e(1 - 3x_e) \qquad (V5.5)$$

If the wave number values of the band centres of the fundamental, $\tilde{\nu}_{0\to1}$, and the first overtone, $\tilde{\nu}_{0\to2}$, are measured the parameters $\tilde{\nu}_e$ and x_e may be calculated. The force constant, f, of the bond may be calculated from the Hooke's law expression

$$\tilde{\nu}_e = \frac{1}{2\pi c}\left(\frac{f}{\mu}\right)^{1/2} \qquad (V5.6)$$

where

$$\frac{1}{\mu} = \frac{1}{m_H} + \frac{1}{m_{Cl}} \qquad (V5.7)$$

where m_H is the mass of hydrogen and m_{Cl} is the mass of chlorine. From the average of the measured separation of the peaks in P and R branches $(2\tilde{B})$ the average moment of inertia may be calculated using equation V5.3. The average internuclear separation, r, may then be calculated from the equation

$$I = \mu r^2 \qquad (V5.8)$$

In the HCl molecule different values for these parameters which are mass dependent may be expected for the isotopes ^{35}Cl and ^{37}Cl.

Theory (b)

If allowance is made for the possibility that the rotational constant $\tilde{B}$ is different in the states $V = 0$ and $V = 1$ the analogous equations to V5.1 and V5.2 become

$$\tilde{\nu}_P = \tilde{\nu}_e(1 - 2x_e) - (\tilde{B}_1 + \tilde{B}_0)J^1 + (\tilde{B}_1 - \tilde{B}_0)J^{12} \qquad (V5.9)$$

$$\tilde{\nu}_R = \tilde{\nu}_e(1 - 2x_e) + 2\tilde{B}_1 + (3\tilde{B}_1 - \tilde{B}_0)J^1 + (\tilde{B}_1 - \tilde{B}_0)J^{12} \qquad (V5.10)$$

The values of $\tilde{B}_0$ and $\tilde{B}_1$ may be calculated by application of the difference formulae obtained by subtraction of equation V5.9 from equation V5.10 for appropriate values of J^1 as follows

$$\tilde{v}_R(J^1) - \tilde{v}_p(J^1) = 4\tilde{B}_1 J^1 + 2\tilde{B}_1 \qquad (V5.11)$$

$$\tilde{v}_R(J^1 - 1) - \tilde{v}_p(J^1 + 1) = 4\tilde{B}_0 J^1 + 2\tilde{B}_0 \qquad (V5.12)$$

Hence these two differences may be plotted against J^1 to give straight lines from which $\tilde{B}_0$ and $\tilde{B}_1$ and hence I_0, I_1, r_0 and r_1 may be calculated using equations corresponding to equations V5.3 and V5.8 for each vibrational level.

Apparatus

Infrared spectrometer (2500–3100 cm^{-1}) with medium resolution; gas cell with 100-mm path-length containing HCl gas at 33 000 N m^{-2} pressure.

Procedure

Determine the optimum conditions by scanning a selected peak in the spectrum of HCl in the region 2500–3100 cm^{-1} for a series of different values of slit-width (around 0·1 mm) at a slow scanning speed and at a high amplifier gain. Choose conditions which give a maximum resolution of the isotopic satellite band (approximately 2 cm^{-1} from the main band). Record the spectrum of HCl between 2500 and 3100 cm^{-1} using the best conditions for resolution consistent with a reasonable signal to noise ratio and time of scan.

Results (a)

Measure the wave number values of the main peaks and the isotope satellite peaks. Tabulate the results as in Table V5(a)

Table V5(a)

Assignment				
Lower	Upper	$\tilde{v}_{H-^{35}Cl}$	$\Delta\tilde{v}_{H-^{35}Cl}$	$\tilde{v}_{H-^{35}Cl} - \tilde{v}_{H-^{37}Cl}$
$J = 10$	$J = 9$			
$J = 9$	$J = 8$			
$J = 8$	$J = 9$			
$J = 9$	$J = 10$			

Calculate the wave number value of the centre of the fundamental band. Given that the centre of the first overtone band is at 5668 cm^{-1}, calculate $\tilde{v}_e$, x_e and f.

Calculate the mean separation of the nineteen peaks listed in the table and calculate average values of $\tilde{B}$, I and r.

Determine whether the subsidiary bands present as shoulders on the low wave number side of the main bands are consistent with ^{37}Cl and ^{35}Cl present in the ratio $1:3$.

From the fundamental band make a scale drawing of the energy level (expressed in cm^{-1}) and assign the set of main peaks to particular transitions. Consider only the transitions involving $J = 0, 1, 2, 3$ and 4.

$$\text{Scale: take } 100 \text{ cm}^{-1} = 1 \text{ inch between } J = 0, 1, 2, 3 \text{ and } 4$$

$$\text{take } 500 \text{ cm}^{-1} = 1 \text{ inch between } V = 0 \text{ and } V = 1.$$

Discussion (a)

1. The degeneracy of a rotational level J is $(2J + 1)$. How does this affect the intensities of the bands?

2. Would you expect similar spectra from the following systems (i) H_2, (ii) Cl_2, (iii) HCl in CCl_4 solution?

3. Compare the spectra of HCl and CO (Fig. 1.6) and predict the ^{13}C isotope effect in the latter.

4. What is the significance of any variation in spacings of the rotational peaks?

Results (b)

Tabulate the parameters in Table V5(b) for the H—^{35}Cl molecule using the data tabulated in (a).

Table V5(b)

J^1	$\tilde{v}_R(J^1)$,	$\tilde{v}_P(J^1)$,	$\tilde{v}_R(J^1)$,- $\tilde{v}_P(J^1)$,	$\tilde{v}_R(J^1-1) - \tilde{v}_P(J^1+1)$
0				
1				
2				

Determine $\tilde{B}_1$ by plotting $\tilde{v}_R(J^1) - \tilde{v}_P(J^1)$ against J^1.

Determine $\tilde{B}_0$ by plotting $\tilde{v}_R(J^1 - 1) - \tilde{v}_P(J^1 + 1)$ against J^1.

Calculate I_0, I_1, r_0 and r_1 for H—^{35}Cl.

Discussion (b)

Explain the significance of differences between r_0 and r_1 in terms of the potential energy diagram of the HCl molecule.

V6. The Vibration–Rotation Spectra of CO_2 and H_2O

Object

To relate the observed infrared spectrum of CO_2 gas and H_2O vapour to fundamental modes of vibration and vibration–rotation interactions.

Theory

CO_2 and H_2O are examples of linear and bent triatomic molecules respectively. The vibrational properties of molecules of these types have been discussed on page 178. The fundamental modes of vibration are as shown in Fig. V6.

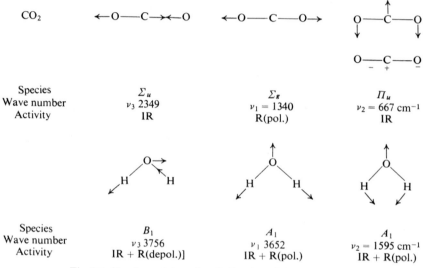

Species	Σ_u	Σ_g	Π_u
Wave number	ν_3 2349	$\nu_1 = 1340$	$\nu_2 = 667$ cm⁻¹
Activity	IR	R(pol.)	IR

Species	B_1	A_1	A_1
Wave number	ν_3 3756	ν_1 3652	$\nu_2 = 1595$ cm⁻¹
Activity	IR + R(depol.)]	IR + R(pol.)	IR + R(pol.)

Fig. V6. Fundamental modes of vibration of CO_2 and H_2O

Aspects of these vibrations in relation to symmetry and infrared and Raman activities, degeneracy of vibrational energy levels, Fermi resonance between vibrational energy levels and rotational structure of infrared bands have been considered earlier.

Apparatus

Infrared spectrometer (650–4000 cm⁻¹) with medium resolution and single beam mode of operation.

Procedure

Record the spectrum of the atmosphere in the optical path of the infrared spectrometer under single-beam conditions. Adjust the amplifier gain in the various regions of the spectrum to obtain suitable presentation of energy of the source with clear display of absorption bands from the atmosphere.

Results and Discussion

1. Select bands due to CO_2 and H_2O and assign them to appropriate modes of vibration.

2. Compare the CO_2 antisymmetric stretching band with the HCl fundamental band (Expt. V5) and/or the CO fundamental band (Fig. 1.6). Explain any similarities and differences in the rotational structure.

3. Note the changes in rotational structure for the CO_2 mode which vibrates *perpendicular* to the molecular axis compared with the mode which vibrates *parallel* to the molecular axis. Suggest a generalization from these observations for linear molecules.

4. Note the complexity of the rotational structure of H_2O. Suggest the basis for this in terms of classification of types of rotor.

5. How can the rule of mutual exclusion be applied to these molecules?

6. Rationalize the existence of four and three fundamental vibrations in CO_2 and H_2O respectively.

V7. Structure of the Nitronium Ion

Object

To observe the Raman spectrum of the nitronium ion and to deduce the structure of the ion.

Theory

The nature of the nitrating species present in a solution of nitric acid in sulphuric acid has been the subject of considerable speculation. Cryoscopic measurements have shown the presence of four species in dilute solution. Detailed considerations of Raman spectra by Ingold *et al.* (1950) suggested the existence of the nitronium ion NO_2^+ in the mixture. The ion is presumably formed by the reaction

$$HNO_3 + 2H_2SO_4 = 2HSO_4^- + NO_2^+ + H_3O^+$$

Raman spectra provide information on molecular vibrations from observations in the visible region of the spectrum. The vibrational spectra of aqueous and corrosive systems which cannot be contained in infrared cells may, therefore, be obtained by the Raman effect using glass cells. In the case of symmetrical molecules and ions the Raman active vibrations may be more significant than the infrared ones. In particular the Raman spectrum may give unequivocal evidence for the symmetry of a species. For example the NO_2^+ ion may either resemble H_2O (bent) or CO_2 (linear). If the ion is linear and centro-symmetric the infrared and Raman spectra will be very similar to those for CO_2 (Expt. V6). Examination of the Raman spectra can, therefore, provide evidence for the structure of the NO_2^+ ion.

Apparatus

Raman spectrograph (photoelectric or photographic recording instrument); colourless and suspension free samples of HNO_3 and H_2SO_4.

Procedure

Make up a solution of 15 % v/v HNO_3 in H_2SO_4 and carefully fill three Raman cells with solvent, solution and solute respectively. Obtain Raman spectra of the solvent, solution and solute under comparable conditions using either a photographic or a photoelectric method. Adjust the conditions under which the spectra are obtained to get optimum performance.

Obtain a spectrum of the excitation source by insertion of a piece of cotton wool in the sample position within the source to reflect the radiation of the source into the spectrometer.

If the spectra are recorded photographically, record the emission spectrum of a suitable calibration arc and superimpose on this a weak spectrum of the excitation source. The solvent, solution, solute, source and calibration spectra may be all recorded on one plate by racking the plate into different positions between exposures. Develop, fix, wash and dry the plate using the standard procedures for the particular plates used.

Results

Examine the spectra of the solvent, solution and solute and by comparison note those lines in the spectrum of the solution which may be due to (a) H_2SO_4, (b) HNO_3, (c) any new species. Measure the Raman wave number value of any new species. If the spectra are photographically recorded use the methods described on page 169 for measurement of the wave number values of lines on the plate.

Discussion

1. Is the Raman spectrum consistent with a linear or bent structure for the NO_2^+ ion?

2. How many valency electrons are there in CO_2 and NO_2^+? Comment on the implications of this comparison.

3. What further information would be obtained by measurement of the polarization of the lines in the Raman spectrum?

4. What are the causes of the differences between the shape of the bands of CO_2 (gas) and NO_2^+ (solution)?

V8. Vibrational Spectra of Triangular, Pyramidal and Tetrahedral Molecules or Ions

Object

To relate the observed spectra of compounds of the type AB_3 and AB_4 to their structure and symmetry.

Theory

The vibrational spectra of triangular and pyramidal molecules and ions of type AB_3 and of tetrahedral molecules and ions of type AB_4 have been discussed on page 138. The forms of the vibration are shown in Fig. 3.9 and the wave number values of particular molecules or ions are summarized in Table V8.

Table V8

(i) AB₃ planar		Point group D_{3h}		
Symmetry species	A_1'	A_2''	E_1'	
Activity	R(pol.)	IR	R(depol.)	+ IR
Wave number NO_3^-	1050	831	1390	720
CO_3^{2-}	1063	879	1415	680

(ii) AB₃ pyramidal		Point group C_{3v}		
Symmetry species	A_1		E	
Activity	R(pol.) + IR		R(depol.) + IR	
Wave number SO_3^{2-}	960	612	925	471

(iii) AB₄ tetrahedral		Point group T_d		
Symmetry species	A_1	E	F_2	
Activity	R(pol.)	R(depol.)	R(depol.) + IR	
Wave number CCl_4	458	218	762⎫ 790⎭	314
SO_4^{2-}	980	450	1100	620

Apparatus

Infrared spectrometer (400–2000 cm^{-1}); Raman spectrograph (photoelectric or photographic recording instrument); BaF_2 plates (or equivalent) for recording infrared spectra of aqueous solutions as thin films; equipment for making halide discs; $NaNO_3$, Na_2CO_3, Na_2SO_3, Na_2SO_4 and CCl_4.

Procedure

Record the infrared spectra of the solids as KBr or KCl discs. Make up a saturated solution (approximately 2 mol dm^{-3}) of the salts in water. Record the infrared spectrum of an aqueous solution of Na_2SO_4 between 800 and 2000 cm^{-1} as a thin film. Record the infrared spectrum of CCl_4 between 400 and 2000 cm^{-1} at 0·1 mm path-length. Repeat the infrared spectrum for any band for which the absorption is intense using less sample in the case of discs and using a thinner film in the case of Na_2SO_4 solution or CCl_4.

Filter each solution into a Raman cell to avoid traces of suspended matter scattering radiation into the spectrometer. Obtain Raman spectra of all solutions and of water following the general procedure outlined in Expt. V7.

Results

Examine the spectra of the solids and solutions, tabulate the wave number values of the various bands and lines and assign these to the vibrations of the various molecules and ions.

Discussion

1. Consider the tabulated wave number values. In which cases do the spectra give positive evidence for structure and symmetry?

2. Compare the infrared spectra of the sulphate ion in the solid state and in solution and comment on changes in the nature of the bands.

3. What effect would be expected in the spectra of the carbonate ion on formation of, (i) monodentate, (ii) bidentate complexes?

4. Observe the effect of Fermi resonance on one of the fundamental vibrations in both the infrared and Raman spectrum of CCl_4. Suggest the cause of this in terms of components involved in the resonance interaction (p. 141).

C. STRUCTURAL CORRELATIONS BY INFRARED SPECTROSCOPY

Characteristic vibrations associated with specific groups are considered in this section. An experiment is provided (V9) to gain familiarity with the empirical basis of group vibrations. The effect of hydrogen bonding on O—H bands in acids is investigated (V10) and the effect of deuterium substitution on X—H vibrations is also examined (V11). Substituent effects on carbonyl and olefinic C—H bands are studied (V12) and the correlation between substitution patterns in aromatic compounds and certain regions of their infrared spectra is also investigated (V13).

V9. Characteristic Infrared Group Vibrations

Object

To record the spectra of selected compounds utilizing a range of sample mounting techniques and to study correlations in group vibrations.

Theory

On page 143 the factors responsible for the existence of characteristic group vibrations were outlined; it was shown that the following modes of vibration gave rise to absorption bands characteristic of particular groups.

Type I   X—H stretch   C—H, N—H, O—H

Type II (a) X=Y stretch   C=C, C=N, C=O

       (b) X≡Y stretch   C≡C, C≡N

It may be shown by the methods of classical dynamics that the above characteristic vibrations occur at relatively high wave numbers, in the region 4000–1500 cm^{-1}. This region may be examined, therefore, for information on which functional groups are present in a molecule. It also follows from classical dynamics that vibrations at low wave number values (1500–100 cm^{-1}) are skeletal modes in which each atom contributes some motion to the mode. In the case of skeletal modes there is no simple relationship between an absorption band and a particular characteristic group. A complex set of bands exist which are unique for a particular structure and are likely to be highly characteristic for a structure rather than a functional group. This leads to the concept of the "fingerprint region" for the region 1500–100 cm^{-1}.

Certain correlations are, however, found in the lower wave number region which suggest some modes are dominated by particular groups in the molecule. If local symmetry elements are present in a molecule (for example, the plane of an ethylenic group or a benzene ring) vibrations are limited to specific regions by symmetry considerations. Thus out-of-plane vibrations of olefins and aromatics have characteristic wave number values. The out-of-plane vibrations do not couple with the in-plane vibrations because of symmetry factors. Other correlations in the 100–1500 cm^{-1} region are found for C—Cl and C—O groups.

Apparatus

Infrared spectrometer (650–4000 cm^{-1}, fast scan). Equipment for preparing samples as films, solutions, mulls and discs. Compounds selected from Table V9.

Procedure

Record the infrared spectra of a selection of compounds from Table V9 under the conditions suggested. It is intended that other members of a group of students run different sets of spectra so that a large collection of spectra is made. When several spectra are to be recorded of the same compound under different conditions superimpose the spectra on a single chart.

Discussion

The various spectra in the suggested set form a basis for discussion of correlations of group vibrations.

1. Compare spectra 1, 2, 3 and 4 at the same path-length. Only one fundamental vibration in carbon tetrachloride is observed above 650 cm^{-1}. This is the antisymmetric C—Cl stretching vibration in Fermi resonance with a non-fundamental vibration. To what extent can analogous modes be picked out in spectra 2, 3 and 4?

2. From the wave number values of the fundamental vibrations of carbon tetrachloride suggest assignments for the bands observed at 1·0 mm path-

Table V9

No.	Sample	State	Path-length (mm)	Range (cm^{-1})
1	Carbon tetrachloride	Liq	0·1	650–4000
			1·0	650–4000
2	Chloroform	Liq	0·1	650–4000
3	n-Propyl chloride	Liq	0·1	650–4000
4	Chlorobenzene	Liq	0·1	650–4000
5	n-Hexane	Liq	Thin film	2500–3500
			0·1	650–4000
6	Toluene	Liq	Thin film	2500–3500
			0·1	650–4000
7	Polystyrene	Solid film		650–4000
8	Benzonitrile	Liq	0·05	650–4000
9	Phenylacetylene	Liq	0·05	650–4000
10	Oct-1-ene	Soln. CCl_4 10% v/v	0·1	650–4000
11	trans-Oct-2-ene	Soln. CCl_4 10% v/v	0·1	650–4000
12	2,4,4-Trimethylpent-1-ene	Soln. CCl_4 10% v/v	0·1	650–4000
13	Styrene	Soln. CCl_4 10% v/v	0·1	650–4000
14	Ethanol	Soln. CCl_4 1% v/v	1·0	2000–4000
			0·1	650–4000
15	n-Hexanol	Soln. CCl_4 1% v/v	1·0	2000–4000
			0·1	650–4000
16	Phenol (on warming)	Liq	Thin film	650–4000
17	Benzyl alcohol	Soln. CCl_4 10% v/v	1·0	2000–4000
			0·1	650–4000
18	n-Propyl aldehyde	Soln. CCl_4 1% v/v	0·1	1600–1800
			0·1	650–4000
19	Benzaldehyde	Soln. CCl_4 1% v/v	0·1	1600–1800

No.	Compound	Preparation	Path length (mm)	Range (cm⁻¹)
21	Methyl n-butyl ketone	Soln. CCl₄ 10% v/v	0·1	2600–3200
				650–4000
		Soln. CCl₄ 1% v/v	0·1	1600–1800
22	Acetophenone	Soln. CCl₄ 10% v/v	0·1	2600–3200
				650–4000
		Soln. CCl₄ 1% v/v	0·1	1600–1800
23	Benzophenone	Soln. CCl₄ 10% v/v	0·1	2600–3200
				650–4000
		Soln. CCl₄ 1% v/v	0·1	1600–1800
24	Ethyl acetate	Soln. CCl₄ 10% v/v	0·1	2600–3200
				650–4000
		Soln. CCl₄ 1% v/v	0·1	1600–1800
25	Ethyl benzoate	Soln. CCl₄ 10% v/v	0·1	2600–3200
				650–4000
		Soln. CCl₄ 1% v/v	0·1	1600–1800
26	Acetic acid	Soln. CCl₄ 10% v/v	0·1	650–4000
		Liq	Thin film	650–4000
27	n-Hexoic acid	Soln. CCl₄ 1% v/v	1·0	1600–4000
		Soln. CCl₄ 10% v/v	0·1	650–4000
28	Benzoic acid	Solid, mull and disc		650–4000
29	Sodium acetate	Solid, mull and disc		650–4000
30	Sodium benzoate	Solid, mull and disc		650–4000
31	Sodium formate	Solid, mull and disc		650–4000
32	Nitrobenzene	Liq	Thin film	650–4000
33	Formamide	Liq	Thin film	650–4000
34	Dimethyl formamide	Liq	Thin film	650–4000
35	Acetamide	Solid, mull and disc		650–4000
36	N-Methyl acetamide	Solid, mull and disc		650–4000
37	Benzamide	Solid, mull and disc		650–4000

length in the regions (a) 800–1600 cm^{-1} (b) 1600–2500 cm^{-1}. What values for path-length of cell may be utilized in the X—H stretching region using, (a) CCl$_4$, (b) CHCl$_3$ as solvents.

3. Note the different positions of bands corresponding to aliphatic, aromatic and olefinic C—H stretching modes in spectra 2, 3, 4, 5, 6 and 10. What are the assignments of the bands between 2500–3500 cm^{-1} in polystyrene (No. 7)?

4. Note any correlations for the phenyl structural unit in spectra 4, 6, 7, 8 and 9. What assignments may be made for the main additional bands in the spectra? Compare patterns in the regions 650–800 cm^{-1} and 1600–2000 cm^{-1} with other types of substituted aromatic compounds. (See Expts V13 and V15.)

5. Assign bands due to the olefinic groups in spectra 10, 11 and 12. Why is the C=C stretching mode considerably weaker in spectrum 11 than in spectrum 12? Assign the strong bands arising from the out-of-plane vibrations

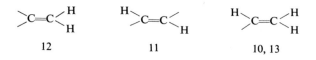

(850–1000 cm^{-1}) of the olefinic C—H group in the structures. Further consideration of these modes is given in experiment V12.

6. Note the bands due to O—H stretching vibrations in spectra 14, 15 and 17 and the changes in, (i) wave number value, (ii) intensity and (iii) band-width with dilution under constant "number of molecules in beam" conditions.

7. Comment on any differences between the wave number value of the bands corresponding to the O—H stretching vibration in spectra 16 and 17.

8. Which bands in spectra 18 and 19 can be assigned to vibrations of the —CH=O group. Note the Fermi resonance splitting (see p. 141) of the band due to the aldehydic C—H stretching mode and suggest the components which contribute to the resonance.

9. Note the intensities of the bands due to the CH$_3$ stretching mode and symmetric CH$_3$ bending mode (1360 cm^{-1}) in spectra 20 and 21. What is the influence of an adjacent carbonyl group on CH$_3$ stretching and bending modes?

10. Measure the wave number of the carbonyl vibration in spectra 24 and 25. What may the strong bands between 1000 and 1300 cm^{-1} be assigned to in these spectra? What classes of compounds would you examine to confirm these assignments?

11. How does the band due to the O—H stretching vibration in acids (27 and 28) differ from that in phenols (16) and alcohols (14, 15 and 17)? Compare the effects of hydrogen bonding in each case on the O—H and also on the C=O band for the acids. Comment on the effect of the change from 10% concentration at 0·1-mm path-length to 1% concentration at 1·0-mm path-

length in each case, and estimate the relative difference in stability of the hydrogen-bonded form between acids and alcohols. How may this be explained? (Experiment V10 extends this study further.)

12. Compare the spectra of acetic acid (26) and benzoic acid (28) with their sodium salts (29 and 30) on the basis of the structure of the carboxylate ion.

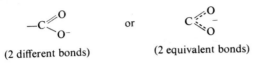

(2 different bonds)   (2 equivalent bonds)

13. Note any similarity between spectra 30 and 32. Do the results suggest closely similar structures for these species? Compare the electronic structures of these species.

14. By comparing the spectra of formamide and dimethyl formamide (33 and 34) and acetamide and N-methyl acetamide (35 and 36) assign bands due to N—H stretch, N—H bend and C=O stretch. An analysis of amide modes is described on page 156. Assign the amide modes in benzamide.

15. Measure the approximate wave number values of carbonyl bands in the following compounds for R = methyl and R = phenyl.

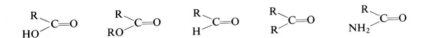

Further consideration of the factors affecting carbonyl vibrations are given in Expt. V12.

VI0. Hydrogen Bonding in Acids

Object

To study the effect of hydrogen bonding on the infrared spectra of carboxylic acids.

Theory

The spectrum of a 1% solution of n-hexoic acid in CCl_4 provides evidence that n-hexoic acid exists mainly in a hydrogen-bonded form. The equilibrium between the bonded and non-bonded forms may be studied as follows

>C=O------H—O⏜	⇌	>C=O	+	H—O⏜
strong	broad strong	strong		sharp, medium bands
1715	2500–3500	1760		3550 cm⁻¹

Hydrogen bonds may be intermolecular, which should be concentration dependent, or intramolecular, which should be invariant with concentration. Examples of inter- and intramolecular hydrogen-bonded systems are as follows

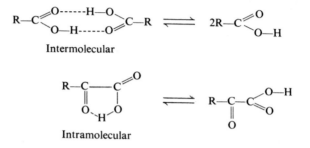

Intermolecular

Intramolecular

Dilution studies on any particular sample should provide information on the type of bonding present, together with a semi-quantitative estimate of the relative amount of each form present at various concentrations.

Apparatus

Infrared spectrometer (650–4000 cm^{-1}); cells and spacers for path-lengths of 0·1, 1·0 and 10·0 mm; acetic acid, pyruvic acid and spectroscopically pure CCl_4.

Procedure

Record the spectra of acetic acid and pyruvic acid as thin films between 650 and 4000 cm^{-1}. Record the spectra of the following solutions using solvent compensation in the reference beam to reduce solvent background.

Path length mm	1·0	10
Concentration in CCl_4 solution % v/v	0·1	0·01

Assign bands in the spectra to O—H and C=O stretching vibrations of bonded and unbonded forms.

Results

By comparison between the spectra which have been recorded and the spectra in the figure, summarize the spectral changes occurring with hydrogen bond formation.

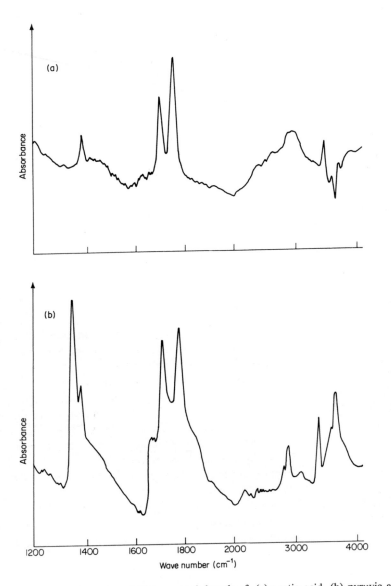

Fig. V10. Infrared spectra of 100-mm path-length of, (a) acetic acid, (b) pyruvic acid (concentration $\approx 0.001\%$ v/v)

Discussion

1. Give reasons for the choice of assignments for the C=O stretching vibration in pyruvic acid.

2. Compare the position of equilibrium between bonded and non-bonded forms of acetic acid with that of isopropanol (Fig. 3.12).

VII. Deuterium Exchange Studies

Object

To identify bands associated with labile hydrogens in X—H groups (X = C, N or O) by comparing spectra of compounds containing X—H and X—D groups.

Theory

It has been shown that to a first approximation the ratio of the wave number values of the bands corresponding to the stretching vibration of an X—D to an X—H group is

$$\frac{\tilde{\nu}_{X-D}}{\tilde{\nu}_{X-H}} = \left(\frac{m_H}{m_D}\right)^{1/2} = 2^{-1/2} \tag{V11.1}$$

Bending modes to which C—H groups contribute some motion will also be sensitive to deuterium substitution. In general a number of bending modes may exist and all will show changes with isotopic substitution. Although it will be expected that the wave numbers of these vibrations will move to lower values on deuterium substitution, changes in coupling conditions may cause certain bands to move to higher wave numbers on substitution. In this case the isotopic substitution has produced a different mode. X—H bending modes do not, therefore, necessarily behave as characteristic group vibrations analogous to X—H stretching vibrations. Nevertheless it is in the case of bending modes that deuterium substitution is particularly useful in making assignments. In certain cases an X—H group may give rise to two pure bending modes. For example, an isolated X—H constrained in a plane will give rise to a high wave number X—H stretch, a low wave number out-of-plane X—H bend and an in-plane bend at an intermediate wave number.

Apparatus

Infrared spectrometer (650–4000 cm^{-1}), dry-box, vacuum line for bulb-to-bulb distillation as Fig. V11, refrigerant (solid CO$_2$); apparatus for making nujol mulls, sodium acetylacetonate, phthalimide sodium α-hydroxyisobutyrate.

To vacuum line

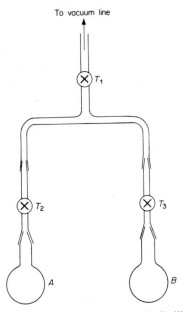

Fig. V11. Apparatus for bulb-to-bulb distillation

Procedure
Select a sample to illustrate one of the exchange processes.

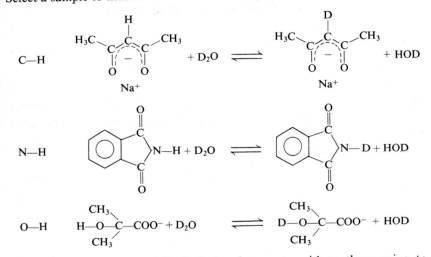

Place about 50 mg of sample in flask *A* and evacuate, with gentle warming, to dry the sample. Close tap T_2, detach at the joint above T_2 and transfer the sample to a dry-box. Detach the flask at the joint below T_2, remove a small amount of sample, prepare a nujol mull and record the infrared spectrum between 650 and 4000 cm^{-1}. Within the dry-box add about 1 cm^3 of D_2O to the sample within flask A, reconnect to T_2 (closed). Withdraw the sample from the dry-box and allow the solid to dissolve in the D_2O by gently warming the flask. Evacuate flask A with gentle warming to remove the D_2O and to dry the sample. (If it is desirable to recover the D_2O, this may be distilled into flask B.) Close tap T_2, detach at the joint above T_2 and transfer the sample to the dry-box.

If time permits add a further 1 cm^3 of D_2O to the sample within the flask and repeat the above procedure.

Prepare a nujol mull of the deuterated sample and record its infrared spectrum between 650 and 4000 cm^{-1}.

Results and Discussion

Compare the spectra of the protonated and deuterated forms of the samples.

1. Comment on the assignment of the X—H stretching and bending modes in the compound studied. Is there evidence of coupling between the X—H modes and other modes?

2. Compare the observed isotopic shift for the X—H stretching mode with the calculated shift using equation V11.1.

V12. Effect of Substituents on Characteristic Group Vibrations
(a) Carbonyl Stretching Modes
(b) Olefinic C—H Out-of-plane Bending Modes

Object

To study the effect of neighbouring substituents on characteristic group vibrations for (a) carbonyl groups, (b) olefinic C—H groups.

Theory (a): Carbonyl Vibrations

The wave number value of a carbonyl vibration is dependent on the chemical environment of the carbonyl group provided that intermolecular effects are absent (these are minimized in dilute solution in CCl_4). Four principal substituent effects may be recognized in adjacent groups:

(i) Electronegative substituent leading to inductive electron withdrawal.

(ii) Substituent with lone-pair electron leading to electron supply by resonance.

(iii) Substituent which is unsaturated or aromatic.

(iv) Substituent involving carbonyl group in strained ring.

An explanation may be provided (p. 150) for the observation that (i) and (iv) lead to a high value of the wave number of the carbonyl vibration whereas (ii) and (iii) lead to a low value.

Apparatus (a)

Infrared spectrometer (1600–1850 cm^{-1}, medium resolution). Cells at 0·1 and 1·0 mm path-length; carbon tetrachloride, acetone.

(i) Acetaldehyde, acetic acid, methyl and phenyl acetate, acetyl chloride.
(ii) Acetamide, dimethylacetamide.
(iii) Mesityl oxide, acetophenone.
(iv) γ-Lactone, cyclopentanone.

Procedure (a)

Prepare a solution of acetone and a selection of the samples listed at concentrations of up to 0·1 % v/v in CCl$_4$. Record the spectra between 1600 and 1850 cm^{-1} at 0·1-mm path-length. If the carbonyl band is too weak to be measured record the spectrum at 1·0-mm path-length. Measure the wave number values of all carbonyl bands.

Results (a)

Tabulate the wave number value of all measured carbonyl bands and tabulate the shift in wave number from the wave number value of the band in acetone.

Discussion (a)

1. Correlate the shift in wave number with the nature of the substituent. Rationalize the effects as far as possible.
2. Predict the wave number values of the carbonyl bands in the following compounds: benzophenone, benzoyl chloride, benzaldehyde, benzamide, phosgene, acrolein, phenylbenzoate, β-lactone.

Compare your predicted values with literature values where possible.

Theory (b): Olefinic C—H Out-of-plane Vibrations

The assignment of the olefinic C—H out-of-plane bending vibrations has been discussed on page 155 for compounds of the type

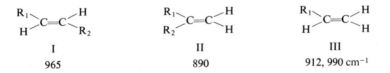

I	II	III
965	890	912, 990 cm^{-1}

The wave number values given are for alkyl substituents. When R is a polar group the wave number values may be modified. Whilst attempts have been made to interpret these changes in terms of resonance and inductive effects (Potts and Nyquist, 1959) the problem is more complex than for the stretching vibration of a single carbonyl group. However, it is worth attempting a rationalization on the basis of additivity of empirically derived shifts. This can allow the prediction of wave number values in unknown compounds.

Apparatus (b)

Infrared spectrometer (800–1000 cm^{-1}, medium resolution). 0·1 mm path-length cell; *trans* $CH_3CH=CHCl$, $CH_3CH=CHC\equiv N$, $CH_2ClCH=CHCl$, $CH_3CH=CHCOOC_2H_5$, $CH_3ClC=CH_2$, $CH_3(C\equiv N)C=CH_2$, $CH_3(CH_2Cl)$-$C=CH_2$, $CH_3(COOC_2H_5)C=CH_2$.

Procedure (b)

Record the infrared spectra of a 5% v/v solution in CCl_4 at 0·1 mm path-length between 820 and 1000 cm^{-1} for a selection of the compounds listed. Assign bands in this region to olefinic C—H out-of-plane vibrations and measure the wave number values.

Results (b)

Tabulate the wave number values of all measured bands and tabulate the shift in wave number from the wave number value of the bands for alkyl substituents.

Discussion (b)

1. Correlate the shift in wave number with the substituent. Suggest quantitative values of shifts induced by various substituents.

2. Assuming additivity of shifts when two substituents effect the wave number value of a particular band, predict the values of the C—H out-of-plane vibration in the compounds I and II for the substituents

$R_1 = R_2 = C\equiv N$ $R_1 = C\equiv N$ $R_2 = Cl$
$R_1 = R_2 = Cl$ $R_1 = C\equiv N$ $R_2 = CH_2Cl$
$R_1 = R_2 = CH_2Cl$ $R_1 = C\equiv N$ $R_2 = COOC_2H_5$
$R_1 = R_2 = COOC_2H_5$ $R_1 = Cl$ $R_2 = COOC_2H_5$

3. Assuming the two out-of-plane modes of the vinyl group (III) at 912 and 990 cm^{-1} are analogous to the modes in II and I respectively, predict the wave number value for the C—H out-of-plane vibrations in the compounds III for the substituents

$$R_1 = C\equiv N, \; Cl, \; CH_2Cl, \; COOC_2H_5, \; C(Cl)=CH_2$$

4. Seek literature values for the wave number values you have predicted and comment on the comparison.

VI3. Correlations of Substitution Patterns in Aromatic Compounds

Object

To study the relationship between the wave number and intensities of bands assigned to out-of-plane vibrations and the pattern of substituents around a benzene ring.

Theory

It has been suggested by Young *et al.* (1951) that each of the various types of substitution in benzene compounds gives rise to a pattern of bands which, within limits, is independent of the nature of the substituents. This pattern of bands occurs between 1650 and 2000 cm^{-1}. It has been shown by Whiffen (1955) that these bands are overtone and combination bands arising from C—H out-of-plane and ring out-of-plane vibrations which occur between 650 and 1000 cm^{-1}. The infrared spectra of a series of methyl substituted benzene compounds is shown in Fig. V13.

Apparatus

Infrared spectrometer (650–2000 cm^{-1}), 0·1-mm path-length cell; cyclohexane and a selection of the following compounds

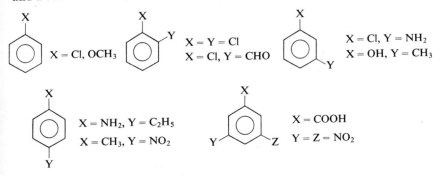

$X = Cl, OCH_3$

$X = Y = Cl$
$X = Cl, Y = CHO$

$X = Cl, Y = NH_2$
$X = OH, Y = CH_3$

$X = NH_2, Y = C_2H_5$
$X = CH_3, Y = NO_2$

$X = COOH$
$Y = Z = NO_2$

Procedure

Record the spectra between 650 and 850 cm^{-1} and between 1650 and 2000 cm^{-1} of a selection of the compounds listed as 10% v/v solutions in cyclohexane and at 0·1-mm path-length and between 1650 and 2000 cm^{-1} as liquid films at 0·1-mm path-length.

Mark bands assigned to aromatic out-of-plane vibrations and list those compounds which, in your opinion, fulfil the predictions of Young, Duvall and Wright. List compounds whose substitution patterns could not be identified from these considerations.

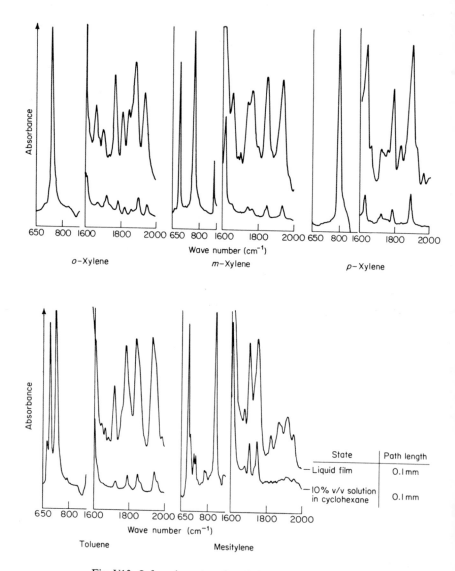

Fig. V13. Infrared spectra of methyl substituted benzenes

Discussion

Indicate which classes of compounds can be profitably examined for type of substitution by this procedure.

D. QUANTITATIVE ANALYSIS BY INFRARED SPECTROSCOPY

Although many competitive techniques exist for quantitative analysis, examples of estimations which can be made, with advantage, by infrared spectroscopy frequently occur. The principles of the method are illustrated with reference to analysis of a two-component mixture of isopropanol and acetone (V14), of a four-component mixture of xylene isomers (V15) and of the various structural repeating units in polychloroprene (V16).

V14. Analysis of a Two-component Mixture

Object

To estimate the amount of acetone present in a mixture of isopropanol and acetone.

Theory

Acetone is a common impurity in isopropanol owing to the facile oxidation of isopropanol:

$$\begin{array}{c} H_3C \\ {} \\ H_3C \end{array}\!\!\!\!> CH(OH) \xrightarrow{\text{[O]}} \begin{array}{c} H_3C \\ {} \\ H_3C \end{array}\!\!\!\!> C{=}O + H_2O$$

Examination of the reference spectra of isopropanol and acetone reveals that the strongest absorption band in acetone, the C$=$O stretching vibration near 1719 cm^{-1}, occurs in a region of the spectrum where isopropanol has very low absorption.

The absorbance, A, of the carbonyl stretching vibration in the mixture may be measured for a solution of known concentration in CCl_4. From a Beer–Lambert law calibration plot, using a series of standard solution, the concentration of acetone in the mixture may be estimated.

A valuable check on the accuracy of the method is to make up a synthetic mixture of acetone and isopropanol in CCl_4. The spectra of this and the original solution should "match" one another to within the transmittance reproducibility of the spectrometer.

Apparatus

Infrared spectrometer (650–4000 cm^{-1}); 0·1 mm path-length cell; acetone, isopropanol; mixture of acetone and isopropanol, CCl_4.

Procedure

Prepare a 10% solution in CCl_4 of the mixture and record the spectrum between 650 and 4000 cm^{-1} at 0·1-mm path-length. If the absorption of the carbonyl stretching vibration occurs outside the range 25–50% of trans-mittance, re-run the spectrum between 1600 and 1900 cm^{-1} using either a 5% or a 20% solution or some other appropriate concentration. Make up a series of solutions of acetone in CCl_4 at 2·0, 1·5, 1·0, 0·5, 0·25% v/v. Record the spectrum between 1600 and 1900 cm^{-1} of each solution, superimposing these spectra on a single chart.

Results

Draw a tangential base line to the carbonyl bands and measure the difference in absorbance units between the peak maxima and the base line at the wave number value corresponding to the peak maxima. Plot the absorbance of each standard solution against concentration and determine the concentration of the unknown solution from the graph. Calculate the extinction coefficient, ϵ, of the carbonyl band in units of $m^2 \, mol^{-1}$ from the slope of the graph. Note: Density of acetone $= 0·790 \, g \, cm^{-3}$.

Further Procedure

Assuming only isopropanol and acetone are present in the mixture, make up a synthetic mixture corresponding to your calculated composition at a total concentration of 10% in CCl_4 (or whatever concentration was used). Record the spectrum and compare with the initial spectrum.

Discussion

1. Briefly comment on the assignment of the functional group vibrations in isopropanol and acetone.

2. Comment on the agreement between your results and the Beer–Lambert law. Would you expect similar agreement if you used the band at 3400 cm^{-1} to estimate the isopropanol content?

3. What alternative techniques would you consider feasible for this estimation?

4. Comment on the relative magnitude of the extinction coefficient of the measured band compared with the extinction coefficient of typical bands in the ultraviolet region of the spectrum.

VI5. Analysis of a Four-Component Mixture

Object

To estimate the amount of *o*, *m* and *p*-xylene and ethyl benzene in a sample of commercial xylene.

Theory

The spectra of o, m and p-xylenes reveal strong bands at 740, 690–770, and 800 cm^{-1} respectively. Cyclohexane has a very low absorbance in this region and is a suitable solvent for quantitative estimations.

Analysis of mixtures of this type can be readily performed by a process of successive approximations by which standards are made up at the calculated composition of the sample. The calculated compositions of successive standard solutions should produce spectra with absorbance values for all components which approach the values for the sample mixture. Any unaccounted difference between the spectra of the sample and the synthetic mixture would arise from a further component. A difference spectrum should reveal bands due to this component which may be identified and estimated in an approximate manner.

Apparatus

Infrared spectrometer (650–4000 cm^{-1}); a pair of matched 0·1-mm path-length cells; commercial xylene (containing o, m and p-xylene and ethyl benzene); o, m and p-xylene, ethyl benzene, cyclohexane.

Procedure

Record the spectra between 650 and 850 cm^{-1} of each of the following solutions in a 0·1-mm cell with a 0·1-mm cell containing cyclohexane in the reference beam.

Commercial xylene 10% v/v. Measure A_o, A_m and A_p in mixture
    o-xylene 1% v/v. Measure A_o and determine $E_o = A_o$
    m-xylene 2% v/v. Measure A_m and determine $E_m = 0·5A_m$
    p-xylene 2% v/v. Measure A_p and determine $E_p = 0·5A_p$

where A_o is the measured absorbance of o-xylene at 740 cm^{-1}, A_m is the measured absorbance of m-xylene at 770, A_p is the measured absorbance of p-xylene at 800 and E is the absorbance of a 1% solution in a 0·1-mm path-length cell.

Estimate, to a first approximation, the concentrations C_o^I C_m^I and C_p^I of the o, m and p isomers in the commercial sample of xylene using the E values obtained above and the relationship $A = EC$.

Note: This treatment ignores, (1) apparent deviations from the Beer–Lambert law, (2) the effect of overlapping absorption. Both these factors can be compensated for by the following procedures.

Prepare three reference solutions at concentration C_o^I, C_m^I and C_p^I and superimpose their spectra on a single chart (650–800 cm^{-1}). Measure the absorbance values of appropriate bands A_o, A_m and A_p and calculate a second set of E values taking overlapping absorption into account.

$$E_o^I = \frac{A_o}{C_o^I}, \quad E_m^I = \frac{A_m}{C_m^I} \quad \text{and} \quad E_p^I = \frac{A_p}{C_p^I}$$

Calculate a second set of concentrations for the components of the mixture in terms of the measured absorbances of the appropriate bands in the spectrum of the mixture using the E^1 values.

Note: In principle it is possible to repeat this procedure until the absorbance values of the sample and reference solutions agree to within the transmittance reproducibility of the instrument.

Finally, make up a synthetic blend of o, m and p-xylenes up to a total concentration of 10% in cyclohexane. Record a difference spectrum by placing the 10% solution of commercial xylene in the sample beam and the 10% solution of the synthetic xylene blend in the reference beam. Compare 'the difference spectrum with that of ethyl benzene. If necessary record the spectrum of this as a solution in cyclohexane.

Results

Tabulate the successive values of A, E and C for the three components. Estimate the approximate amount of ethyl benzene in the sample of commercial xylene.

Discussion

1. Comment on the comparison between the "matched" spectra.

2. What reservations should be made when examining spectra obtained by solvent cancellation? Are any curious effects observed in your spectra?

3. Suggest which is the most suitable method for this analysis: (a) mass spectrometry, (b) ultraviolet-absorption spectroscopy, (c) gas chromatography, (d) infrared-absorption spectroscopy.

VI6. Stucture of Polychloroprene

Object

To estimate the relative amounts of the different types of repeating units in a sample of polychloroprene.

Theory

Chloroprene may polymerize leading to four possible types of repeating units (Fig. V16). Consideration of infrared absorption bands due to olefinic groups and the effect of chlorine substitution on the wave number value of these bands suggest that specific bands should occur for each of the units: *trans* 1,4; *cis* 1,4; 1,2 and 3,4. If suitable model compounds are available containing groupings of these types the relative amount of each repeating unit can be estimated in terms of a model compound as a standard. Methods have been described for the analysis based on monomeric model compounds (Maynard

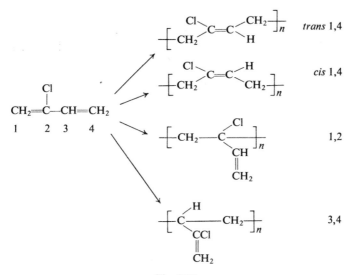

Fig. V16

and Mochel, 1954) and on polymeric model compounds prepared by stereo-specific methods (Ferguson, 1964).

The compounds and the associated modes of vibration listed in Table V16 provide suitable standards for the analysis of polychloroprene.

The relative amount of *trans* 1,4 and *cis* 1,4 addition product may be obtained by application of the Beer–Lambert law taking account of the effect of over-lapping absorptions of bands listed in the table.

$$A_1 = \epsilon_{1T} C_T l + \epsilon_{1C} C_C l \tag{V16.1}$$

$$A_2 = \epsilon_{2T} C_T l + \epsilon_{2C} C_C l \tag{V16.2}$$

where A is the absorbance, ϵ is the extinction coefficient at the wave number value at which A is measured, C is the mole fraction of repeating unit and l is the path-length. The concentration is expressed as a mole fraction because of the necessity of taking the concentration of the other isomer unit into account. The subscripts 1 and 2 refer to measurements made at the wave number value corresponding to the peak absorbance of the bands which occur at 1661·5 and 1652·5 cm^{-1} respectively in the model compounds. The subscripts T and C refer to the *trans* 1,4 and *cis* 1,4 units respectively.

From a set of solutions in which the mole fraction of *cis* 1, 4 units had values between 0 and 0·20, the extinction coefficients were estimated as follows (for path length in units of mm).

ϵ_{1T}	ϵ_{2T}	ϵ_{1C}	ϵ_{2C}
0·642	0·265	0·194	0·0797 mm^{-1}

Table V16

Repeating units	Model	Mode of vibration — form	Mode of vibration — description	Wave number (cm^{-1})
trans 1,4	*trans*-4-Cl-octene-4	$\mathrm{Cl}-\overset{\downarrow}{C}=\underset{\mathrm{H}}{\overset{\uparrow}{C}}$	C=C stretch	1661·5
cis 1,4	*cis*-4-Cl-octene-4	$\mathrm{Cl}-\overset{\downarrow}{C}=\overset{\mathrm{H}}{\overset{\uparrow}{C}}$	C=C stretch	1652·5
1,2	3-Cl-butene-	$\mathrm{CH_3-CH}-\overset{\mathrm{Cl}}{\underset{\,}{\mathrm{CH}}}=\mathrm{C}\begin{smallmatrix}\mathrm{H+}\\ \mathrm{+H}\end{smallmatrix}$	Olefinic C—H out-of-plane bend	926
3,4	Substituted 2-Cl-butene-1 (R = alkyl)	$\mathrm{CH_3-CH}\overset{\mathrm{R}}{\underset{\,}{}}\ \mathrm{CCl}=\mathrm{C}\begin{smallmatrix}\mathrm{H+}\\ \mathrm{+H}\end{smallmatrix}$	Olefinic C—H out-of-plane bend	884

The relative amounts of *trans* 1,4 and *cis* 1,4 repeating units in a solution of polychloroprene may be estimated by measurement of the absorbance of the bands corresponding to those at $1661 \cdot 5$ cm^{-1} and $1652 \cdot 5$ cm^{-1} and substitution of the above ϵ values in equations 1 and 2 which may be solved for C_T and C_C. These values are likely to be in error because the assumption of the validity of the use of model compounds is unlikely to be very accurate. However, the relative concentrations of the two repeating units is likely to have more significance and the results are, therefore, expressed as a ratio

$$C:T = \frac{C_C}{C_T + C_C} : \frac{C_T}{C_T + C_C} \qquad (V16.3)$$

The concentrations of 1,2 and 3,4 repeating units are usually low. Also the bands corresponding to those at 926 and 884 cm^{-1} in the model compounds are not overlapped. The concentrations of the 1,2 and 3,4 repeating units ($C_{1,2}$ and $C_{3,4}$) may be estimated from the Beer–Lambert law

$$C_{1,2} = \frac{A_{1,2}}{\epsilon_{12} l} \qquad C_{3,4} = \frac{A_{3,4}}{\epsilon_{3,4} l} \qquad (V16.4)$$

where A is the absorbance of the appropriate band, l is the path-length and ϵ is the appropriate extinction coefficient. Since the bands are well separated the concentrations can be expressed in mol dm^{-3} for each compound. If the path length is in units of mm the extinction coefficients derived from model compounds have the values

$\epsilon_{1,2}$	$\epsilon_{3,4}$
$11 \cdot 5$	$18 \cdot 4$ m^2 mol^{-1}

Apparatus

Infrared spectrometer (800–2000 cm^{-1}) medium resolution; pair of matched cells with $1 \cdot 0$-mm path-length (accurately known); CS$_2$, sample of polychloroprene which is additive free and soluble in CS$_2$.

Procedure

Make up a solution of polychloroprene in CS$_2$ at a known concentration of about 3 g dm^{-3}. Record the spectrum of this solution between 800 and 2000 cm^{-1} using a $1 \cdot 0$-mm path-length cell and a matching cell containing CS$_2$ in the reference beam. Confirm that the calibration, resolution and transmittance reproducibility of the instrument is up to specification in the region 1500–1700 cm^{-1} by scanning the spectrum of a standard polystyrene film three times in this region using expansion of the wave number scale and suitable values of scan time and gain setting. Record the spectrum of the solution of polychloroprene between 1500 and 1800 cm^{-1} under the instrumental conditions used for polystyrene at $1 \cdot 0$-mm path-length and with a matching cell containing CS$_2$ in the reference beam.

Results

Construct the base line to the pair of bands near 1650 cm^{-1} by drawing a line parallel to the zero-absorbance value so that the line passes through the minimum point of the spectrum between 1700 and 1800 cm^{-1}.

Note: The base line is constructed in this manner since the extinction coefficients of the standards were measured in this way. Select the peak corresponding to the *trans* 1,4 addition product and select the shoulder on the low wave number side corresponding to the *cis* 1,4 addition product.

Note: The wave number values of the bands due to *trans* 1,4 and *cis* 1,4 C=C stretching vibrations have been measured at 1660·2 and 1652·2 cm^{-1} respectively in standard polychloroprene samples prepared by stereospecific methods (Ferguson, 1964). If the position of the shoulder is ill-defined measure 8 cm^{-1} to lower wave number value of the peak maxima of the main band.

Measure the absorbance values at the two selected wave number values and calculate C_T and C_C from equations V16.1 and V16.2. Calculate the ratio $C:T$ from equation V16.3.

Select the bands due to 1,2 and 3,4 repeating units and draw a tangential base line to each band. Calculate $C_{1,2}$ and $C_{3,4}$ from equation V16.4. Convert these concentrations into percentages of each repeating unit present in the original sample.

Tabulate the proportions of each type of repeating unit present in the original sample.

Discussion

1. If a single-beam infrared spectrum of water vapour is available examine the region between 1500 and 1700 cm^{-1} for specific bands which are likely to cause serious interference with this estimation. How can this interference be minimized?

2. If the half-band width of the bands assigned to overlapping C=C stretching vibrations is about 15 cm^{-1}, how narrow should the spectral band-width of the instrument be?

3. Consider briefly possible random errors due to instrumental variations, handling technique, sample inhomogeneity and error in absorbance measurement. Consider also possible systematic errors such as measurement of background absorption, interference from oxidation products or other contaminants, difference in either half-band width or extinction coefficients between sample and model compound. Which sources of error do you consider to be the most significant?

REFERENCES

Badger, R. M., and Zumwalt, L. R. (1938). *J. Chem. Phys.* **6**, 711.

Bellamy, L. J. (1968). "Advances in Infrared Group Frequencies". Methuen, London.

Ferguson, R. C. (1964). *Analyt. Chem.* **36**, 2204.

Herzberg, H. (1945). "Infrared and Raman Spectra". Van Nostrand, New York.

IUPAC. (1961). "Tables of Wave Numbers for the Calibration of Infrared Spectrometers". Butterworths, London.

Ingold, C. K., Millen, D. J., and Poole, H. G. (1950). *J. Chem. Soc.* 2576.

Maynard, J. T., and Mochel, W. E. (1954). *J. Polym. Sci.* **13**, 251.

Miller, R. G. J. (ed.) (1965). "Laboratory Methods in Infrared Spectroscopy". Heyden, London.

Miyazawa, T., Shimanouchi, T., and Mizushima, S. (1958). *J. Chem. Phys.* **29**, 611.

Placzek, G. (1934). "Handbuch der Radiologie", Vol. 2, p. 209. Translation by U.S. Atomic Energy Commission, UCRL Trans. No. 526(L).

Potts, W. J., and Nyquist, R. A. (1959). *Spectrochim. Acta*, **15**, 679.

Rank, D. H., and Kagarise, R. E. (1950). *J. opt. Soc. Am.* **40**, 89.

Sawyer, R. A. (1963). "Experimental Spectroscopy". Dover, New York.

Scherer, J. R., and Potts, W. J. (1959). *J. Chem. Phys.* **30**, 1527.

Wilson, E. B., Decius, J. C., and Cross, P. C. (1955). "Molecular Vibrations". McGraw-Hill, New York.

Whiffen, D. H. (1955). *Spectrochim. Acta*, **7**, 253.

Young, C. W., Duvall, R. B., and Wright, N. (1951). *Analyt. Chem.* **23**, 709.

8

Chapter 4

Nuclear Magnetic Resonance

I. PRINCIPLES OF NUCLEAR MAGNETIC RESONANCE

A. GENERAL PRINCIPLES

Nuclear magnetic resonance (n.m.r.) is a term used to describe the interaction which can occur between an oscillating magnetic field and certain types of nuclei when these nuclei are sited in a static magnetic field. The interaction can only occur if the former field is oscillating at specific frequencies which are related to the strength of the static magnetic field. The field strengths commonly used are such that the frequencies at which interaction can occur correspond to values in the radio-frequency region of the electromagnetic spectrum.

The interaction between an oscillating magnetic field and nuclei located in a static magnetic field can be discussed in both classical and quantum-mechanical terms.

I. Classical Description of N.M.R.

The properties of the nuclei of many elements indicate that the nuclei spin about an axis and that they possess spin angular momentum. Nuclei possess a positive charge and just as a circulating negative charge in a coil sets up a magnetic field, as for instance in an electromagnet (Fig. 4.1a), so the spinning positive charge associated with the nucleus gives rise to a magnetic field. The magnetic field will be parallel to the axis of spin and the nucleus will possess a magnetic moment, μ, along this axis. The nucleus can be considered as a minute bar magnet whose axis coincides with the axis of spin (Fig. 4.1b).

When a nucleus is located in an applied static magnetic field, B_0, then it will in general be shielded by the secondary field set up by the electrons surrounding the nucleus. Thus the local field, B, experienced at the nucleus is generally less than the applied field. This can be expressed in the form

$$B = B_0(1 - \sigma) \qquad (4.1)$$

where σ is the shielding constant for the nucleus.

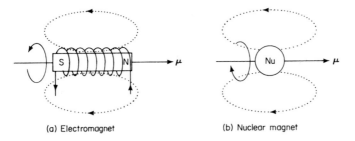

(a) Electromagnet   (b) Nuclear magnet

Fig. 4.1. Fields arising from circulation of charge: (a) negative charge (b) positive charge

The local field experienced by the nucleus exerts a torque tending to align the magnetic moment of the nucleus with the applied field. If the nucleus is spinning the effect of this torque is to cause the magnetic moment of the nucleus to precess around the axis of the applied field B_0 (Fig. 4.2a). The precessing magnetic moment, μ, can be resolved into two components: μ_z which is parallel to the axis of the applied field and is static, and μ_x which is perpendicular to the axis of the field and which rotates at the same angular velocity as the precessing magnetic moment. The angular velocity, ω, of the precessional motion is given, in radians per second, by

$$\omega = \frac{\mu}{I} B \qquad (4.2)$$

or

$$\omega = \gamma B \qquad (4.3)$$

where γ is a constant termed the "gyromagnetic (or magnetogyric) ratio" and is given by the ratio μ/I where I is the spin angular momentum of the nucleus. Equation 4.2 is known as the Larmor equation and can be derived from classical magnetodynamics.

If the Larmor equation is expressed in Hertz instead of radians per second then the following expressions for the frequency, v, of the precessional motion are obtained

$$v = \frac{\mu}{2\pi I} B \tag{4.4}$$

$$v = \frac{\gamma}{2\pi} B \tag{4.5}$$

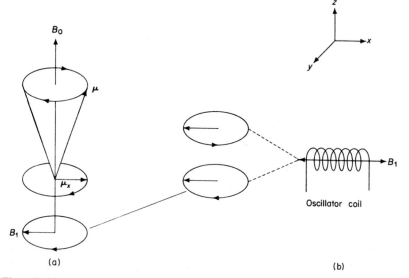

Fig. 4.2. (a) Precessional motion of a nuclear magnetic moment μ around the applied field B_0. (b) Representation of a linear oscillating field as two circularly polarized components

In the n.m.r. experiment an oscillating field, B_1, supplied from an oscillator coil is applied along an axis perpendicular to the axis of the field B_0. The oscillating field can be considered in terms of two circularly polarized component fields which rotate in opposite directions (Fig. 4.2b). In the situation shown in Fig. 4.2b the resultant of the component fields lies along the negative x direction and is of maximum value equal to the sum of the components. As the components rotate, the resultant along the x axis decreases and reaches zero when the components are perpendicular to the x axis, and then rises to a maximum when both components point along the positive x direction. Thus as the components rotate, the magnitude of the resultant field oscillates between the two maximum values.

One way of performing the n.m.r. experiment is to vary the rotational frequency of the circularly polarized components (by varying the oscillation frequency of the coil) until the frequency is the same as the precessional frequency, v, of the nucleus. If the component of the field which is rotating in the same direction as the nucleus has the same precessional frequency as the nucleus, it will remain in phase with the nucleus and will exert a magnetic torque which will tend to change the orientation of the magnetic moment, μ, from the positive z direction (as in Fig. 4.2a) to the negative z direction. When this occurs the rotating component of the field B_1 and the precessing nucleus are in *resonance* and there is an uptake of energy from the oscillator producing the rotating field.

2. Quantum Description of N.M.R.

According to quantum theory a spinning nucleus can only have values for the spin angular momentum I given by the equation

$$I = [I(I+1)]^{1/2} \frac{h}{2\pi} \tag{4.6}$$

where h is Planck's constant and I is the spin quantum number of the nucleus.

Since the magnetic moment, μ, of the nucleus is given by γI it follows that

$$\mu = \gamma [I(I+1)]^{1/2} \frac{h}{2\pi} \tag{4.7}$$

If the nuclear system is placed in a magnetic field then according to quantum theory the component, μ_z, of the magnetic moment along the field direction can only have the specific values given by the equation

$$\mu_z = m_I \gamma \frac{h}{2\pi} \tag{4.8}$$

where m_I is a magnetic quantum number which can have $2I+1$ values, viz. $I, I-1, I-2, \ldots -I$.

As a consequence of equation 4.8 the orientations of the magnetic moment, μ, of the nucleus with respect to the magnetic field can only be such that the components μ_z in the field direction have the specific values given by the equation. For example, if a nucleus has a spin quantum number equal to one-half, then the possible values for μ_z are $+\frac{1}{2}(\gamma h/2\pi)$ and $-\frac{1}{2}(\gamma h/2\pi)$, and the magnetic moment can only adopt the two orientations in the magnetic field which give these components in the field direction (Fig. 4.3a). In classical terms this corresponds to orientations of the magnetic moment where the components are parallel ($m_I = +\frac{1}{2}$) and anti-parallel ($m_I = -\frac{1}{2}$) to the direction of the applied magnetic field.

When a nucleus possessing a magnetic moment is introduced into a magnetic field, interaction will occur between the magnetic field and the component of

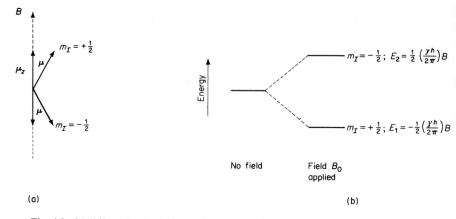

Fig. 4.3. (a) Allowed orientations of the magnetic moment, μ, of a nucleus with $I = \frac{1}{2}$ in a magnetic field. (b) Energy levels of a nucleus with $I = \frac{1}{2}$ in the absence and in the presence of a magnetic field

the magnetic moment aligned in the field direction, and the interaction will alter the energy of the nucleus. The energy, E, of the nucleus relative to that in zero field is given by

$$E = -\mu_z B \qquad (4.9)$$

Since μ_z is a function of the magnetic quantum number m_I, equation 4.9 can be written as

$$E = -m_I \frac{\gamma h}{2\pi} B \qquad (4.10)$$

Thus for a nucleus with $I = \frac{1}{2}$, the energies E_1 and E_2 for the two states with $m_I = +\frac{1}{2}$ and $m_I = -\frac{1}{2}$ respectively are

$$E_1 = -\frac{1}{2}\left(\frac{\gamma h}{2\pi}\right) B \qquad (4.11)$$

$$E_2 = +\frac{1}{2}\left(\frac{\gamma h}{2\pi}\right) B \qquad (4.12)$$

The energies of these two states are represented in Fig. 4.3b.

The nucleus can be promoted from the lower energy state E_1 to the higher energy state E_2 by absorption of energy, ΔE, equal to the energy difference $E_2 - E_1$. That is, the absorption of energy ΔE causes the magnetic moment of the nucleus to change from the parallel state ($m_I = +\frac{1}{2}$) to the anti-parallel state ($m_I = -\frac{1}{2}$). If the nucleus is in the upper energy state E_2 and radiation of energy ΔE is incident upon the system then the nucleus will spontaneously emit energy corresponding to ΔE and transfer to the lower energy state E_1.

From the Bohr relationship ($\Delta E = h\nu$), the frequency ν at which energy is absorbed or emitted is given by the equations

$$\nu = \frac{E_2 - E_1}{h} \tag{4.13}$$

$$\nu = \frac{\frac{1}{2}(\gamma h/2\pi)\, B + \frac{1}{2}(\gamma h/2\pi)\, B}{h} \tag{4.14}$$

$$\nu = \frac{\gamma}{2\pi} B \tag{4.15}$$

Equation 4.15 relating the magnitude of the field B experienced by the nucleus to the frequency at which energy is absorbed is the fundamental equation of n.m.r. It can be seen that this equation is the same as equation 4.5 which was obtained on the basis of the classical description of the n.m.r. phenomenon.

3. Experimental Method

Equation 4.15 shows that nuclei which experience a magnetic field of strength B will come into resonance, i.e. be induced to transfer from a lower to a higher energy state and vice versa, when the nuclei are subjected to radiation of frequency ν. For a nucleus which has a spin quantum number of one-half there are two energy levels for the resonance to occur between. Although both of these levels will be populated, the population of nuclei in the lower energy state will be slightly greater than that in the higher energy state. (The Maxwell–Boltzmann distribution law predicts an excess relative population of six in every million at a temperature of 300° K.) Consequently if an oscillating field B_1 of frequency ν is applied to the system the probability of transitions to the upper energy state (absorption of energy) is slightly greater than that of transitions to the lower energy state (emission of energy) and there will be a net absorption of energy from the field B_1. It is this absorption of energy which is detected in an n.m.r. experiment.

The value for the gyromagnetic ratio, γ, for a proton nucleus is such that when the nucleus is located in magnetic fields of 1·4092 or 2·3487 Tesla the frequencies at which resonance occurs, as given by equation 4.15, are 60 and 100 MHz respectively. The precise frequency at which resonance occurs depends upon the environment of the nucleus (see later) but for most protons the resonance positions lie within a range of 1000 Hz when the frequency of the field B_1 is 60 MHz. The resonance frequency can be found by either applying a constant field B_0 and varying the frequency of the oscillating field B_1 (frequency-sweep method), or by keeping the frequency of the oscillating field B_1 constant and varying the magnetic field B_0 (field-sweep method). The latter method is the more convenient experimentally and in an n.m.r. experiment the energy absorbed by nuclei is monitored as the magnetic field B_0 is varied. As the field B_0 is increased so the precessional frequency of the nuclei increases, and when this frequency is equal to the frequency of oscillation of field B_1, transitions are induced between nuclear energy states. The energy absorbed in this process produces a signal at the detector and this signal is

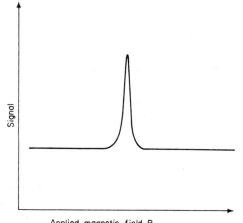

Fig. 4.4. An absorption spectrum in n.m.r. spectroscopy

amplified and recorded as a band in the spectrum. An n.m.r. spectrum is generally presented as a plot of signal at the detector against the strength of the applied magnetic field B_0 (Fig. 4.4). The x axis of the charts used for recording n.m.r. spectra is normally calibrated in units of frequency rather than in units of magnetic field strength. The relationship between the strength of the magnetic field and the frequency is given by equation 4.15.

4. Magnetic Properties of Nuclei

The n.m.r. effect can only be observed for nuclei which have a non-zero spin quantum number. Those nuclei with a spin quantum number equal to zero have no spin angular momentum and therefore no magnetic moment. The value of the spin quantum number is related to the atomic number and the mass number of a nucleus as shown in Table 4.1 ˙

The spin quantum numbers and the magnetic properties of some nuclei are given in Table 4.2. The frequency at which n.m.r. occurs is related to the field experienced by the nucleus and to the gyromagnetic ratio of the nucleus

Table 4.1

Atomic number	Mass number	Spin quantum number
Even	Even	0
Even or odd	Odd	$\frac{1}{2}, \frac{3}{2}, \frac{5}{2}$, etc.
Odd	Even	1, 2, 3, etc.

Table 4.2. Magnetic properties of selected nuclei

Nucleus	Nuclear spin quantum no., I	Magnetic moment μ (ampere square metre $\times 10^{27}$)	Resonance frequency in MHz for magnetic field of 1·4092 Tesla
^{1}H	$\frac{1}{2}$	14·09	60·000
^{2}H	1	4·34	9·211
^{4}He	0	—	—
^{10}B	3	9·09	6·447
^{11}B	$\frac{3}{2}$	13·59	19·250
^{12}C	0	—	—
^{13}C	$\frac{1}{2}$	3·53	15·085
^{14}N	1	2·02	4·335
^{15}N	$\frac{1}{2}$	−1·14	6·081
^{16}O	0	—	—
^{17}O	$\frac{5}{2}$	−9·55	8·134
^{19}F	$\frac{1}{2}$	13·28	56·446
^{31}P	$\frac{1}{2}$	5·71	24·288

(equation 4.15). Since different nuclei have different values for the gyromagnetic ratio, the resonance frequency for a fixed value of the magnetic field will vary for different nuclei. The resonance frequencies for nuclei when in a magnetic field of 1·4092 Tesla are given in Table 4.2 (a field of 1·4092 Tesla is commonly used in commercial spectrometers).

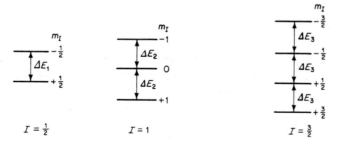

Fig. 4.5. Energy levels and allowed transitions between levels for nuclei with $I = \frac{1}{2}$, 1 and $\frac{3}{2}$

The magnetic quantum number, m_I, of a nucleus can have $2I + 1$ values and according to equation 4.10 there will be $2I + 1$ possible energy states for nuclei with $I = \frac{1}{2}$ (^{1}H, ^{19}F), three for nuclei with $I = 1$ (^{2}H, ^{14}N), and four for nuclei with $I = \frac{3}{2}$ (^{11}B). These energy states are represented in Fig. 4.5. As can be seen from the figure the number of allowed transitions (i.e. those for which $\Delta m_I = \pm 1$) increases as the spin quantum number of the nucleus increases.

B. CHEMICAL SHIFT

The frequency at which a nucleus comes into resonance in a magnetic field is given by equation 4.15, and in principle this frequency can be calculated if the value of the field at which resonance occurs could be derived from the n.m.r. spectrum. However it is difficult to measure accurately the absolute value of the field corresponding to a peak in a spectrum, and in practice the difference in frequency between peaks is measured. The resonance frequencies of nuclei in a sample are measured relative to the resonance frequency of a nucleus in a reference compound, and the frequencies are quoted relative to the reference frequency. The normal reference compound used for ^{1}H nuclei in organic compounds is tetramethylsilane (TMS). The protons in TMS are all equivalent and the n.m.r. spectrum consists of a single band.

The n.m.r. spectrum of a compound such as p-xylene which contains two sets of protons in different environments consists of two bands, and if TMS is present as reference, three bands will be observed in the spectrum (Fig. 4.6).

One important feature of n.m.r. spectra is that under the correct operating conditions of the spectrometer, the areas of the bands, i.e. the integrated intensities, are in approximately the same ratio as the number of protons responsible for the resonance. (In Fig. 4.6 the relative areas are $A:B = 2:3$.) Consequently if the total number of protons in the molecule is known the number of protons in each particular environment in the molecule can often be found.

The position of the peaks in a spectrum relative to the reference peak can be quoted in terms of the *chemical shift*, δ, as

$$\delta = \frac{\Delta \nu}{\nu} \times 10^6 \tag{4.16}$$

where $\Delta \nu$ is the difference in frequency between the position of a peak and that of the reference peak and ν is the frequency of the oscillating field B_1. The chemical shift, δ, is dimensionless and is expressed in parts per million (p.p.m.) on account of the factor 10^6 which is included in the definition in order to avoid having to quote very small values.

If the spectrum shown in Fig. 4.6 had been obtained using a frequency of 60 MHz and if the values of $\Delta \nu'$ and $\Delta \nu''$ were 134 and 421 Hz respectively, then δ_A would equal 2·24 p.p.m. and δ_B would equal 7·77 p.p.m. The difference in resonance positions of two nuclei when expressed in the dimensionless form of chemical shift will be a constant independent of both the strength of the static field B_0 and the frequency of the oscillating field B_1. However, as can be seen from equation 4.16, if the δ value is constant, then as the frequency ν of the field B_1 is increased the value of $\Delta\nu$ will also increase. That is, as the operating frequency of the spectrometer is increased the peaks in the spectrum become more widely spaced. It is obviously an advantage to operate at high

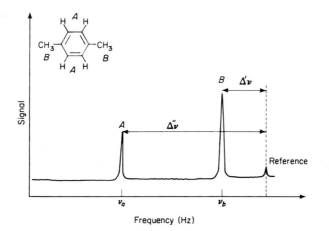

Fig. 4.6. N.m.r. spectrum of p-xylene plus TMS reference showing relative positions of peaks

frequencies because bands which are superimposed at low operating frequencies may be separated at high frequencies.

An alternative system which is commonly used for defining the position of resonance relative to a reference is the tau (τ) scale. On this scale the reference, TMS, is assigned the arbitrary position of 10 and the τ values of other resonances are given by $10 - \delta$ where δ has the same significance as above. Whereas the τ scale refers specifically to TMS as reference, the chemical shift can be measured relative to any reference by use of equation 4.16.

The chemical-shift or τ value of a resonance band in n.m.r. spectroscopy is analogous to the wave-number value of an absorption band in electronic, vibrational and rotational spectroscopy. Just as in other branches of spectroscopy absorption bands arise at different wave-number values, so in n.m.r. spectroscopy bands are observed at different chemical-shift or τ values. This difference arises because different nuclei in a molecule experience different magnetic fields as a result of the secondary magnetic fields associated with the molecule. These secondary fields arise from the induced circulation of electrons in the molecule under the influence of the applied field B_0. The secondary fields set up by the circulating electrons (specific cases will be considered later) may either oppose the applied field at a particular nucleus in the molecule or else reinforce the applied field; that is the nucleus is either positively shielded or negatively shielded by the secondary field. In the former case the effective field experienced by the nucleus is less than the applied field and the value of the applied field necessary to bring the nucleus into resonance will be greater than that if there were no secondary opposing field. Thus when the nucleus is shielded the resonance position moves upfield. When the nucleus is deshielded the resonance position moves downfield (Fig. 4.7).

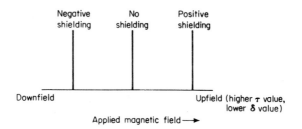

Fig. 4.7. Effect of shielding and deshielding on position of resonance of nuclei

The field experienced by a nucleus may be modified by fields set up by induced circulation of electrons localized on that nucleus (local shielding) or by fields set up by induced circulation of electrons within neighbouring groups or over the entire molecule (long-range shielding). The two types of shielding are considered in the following sections.

I. Local Shielding

Consider the proton nucleus. When the nucleus is subjected to a magnetic field B_0, the two $1s$ electrons are induced to circulate. The circulatory motion is perpendicular to the direction of B_0 (Fig. 4.8) and the secondary field set up at the nucleus by the circulation of electrons is opposed to the applied field, i.e. the circulation is diamagnetic. The effective field experienced by the nucleus is less than the applied field B_0 and the nucleus is positively shielded.

The greater the electron density around the nucleus the greater will be the opposing secondary field and the further upfield the resonance position will be. For instance, the electron density around the protons in methyl chloride is lower than that around the protons in TMS because of the electronegativity of the chlorine atom, and the resonance position of the methyl protons in methyl chloride is downfield relative to that of the protons in TMS. The protons in TMS are highly shielded and the position of resonance is at higher field than that of protons in most compounds.

The secondary field set up by the induced circulation of electrons around the proton nucleus is spherically symmetrical (i.e. the field is isotropic), and for a system in solution the random tumbling of the nucleus and hence the random orientation of the nucleus to the applied field results in the effect of the secondary field of neighbouring nuclei being averaged to zero. Thus the effect of the induced field is localized to the nucleus with which the electrons are associated.

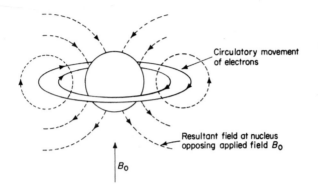

Fig. 4.8. Induced diamagnetic circulation of electrons around a proton nucleus

2. Long-range Shielding

In aromatic compounds the secondary fields set up by the induced circulation of π electrons often influence the fields experienced by nuclei not directly associated with the π-electron system. For example, the induced field in benzene, arising from circulation of the π electrons above and below the plane of the benzene ring, is experienced by the proton nuclei. The direction of the lines of force of this field with respect the direction of the applied field B_0 is shown in Fig. 4.9a for the situation where the plane of the benzene ring is at right angles to the direction of the applied field. It can be seen from Fig. 4.9a that the protons lie in a region where the induced field reinforces the applied field. Consequently the proton nuclei experience a deshielding effect and the resonance position of the proton nuclei is shifted to lower field than for the case if there were no shielding effect. The position of proton resonance in benzene ($2 \cdot 73 \ \tau$) can be compared with that in cyclohexane ($8 \cdot 57 \ \tau$) where there are no secondary fields arising from "ring currents".

In the liquid state or in solution the benzene molecules are in random motion and the orientation of the plane of the ring with respect to the applied field is continuously changing. Thus there will be a continuous change in the magnitude of the component of B_0 at right angles to the ring, and a continuous

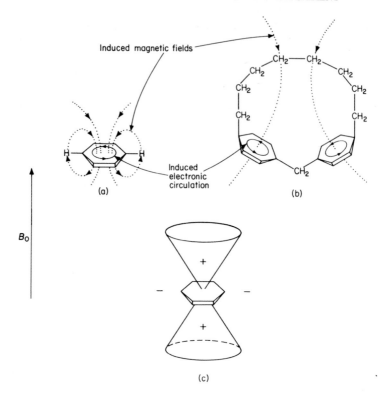

Fig. 4.9. The deshielding and shielding of protons associated with aromatic compounds. (a) Deshielding of protons in benzene. (b) Shielding of aliphatic protons in [1,8] paracyclophane. (c) Shielding envelope associated with the benzene ring

change in the magnitude of the induced field. Nevertheless the direction of the lines of force of the field with respect to the component of B_0 will not change, and the net effect of the induced field averaged over many rotations is to deshield the proton nuclei. In general terms, since the induced field in benzene is anisotropic, i.e. not spherically symmetrical, the effect of the field at any point in surrounding space does not average to zero for all possible orientations of the ring with respect to the applied field.

The position of nuclei within the induced field determines whether they are in a shielded or deshielded environment. For example, the central methylene protons in 1,8-paracyclophane are located over the benzene ring and are in a shielded environment (Fig. 4.9b). The result is that the resonance position of the central methylene protons in the aliphatic chain is at higher field ($9\cdot70\,\tau$) than that of methylene protons in an aliphatic hydrocarbon ($\sim8\cdot7\,\tau$). Similarly in the complex formed between benzene and chloroform, the proton in the

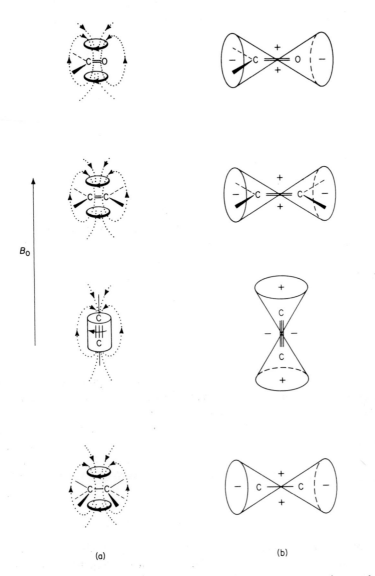

(a) (b)

Fig. 4.10. (a) Induced circulation of electrons in various common groupings and (b) corresponding shielding envelopes

chloroform molecule is sited above the ring and the resonance position is at higher field than that for chloroform in cyclohexane (see Expt. N11). The regions of shielding (+) and deshielding (−) associated with the induced field in benzene are shown in Fig. 4.9c.

The induced fields in some common anisotropic groupings are shown in Fig. 4.10a for a particular orientation between the group and the applied field B_0. The corresponding shielding envelopes shown in Fig. 4.10b represent the average effect of the induced field for all mutual orientations between the groups and the applied field.

It can be seen from Fig. 4.10 that the protons on an acetylenic linkage lie in a shielded region whereas those on a carbonyl group or ethylenic linkage lie in a deshielded region. This accounts for the observation that acetylenic protons resonate at relatively high field (7–8 τ) as compared to protons on a carbonyl group (0–0·7 τ) and ethylenic protons (3·5–5·5 τ).

The long-range shielding effects are not so marked for saturated groups as for unsaturated groups. Nevertheless the effect is appreciable and accounts for the fact that the resonance position of the protons shown in the alkanes I, II and III moves downfield on increasing substitution even although the expected shift on the basis of increasing electron density would be upfield.

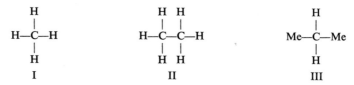

The chemical-shift value for protons can often be predicted with the aid of tables which list data on the resonance position of protons in common structures and the effect of substituents on the position of resonance. Correlation tables are given on pages 250 and 256 and examples on the use of these tables for calculation of chemical-shift values for protons in selected molecules are given in Expts N2 and N10. A more comprehensive set of correlation tables are given in texts by Chapman and Magnus (1966) and Matheson (1967).

C. SPIN–SPIN COUPLING

The interaction between the spins of neighbouring nuclei in a molecule may result in the resonance for a particular nucleus appearing in the n.m.r. spectrum as a set of bands rather than as a single band. For instance, the spectrum of 1,1-dichloroethane exhibits two sets of bands; the one centred at 4·15 τ due to the methine proton is split into a multiplet of four bands while the one centred at 7·95 τ due to the methyl protons appears as a doublet (Fig. 4.11). The mechanism for the splitting of resonances into sets of bands is considered

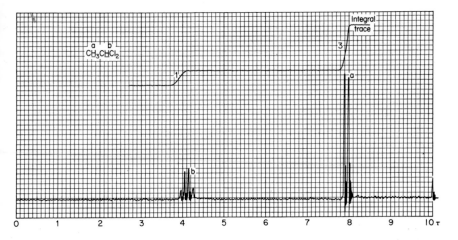

Fig. 4.11. N.m.r. spectrum of 1,1-dichloroethane

in the following section. Although the reasonance for a particular group of nuclei appears as a set of bands, the integrated area of the multiplet bands is still proportional to the number of protons giving rise to the absorption.

I. Mechanism of Spin–Spin Coupling

The fact that an individual nucleus can give rise to a set of bands in a spectrum arises from the effect of the spin associated with neighbouring nuclei on the field experienced by the nucleus undergoing resonance. The effect of the spin of neighbouring nuclei on a particular nucleus is transmitted via the electrons in the intervening bonds.

Consider the hypothetical case where two protons H_a and H_b in a molecule are separated by a single bond and are in markedly different chemical environments. In an applied magnetic field B_0 the components of the magnetic moment of the nuclei in the field direction may be either parallel ($m_I = +\frac{1}{2}$) or anti-parallel ($m_I = -\frac{1}{2}$) to the applied field. The former case is designated as the α spin state and the latter as the β spin state. When the nucleus is in a particular spin state the electrons are polarized so that adjacent nuclear and electron spins are paired. If the proton H_a is in the α spin state then the two electron spins will be polarized as in structure (a) of Fig. 4.12. Since the field associated with the electron adjacent to nucleus H_b will be in the same direction as the applied field B_0, the nucleus H_b will sense a net field which will support the applied field. When nucleus H_a is in the β spin state nucleus H_b will experience a net field which opposes the field B_0. Conversely, depending upon whether nucleus H_b is in the α or β spin state, nucleus H_a will experience a net field supporting or opposing the applied field. The overall result is that there are

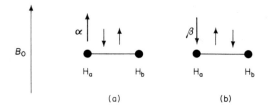

Fig. 4.12. Coupling of electron and nuclear spin states. (a) Polarization of electron spins when nucleus H_a is in α-spin state. (b) Polarization of electron spins when nucleus H_a is in β-spin state

two values for the total effective field sensed by nuclei H_a and H_b, and hence two values for the frequency at which each of these nuclei come into resonance. If it is assumed that the chemical shift values for the nuclei are very different, the n.m.r. spectrum for the H_a—H_b system in the molecule would then be as shown in Fig. 4.13. The resonance due to each proton appears as a doublet, and since a nucleus has almost equal probability of being in the α or β spin state, the intensities of each component of the doublet are equal. Since the coupling effect of H_a on H_b is equal to that of H_b on H_a the splitting between the peaks in each doublet is the same. (The frequency difference between the peaks within a multiplet pattern is known as the coupling constant J.) As the coupling interaction between the nuclei is an intramolecular effect the magnitude of the coupling, i.e. the value of J, is not affected by the strength of the applied field B_0.

Applied field B_0

Fig. 4.13. Spectrum of a system H_a—H_b showing effect of coupling between H_a and H_b

The possible arrangements giving rise to full correlation between nuclear and electron spins in the structures H_a—C—H_b and H_a—C—C—H_b are shown in Fig. 4.14. In each structure the two electrons associated with each carbon atom have their spins in the same direction. This arises because these electrons occupy degenerate orbitals of the atom and according to Hund's principle will tend to have parallel spins.

It can be seen from Fig. 4.14 that the two states with full spin correlation for the structure H_a—C—H_b have the proton nuclear spins unpaired ($\alpha\alpha$ and $\beta\beta$)

whereas for the structure H_a—C—C—H_b the proton nuclear spins are paired ($\alpha\beta$ and $\beta\alpha$). If the state with unpaired spins is of lower energy than the corresponding state in the absence of coupling then the coupling constant J is defined as being negative, while in the reverse situation the coupling constant is defined as being positive. If the state with paired nuclear spins is of lower energy than the corresponding state in the absence of coupling then J is defined as positive, while in the reverse case J is negative.

The effect of spin–spin coupling in structures H_a—C—H_b and H_a—C—C—H_b is the same as in the hypothetical structure H_a—H_b in that nuclei

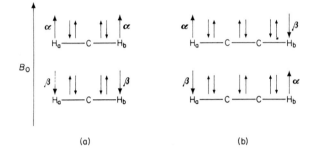

(a)   (b)

Fig. 4.14. Correlation of electron and nuclear spins in (a) the structure H—C—H and (b) the structure H—C—C—H

H_a and H_b experience two magnetic fields according to whether the other proton is in the α or β spin state. If the nuclei are in markedly different environments then the spectrum for each structure will be a symmetrical pair of doublets as in Fig. 4.13. Quantum-mechanical selection rules are such that if H_a and H_b are in equivalent environments then spin–spin coupling between the nuclei is not observed and only a single band is shown in the spectrum. The equivalence and non-equivalence of nuclei is discussed in the next section.

2. Equivalent and non-equivalent nuclei

Nuclei may be equivalent in one of three ways:

(a) Chemical equivalence

Nuclei which have the same value for the chemical shift are chemically equivalent, e.g. the protons in 1,2-dichloroethane.

(b) Symmetrical equivalence

Nuclei which interchange positions on application of a symmetry operation are symmetrically equivalent, e.g. the protons in 1,1-dichloroethylene. Nuclei which possess symmetrical equivalence are of necessity chemically equivalent.

(c) Magnetic equivalence

Nuclei which have the same chemical shift and which couple to the same extent with all other nuclei in the molecule are magnetically equivalent, e.g. the methyl protons in 1,1-dichloroethane. Coupling between magnetically equivalent nuclei has no effect on the observed n.m.r. spectrum.

The stereochemistry of a molecular structure has often to be taken into account when considering the equivalence and non-equivalence of nuclei. Consider the three rotamers of 1,2-dibromo-2,2-dichloroethane in Fig. 4.15. If rotamer (a) is viewed along the C—C bond it can be seen that protons H_a and H_b are both equally sited between a Cl and a Br atom on the neighbouring

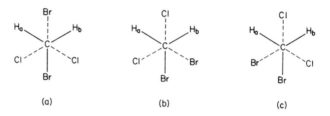

(a)   (b)   (c)

Fig. 4.15. Showing the environments of the protons in the rotameric forms of 1,2-dibromo-2,2-dichloroethane

carbon atom. Since the chemical environments of nuclei H_a and H_b are the same they are chemically equivalent. They are also magnetically equivalent since the coupling with all the other nuclei in the molecule is zero. The spectrum corresponding to rotameric form (a) would consist of a single resonance band.

Inspection of the spatial arrangements of the atoms in rotamers (b) and (c) reveals that the environments of H_a and H_b within the individual rotamers are different and that the environment of H_a in each rotamer is equivalent to the environment of H_b in the other rotameric form. Thus H_a and H_b are chemically non-equivalent in rotamers (b) and (c), and coupling will occur between the nuclei. The spectrum for each rotamer would appear as a pair of doublets; the position of the doublets being the same for each rotamer. If rotation about the C—C bond is rapid, as is the case at room temperature, then rapid interchange will occur between the three rotameric forms and the average environments of H_a and H_b will be the same. In such a case the spectrum would appear as a single band and the position of the band would correspond to the average of the environments of H_a and H_b in the three rotameric forms. The rate of rotation about the C—C bond slows down as the temperature of the system is lowered and at sufficiently low temperatures the spectrum will exhibit bands arising from rotamers (a), (b) and (c).

3. Some Spin–Spin Coupling Patterns

The n.m.r. spectrum of 1,1-dichloroethane (Fig. 4.11) can be explained by considering the possible arrangements of the spin states of the proton nuclei. As can be seen from Fig. 4.16 there are four ways in which the spin states of the methyl protons can be arranged. Consequently the neighbouring methine proton experiences four different fields and the nucleus will come into resonance at four different frequencies. Thus as shown in Fig. 4.11 the resonance of the methine proton appears as a quartet of bands. The intensities of the bands are in the ratio $1:3:3:1$ which is equal to the relative probabilities of the different arrangements of the spin states of the methyl protons. The resonance of the methyl protons is a doublet corresponding to the adjacent methine proton

Total spin	CH$_3$ Protons	CH Proton
$-\frac{3}{2}$	$\beta\beta\beta$	
$-\frac{1}{2}$	$\alpha\beta\beta$ $\beta\alpha\beta$ $\beta\beta\alpha$	β
$+\frac{1}{2}$	$\alpha\alpha\beta$ $\alpha\beta\alpha$ $\beta\alpha\alpha$	α
$+\frac{3}{2}$	$\alpha\alpha\alpha$	

Fig. 4.16. Possible arrangements of spin states of protons in the group CH$_3$CH—

being in either the α or β spin state. The probability of being in either of these states is equal and so the intensities of the bands in the doublet are the same. The coupling between the methyl protons and the methine proton is mutual and the spacings between the peaks in the doublet and quartet are the same.

Examples of the spectra of some molecules are given in Fig. 4.17. It can be readily confirmed that the number of ways in which the nuclear spin states can be arranged to give the same value for the total spin correspond to the relative intensities of the bands for the neighbouring protons.

The multiplet patterns shown in the spectra of Fig. 4.17 are in accord with the following rules:

(1) If a proton has n chemically equivalent neighbouring protons then the number of bands in the multiplet is $(n + 1)$.

(2) The relative intensities of the bands in a multiplet are given by the coefficients of the terms in the expansion of the expression $(x + y)^n$.

Rule 2 is only applicable when the ratio $(\nu_a - \nu_b)/J_{ab}$ for the interacting nuclei a and b is greater than about ten, in which case the spectrum is termed "first order". Rule 1 is only applicable when the neighbouring protons giving rise to the multiplet pattern are chemically equivalent. As discussed in the following, the non-equivalence of neighbouring protons may result in a larger number of bands appearing in the spectrum than would be predicted by the rule.

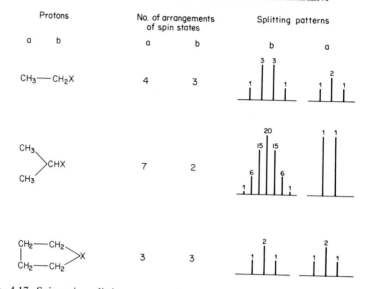

Fig. 4.17. Spin–spin splitting patterns in selected structures. (Peak heights and the separation between peaks are not drawn to scale)

Consider the molecules

where R is an electropositive group and R′ an electronegative group. In structure IV, J_{ab} will equal J_{bc} since protons a and c are in chemically equivalent environments, whereas in structure V, J_{xy} and J_{yz} will probably not be equal since protons x and y are not in chemically equivalent environments.

The splitting pattern for protons b may be determined by considering the splitting due to protons a and c in subsequent steps (Fig. 4.18a). The origin line will be split into a doublet by the influence of proton a, and subsequently each line of the doublet will be further split into two lines by the effect of proton c. Since $J_{ab} = J_{bc}$ the middle line in the final splitting pattern for protons b will consist of two superimposed lines, and as a consequence the intensities of the peaks in the final pattern will be in the ratio 1:2:1. In the case of proton y, since $J_{xy} \neq J_{yz}$ the subsequent splitting of the doublet arising from the effect of proton x will not result in the superimposition of the central lines and a quartet of lines of equal intensity will be obtained (Fig. 4.18b). It should be noted that the same splitting pattern would be obtained if the splitting of the origin lines of protons b or y by protons c and z respectively were first considered, followed by splitting by protons a and x respectively.

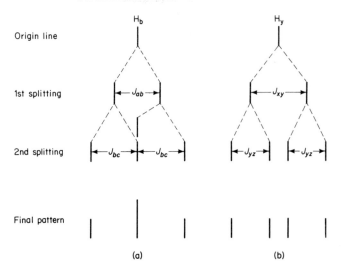

Fig. 4.18. Splitting patterns for a —CH₂— group with (a) two equivalent protons as nearest neighbours and (b) two non-equivalent protons as nearest neighbours and assuming $J_{xy} > J_{yz}$

4. Relative Intensities of Bands and Classification of Systems

The relative intensities of peaks in a multiplet depend upon the ratio $(\nu_a - \nu_b)/J_{ab}$ where ν_a, ν_b represent the resonance frequencies of the interacting protons and J_{ab} the coupling constant. If the groups R and R′ in the molecule

$$\underset{R}{\overset{R}{>}}CH\underset{a}{\underset{}{—}}CH\underset{b}{\overset{}{<}}\overset{R'}{\underset{R'}{}}$$

VI

are electropositive and electronegative respectively, then protons a and b will be in markedly different environments and $\nu_a - \nu_b$ will be large. J_{ab} is normally small compared to the difference in frequencies and consequently the ratio $(\nu_a - \nu_b)/J_{ab}$ may be large. In such a case the spectrum will be "first order" and will consist of a symmetrical pair of doublets (Fig. 4.19a). When the interacting protons are in very different environments the system is designated by letters at the opposite end of the alphabet. The example under discussion would be termed an AX system. The classification of some common groupings are given in Table 4.3, where it is assumed that the substituents R and R′ are such that the ratio $(\nu_a - \nu_b)/J_{ab}$ is large.

If in structure VI the electron donating or accepting properties of the groups R and R′ are not very different then protons a and b may be in similar environments and the difference between ν_a and ν_b will be small. The ratio $(\nu_a - \nu_b)/J_{ab}$

Table 4.3

Structure	Classification
R\ /R' CH—CH R/ \R'	AX
CH$_3$—CH$\diagup$R'$\diagdown$R'	A$_3$X
R | CH$_3$—CH—CH$_3$	AX$_6$
CH$_3$—CH$_2$—R	A$_2$X$_3$
R—CH$_2$—CH$_2$—R'	A$_2$X$_2$

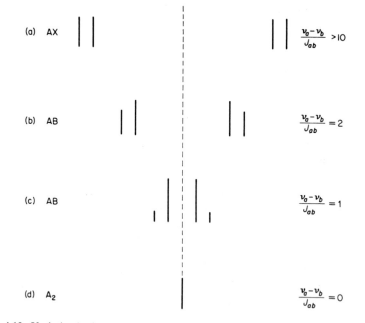

Fig. 4.19. Variation in the spectrum of a system of coupled protons H$_a$ and H$_b$ as the ratio $(\nu_a - \nu_b)/J_{ab}$ changes

will also be correspondingly small. The spectrum will now be "second order" and will consist of a pair of non-symmetrical doublets such as shown in Fig. 4.19b. The system is designated as an AB system to denote that protons a and b are in similar environments.

As the ratio $(\nu_a - \nu_b)/J_{ab}$ decreases the positions of the doublets come closer together and also the intensities of the inside lines of the doublet increase while the intensities of the outside lines decrease. When $R = R'$, protons a and b are in chemically equivalent environments and a single line spectrum is obtained (Fig. 4.19d). In this case the system is designated as A_2.

5. Spin–Spin Coupling in Relation to Energy Levels of Nuclear Spin States

An AX system consisting of protons H_a and H_b can be considered in terms of the total energy of the two spin system. The nuclei H_a and H_b can each occupy the α or β spin states and the possible combinations of spin states for the two nuclei are $\alpha\alpha$, $\alpha\beta$, $\beta\alpha$ and $\beta\beta$. These combinations of spin states are represented on the energy diagram in Fig. 4.20a for the situation where there is no coupling between H_a and H_b $(J_{ab} = 0)$. In the n.m.r. experiment the absorption of energy from the radio-frequency field B_1 causes a change in the spin state $(\alpha \to \beta, \beta \to \alpha)$ of one of the nuclei. The transitions involving change in spin states (spin flip) of nuclei H_a and H_b are shown by arrows in Fig. 4.20a. There are two possible transitions involving each nucleus but since each transition corresponds to spin flip of the same nucleus the energies of the two transitions are the same. As shown in the figure, the transition involving H_a requires greater energy than that involving H_b and a two line spectrum would be obtained for this system. If the frequency of the field B_1 is kept constant then the energy, $\Delta E = h\nu$, available for inducing transitions is constant, and transitions involving either H_a or H_b can only occur when the energy gap between the appropriate levels is equal to ΔE. Consequently if the spectrum is recorded by varying the applied field B_0 the field value at which transitions involving H_a occur will be less than those involving H_b and the spectrum will be as shown in Fig. 4.20a.

If H_a and H_b are coupled and if the sign of the coupling constant is positive, then as we have seen previously for a system with three bonds between the interacting nuclei (page 231) the states where the nuclear spins are paired $(\alpha\beta$ or $\beta\alpha)$ are more stable than the corresponding states in the absence of coupling. On the other hand, the energy of the states where the nuclear spins are not paired $(\alpha\alpha$ or $\beta\beta)$ is greater than that of the corresponding states in the absence of coupling. The energy levels for the system are then as shown in Fig. 4.20b. If the energy change for the individual levels is assigned a value of $J/4$ then, as can be seen from the figure, the energy of each transition is changed by $J/2$ relative to the energy of the corresponding transition when there is no coupling between the nuclei. Two lines now appear in the spectrum for each nucleus and since the probabilities of the transitions are the same, the lines are of equal intensity. Since each line of the doublet has shifted by the same amount ($J/2$) from the position of resonance in the absence of coupling the mid-point between the lines in the doublets gives the values of ν_a and ν_b for protons H_a and H_b respectively.

If the coupling constant for the interaction between H_a and H_b is negative then each of the energy levels in the coupled system would be decreased by an amount $J/4$ relative to the energy levels in the uncoupled system. However

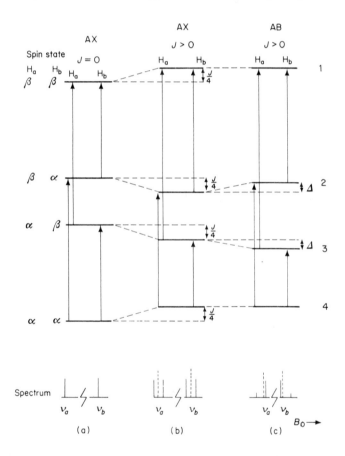

Fig. 4.20. Energy-level diagram for two spin system H_a, H_b. The diagram is not drawn to scale—the separation between the $\alpha\beta$ and $\beta\alpha$ states in frequency units is equal to $\nu_a - \nu_b$ (of the order of 100 Hz) whereas the separation between the states connected by vertical lines is close to ν_a or ν_b (of the order of 100 MHz)

the differences in energy between the levels in the coupled system would be the same as in Fig. 4.20b and the spectrum would be identical to that for the case where the coupling constant is positive.

It should be noted that the difference in energy between the levels representing the $\alpha\beta$ and $\beta\alpha$ combinations of the spin states of the nuclei corresponds to the chemical-shift difference between nuclei H_a and H_b. For an AX system the difference in chemical shift is relatively large and there will be no interaction between these levels. However for an AB system the difference between the chemical shifts of the two nuclei is small and the energies of the levels are close to each other. The net result is that there is a "repulsion" effect between the two spin states and the upper state is increased in energy by an amount Δ while the lower state is decreased in energy by the same amount (Fig. 4.20c). Thus the transitions involving spin flip of nucleus H_a are increased in energy by an amount Δ while those involving spin flip of nucleus H_b are decreased by an amount Δ. Consequently the doublets representing these transitions in H_a and H_b are shifted downfield and upfield respectively from the unperturbed origin positions ν_a and ν_b. Since the doublets are no longer symmetrical about the origin position the values for ν_a and ν_b cannot be obtained directly from the spectrum but have to be calculated from the spectral data (see next section). As noted previously the intensities of the inner lines of the doublets increase while those of the outer lines decrease on changing from an AX to an AB system.

D. ANALYSIS OF SPECTRA

The values for the chemical shift of nuclei and for the coupling constants between nuclei give valuable information for the elucidation of structures of unknown molecules. In many instances these values cannot be obtained directly from the n.m.r. spectrum and have to be determined by inserting spectral data into standard formulae which are derived from a quantum-mechanical consideration of nuclear spin states and transitions. The formulae used for the analysis of some common systems are given in the following sections. The frequencies of the lines shown in the diagram are considered to be relative to that of TMS and hence the calculated values of the chemical shift are relative to TMS.

I. AX System

The spectrum of an AX system consists of a symmetrical pair of doublets and the values of ν_a, ν_x and J_{ax} can be read directly from the spectrum (Fig. 4.21).

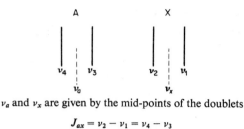

v_a and v_x are given by the mid-points of the doublets

$$J_{ax} = v_2 - v_1 = v_4 - v_3$$

Fig. 4.21

2. AB System

The spectrum of an AB system consists of an asymmetrical pair of doublets (Fig. 4.22) and the values of v_a, v_b and J_{ab} can be calculated using the formulae given below.

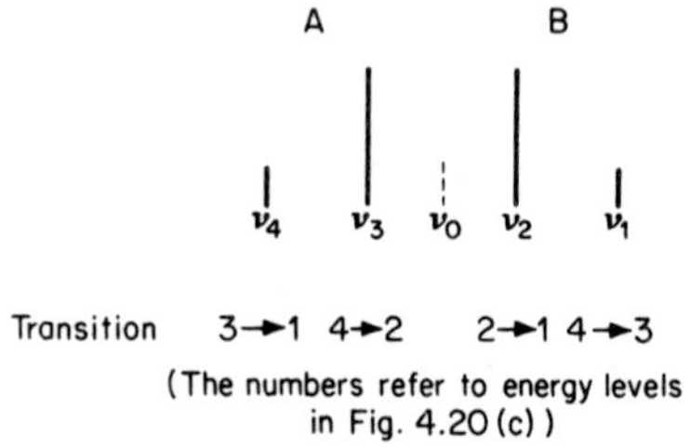

Transition $3 \to 1$ $4 \to 2$ $2 \to 1$ $4 \to 3$
(The numbers refer to energy levels
in Fig. 4.20 (c))

$$J_{ab} = v_2 - v_1 = v_4 - v_3$$
$$v_a - v_b = [(v_4 - v_1)(v_3 - v_2)]^{1/2}$$
$$v_a = v_0 + \tfrac{1}{2}[(v_4 - v_1)(v_3 - v_2)]^{1/2}$$
$$v_b = v_0 - \tfrac{1}{2}[(v_4 - v_1)(v_3 - v_2)]^{1/2}$$

where v_0 is the mid-point of the set of peaks

Fig. 4.22

Table 4.4. Positions and relative intensities of bands in an AB spectrum

Transition[a]	Position[b]	Relative intensity[c]
$3 \to 1$	$\tfrac{1}{2}[J^2 + (v_a - v_b)^2]^{1/2} + \tfrac{1}{2}J$	$1 - \sin 2\theta$
$4 \to 2$	$\tfrac{1}{2}[J^2 + (v_a - v_b)^2]^{1/2} - \tfrac{1}{2}J$	$1 - \sin 2\theta$
$2 \to 1$	$-\tfrac{1}{2}[J^2 + (v_a - v_b)^2]^{1/2} + \tfrac{1}{2}J$	$1 + \sin 2\theta$
$4 \to 3$	$-\tfrac{1}{2}[J^2 + (v_a - v_b)^2]^{1/2} - \tfrac{1}{2}J$	$1 - \sin 2\theta$

[a] The numbers refer to energy levels in Fig. 4.20c
[b] Positions relative to v_0
[c] $\sin 2\theta = J/[J^2 + (v_a - v_b)^2]^{1/2}$

If a quartet of lines in a spectrum has been assigned to an AB system within a molecule then the assignment may be checked using the relationships given in Table 4.4.

3. AX$_2$ System

The spectrum of an AX$_2$ system (Fig. 4.23) consists of five lines and the origin positions and coupling constant can be obtained directly from the spectrum.

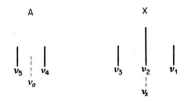

$$J_{ax} = \nu_2 - \nu_1 = \nu_4 - \nu_3 = \nu_5 - \nu_4$$

ν_a and ν_x are given by the mid-points of the multiplets

Fig. 4.23

4. AB$_2$ System

The spectrum of an AB$_2$ system consists of nine lines but one of the lines may be weak or undetectable. As shown in Fig. 4.24 the appearance of the spectrum depends upon the magnitude of the ratio $\nu_a - \nu_b/J_{ab}$. In certain cases lines 5

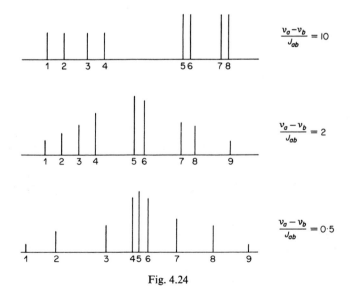

Fig. 4.24

and 6 may coalesce to give a slightly broadened singlet and the spectrum may contain seven or eight lines. The spectrum is readily analysed using the following rules:

(1) ν_b is given by the mid-point between lines 5 and 7
(2) ν_a is given by the position of line 3
(3) J_{ab} is equal to $\frac{1}{3}(\nu_1 + \nu_6 - \nu_4 - \nu_8)$.

5. AMX System

The nomenclature AMX indicates that the M proton has a chemical shift value intermediate between the A and X protons. As for the AX case the spectrum is first order and as can be seen from the splitting diagram (Fig. 4.25) the spectrum will consist of twelve bands of equal intensity when $J_{ax} \neq J_{mx} \neq J_{am}$. The twelve bands in the spectrum will be grouped in three well separated, symmetrical quartets.

The values of ν_a, ν_m, and ν_x are given by the mid-points of the quartets. Three values for coupling constants can be read directly from the spectrum and the values so obtained can be assigned to the appropriate nuclei by consideration of the structure of the molecule.

If two of the coupling constants are equal then the spectrum will consist of eleven lines. For example, if $J_{am} = J_{ax}$ then the resonance due to H_a would appear as a $1:2:1$ triplet instead of a quartet. If all three coupling constants are equal the spectrum will consist of nine lines and will appear as three sets of $1:2:1$ triplets.

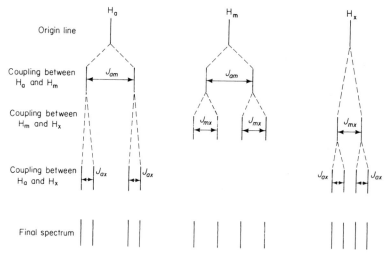

Fig. 4.25

6. ABX System

The spectrum of an ABX system often consists of twelve lines although it may contain up to fourteen lines if the two possible combination lines are present, or less than twelve lines if some of the lines are superimposed. The spectrum of an ABX system may be distinguished from that of an AMX system, which may also contain twelve lines, by the distortion of line intensities in the ABX

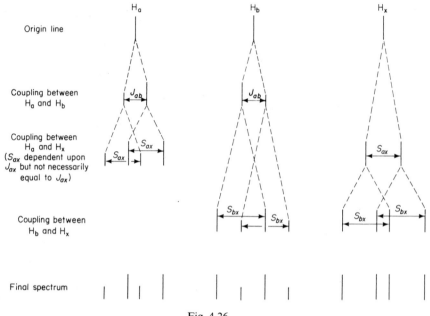

Fig. 4.26

spectrum. A first-order approach to an ABX system predicts a twelve line spectrum with the X region consisting of a symmetrical quartet and the AB region of two asymmetrical quartets (see Fig. 4.26).

It must be emphasized that this pattern of lines for an ABX spectrum is only one of many possible patterns. There may be considerable overlapping and interleaving of lines in the AB region with a consequent change in the number of lines and intensity distribution of the lines in the AB pattern. There may also be a change in position and in the number of lines in the X region. The spectral parameters for an ABX system cannot be obtained directly from the spectrum and each spectrum has to be analysed individually utilizing standard rules (Expt. N15).

For the analysis of the spectra of more complex systems such as ABC, A_2B_2, the reader is referred to Pople et al. (1959) and Emsley et al. (1965).

E. APPLICATION OF N.M.R. TO QUANTITATIVE MEASUREMENTS

I. Band Areas

Analysis of n.m.r. spectra to give quantitative information on the concentration of absorbing species generally involves measurement of band areas. Band areas are used in preference to band heights since the area is directly proportional to the number of absorbing nuclei, while the band height may not remain proportional to the number of species as the band shape changes.

Although in principle the area of a band will be proportional to the number of absorbing nuclei, in practice the area of a band will only be an accurate measure of the number of absorbing species under specific instrumental operating conditions. It is thus often convenient to construct a calibration graph of band areas as a function of the concentration of absorbing species in standard samples and to use this as a reference. In such cases the only instrumental criterion is that the operating conditions are the same for the standard and unknown samples. If the relative concentrations of absorbing species in the same sample are to be determined then the band areas may be used directly as a measure of the number or concentration of absorbing species (Expts N7 and N9).

2. Exchange Effects

The width of an absorption band in an n.m.r. spectrum is dependent upon the physical state of the sample and upon the residence time of the nucleus in a particular environment. The width of absorption bands in the spectra of liquids is small, generally of the order of 2–3 Hz, and when broad bands are observed (10–100 Hz) these can normally be accounted for in terms of exchange effects.

Exchange effects can be discussed in relation to the equilibrium between pyruvic acid and 2,2-dihydroxypropanoic acid:

$$C\underline{H}_3COCOOH + H_2O \ \rightleftharpoons \ C\underline{H}_3C(OH)_2COOH$$
$$\text{a} \qquad\qquad\qquad\qquad \text{b}$$

In this example the chemical environment of the methyl protons is being continually changed from A to B. If the mean lifetime, τ, of the nucleus in each environment is long compared to the time required for the transition between the lower and upper magnetic energy states ($\sim 10^{-3}$ second) then the spectrum for the methyl protons will consist of two sharp bands. However if the mean lifetime is short compared with the transition time between magnetic states then the nucleus will change environments in the time required for the transition and a single sharp band is observed in the spectrum at a position corresponding to the average environment of the nucleus. At intermediate lifetimes either one or two broad resonances may be observed. The appearance of the spectrum depends upon the relative magnitude of the rate of exchange (equal to $1/\tau$) and the difference in chemical shift $(\nu_a - \nu_b)$ between the nuclei in the two environments (Fig. 4.27).

The relationship between the width of individual bands in the spectrum and the mean lifetime, τ, of the protons in each environment is derived from the Heisenberg Uncertainty Principle. This principle states that the shorter the lifetime of a state of a system, the less precisely the energy of the system in that state is known. The principle may be expressed in the form

$$\Delta E . \Delta t = h/2\pi \qquad (4.17)$$

where Δt is the lifetime of the state, ΔE is the uncertainty in the measurement of the energy of the state and h is Planck's constant. In the situation under

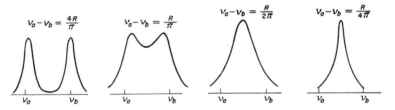

Fig. 4.27. Change in appearance of the spectrum for two exchanging nuclei as the rate of exchange, R, is varied

discussion the lifetime of a state is approximately equal to the mean lifetime of the protons in a particular environment. Therefore equation 4.17 can be written as

$$\Delta E . \tau = h/2\pi \qquad (4.18)$$

The uncertainty in the measurement of the energy of a magnetic state means that there is a similar uncertainty in the measurement of the position of a line arising from the absorption of energy by this state. Consequently a band, which extends over a range of energy, is observed in an n.m.r. spectrum instead of a line.

Dividing equation 4.18 by $h\tau$ and using the Bohr relationship $E = h\nu$, gives

$$\frac{\Delta E}{h} = \Delta \nu = \frac{1}{2\pi\tau} \qquad (4.19)$$

where $\Delta \nu$ is the uncertainty in the measurement of the frequency corresponding to the position of the line. This uncertainty is normally measured as the width of the band at half the band height (half-width). Now

$$\Delta \nu = \Delta \omega / 2\pi \qquad (4.20)$$

9

where $\Delta\omega$ is the uncertainty in the angular frequency of the precessing nuclei. Therefore equation 4.19 can take the form

$$\Delta\omega = \frac{1}{\tau} \qquad (4.21)$$

Each band in an n.m.r. spectrum has a "natural" width and for a system where exchange processes are not occurring, the width of each band is equal to the "natural" width. If exchange processes involving certain nuclei occur, then the bands arising from these nuclei are further broadened and, as shown by equation 4.21, the extent of this broadening, $\Delta\omega$, is proportional to the reciprocal of the mean lifetime of the protons in a particular environment. The expression for the half-width of the exchange broadened band has been shown to be

$$\frac{1}{T_2'} = \frac{1}{T_2} + \frac{1}{\tau} \qquad (4.22)$$

where $2/T_2'$ is the half-width of the exchange broadened band and $1/T_2$ is the half-width of the band in the absence of exchange.

It can be seen from equation 4.22 that the mean lifetime τ of nuclei in a particular environment can be obtained by measuring the half-width of the appropriate bands in the spectrum in the presence and absence of exchange processes. In the case of the equilibrium between pyruvic acid and 2,2-dihydroxypropanoic acid the exchange rate is relatively slow, i.e. the mean lifetime of the protons in each environment is long, and bands corresponding to protons in environments a and b can be observed under suitable conditions. The mean lifetime of the protons in each environment can therefore be calculated (Expt. N13).

Protons involved in hydrogen bonding between acetone and chloroform exchange between the environments of "free" and "bonded" chloroform

$$(CH_3)_2C{=}O + HCCl_3 \rightleftharpoons (CH_3)_2C{=}O----HCCl_3$$

Since the rate of exchange is rapid compared to the time scale of the n.m.r. process the resonance for the chloroform proton appears as a single band. The position of the band is dependent upon the relative proportion of chloroform protons in each environment and is determined by the "weighted" average of the environments. If the equilibrium is displaced to the left then the band position for the proton would move towards that of the chloroform proton in the "free" state. By analysing the change in band position as the equilibrium is displaced, the relative proportions of nuclei in the complexed and uncomplexed states may be calculated and the equilibrium constant determined (Expt. N10).

Exchange processes not only affect the positions of bands in a spectrum but can also affect the multiplicity of the bands. For example, in purified ethanol the rate of exchange of the hydroxyl proton between different alcohol molecules is slow and the hydroxyl resonance appears as a triplet, but if a trace of acid is added the rate of exchange becomes rapid and the hydroxyl resonance appears as a singlet. Also in the former case the methylene resonance appears as a quartet of doublets whereas in the latter case it is a quartet. These observations can be explained in the following way. If the mean lifetime of the hydroxyl proton in each environment is long it will experience each field associated with the different spin arrangements of the neighbouring methylene protons and the corresponding resonance will be a triplet. The methylene protons must also experience the fields associated with the two spin states of the hydroxyl proton and the methylene resonance will consist of a quartet (coupling with the methyl group) with each component of the quartet split into a doublet. If the mean lifetime of the hydroxyl proton in each environment is short the methylene protons will experience a rapidly changing succession of hydroxyl protons. Since the exchanging protons can be in either the α or β spin state the net field experienced by the methylene protons is averaged to zero in the time required for an n.m.r. transition. Consequently the methylene resonance will not show coupling to the hydroxyl proton and will appear as a quartet. Likewise the hydroxyl resonance will not show coupling and will be a singlet.

F. INTERPRETATION OF SPECTRA

It is often helpful when attempting to interpret an n.m.r. spectrum to note the following parameters for each band or multiplet of bands in the spectrum:

1. Integrated intensity
2. Chemical shift (δ or τ value)
3. Multiplicity and spin–spin splitting (J value).

The integral trace on a spectrum may be used to determine the number of protons responsible for the individual bands or multiplets if the number of protons in the molecule is known. This information is best gained by dividing the total integrated height over the whole spectrum by the number of protons in the molecule to obtain the integral height per proton. The integral height for individual bands or multiplets will then indicate the number of protons giving rise to that signal. If the number of protons in the molecule is not known then the individual integrated heights may be used to give the relative number of protons responsible for the signals. Owing to instrumental and other factors the integrated heights may not correspond exactly to an integral number of protons. However the value is generally close enough to an integral value to enable a decision to be made as to the number of protons giving the signal.

The chemical shift of a band or multiplet of bands in the spectrum can yield information regarding the environment of the nucleus or nuclei responsible for the signal, and when used in conjunction with correlation tables can often lead to the assignment of groups in the molecule. The chemical shift may also be used to check whether or not a proposed structure fits the observed spectrum. The chemical shift for nuclei in a proposed structure can often be predicted with the aid of correlation tables and if the structure is correct then the predicted and observed values should be in reasonable agreement.

The multiplicity of a resonance representing a proton or a group of protons and the spin–spin splitting between the peaks in the multiplet can often give information as to the number of neighbouring protons and to the number of

Fig. 4.28

bonds between these neighbours and the group under examination. The following points should be borne in mind when analysing spin–spin splitting patterns: (a) coupling between nuclei is mutual and if, for example, a structure $R—CH_2—CH_2—R'$ is examined, then the splitting between the peaks in the two multiplets for the methylene protons is exactly the same, and (b) multiplets which are associated with each other often have the inner lines of greater intensity than the outer lines. This latter point leads to a "roofing effect" (see Fig. 4.28) which can be useful in indicating systems which are coupled to each other.

I. Examples of Spectrum Interpretation

The conditions used for running Spectra 1–4 discussed in the following sections were as follows:

Operating frequency: 60 MHz
Scale factor: one graticule on chart represents 6 Hz
Reference: internal TMS
Solvent: carbon tetrachloride or deuterochloroform

As the solvents do not contain any protons all the bands in the spectra arise from protons in the solute. The resonance due to TMS is positioned at 10 on the chart in each spectrum and the τ values for the resonances can be read off directly from the calibrated chart.

(a) Spectrum I

The spectrum exhibits two sets of triplets which suggests that the protons giving rise to each triplet have two neighbouring protons in equivalent environments. The integral trace indicates that the triplet centred at $9 \cdot 08 \tau$ arises from three protons, and this observation, combined with the above information, points to a partial structure CH_3—CH_2—. This leaves a fragment CH_3O to be assigned and since it appears that there are two CH_2 groups in the molecule, the structure for this fragment will be —CH_2—OH. The combined structure will then be that of n-propanol

$$CH_3\text{---}CH_2\text{---}CH_2\text{---}OH$$
$$a \qquad b \qquad c \qquad d$$

The integral trace shows that the band at $4 \cdot 32 \tau$ arises from a single proton. This band must therefore represent the resonance of the hydroxylic proton. The complex multiplet centred at $8 \cdot 5 \tau$ results from the methylene group b. The origin line for this group will be split into four by coupling with the methyl group, and each line will be subsequently split into three through coupling with methylene group c. Not all the bands are observed in the spectrum owing to overlap. The magnitude of the couplings between groups a and b and groups b and c would be expected to be very similar but not identical since groups a and c are different. If the coupling constants J_{ab} and J_{bc} were identical a multiplet of six bands would be observed for the resonance of group b.

The complete assignment of the spectrum is summarized in Table 4.5 along with the predicted resonance positions given by Table 4.6.

Table 4.5

Pattern	Position τ	Splitting Hz	Protons	Assignment	Predicted position τ
Triplet	9·08	7	3	CH_3— (a)	~9·12
Complex	~8·5	7	2	—CH_2—(b)	~8·75
Triplet	6·46	6	2	—CH_2—(c)	6·42
Singlet	4·32	—	1	—OH (d)	—

Table 4.6. τ values of the protons of CH_3, CH_2 and CH group in acyclic molecules (R is an alkyl group)[a]

X	CH_3X	$R'CH_2X$	$R'R''CHX$
—MgI	11·3	10·62, 10·99	10·20
—Li			
—CH₃	9·12	8·75	8·50
RCH=CHCO—	8·14		
—COOEt	8·05	7·80	
—COOR	8·00	7·80	
—CONH₂	8·00	7·52	7·3
—CN	8·00		
—COOCH₃	8·00		
—CONHR	7·95		
—SCH₃	7·94		
—SH	7·92	7·60	6·90
—COOH	7·90	7·64	7·42
—COR	7·9	7·6	7·42
—NH₂	7·85	7·50	7·13
—N(Me)₂	7·84	7·58	
—CHO	7·83	7·80	7·62
—NMeCH₂Ph	7·83		
—I	7·83	6·88	5·76
—NR₂	7·81	7·50	7·12
—NMeCH₂CH₂OH	7·73		
—Ph	7·66	7·38	7·13
—COSH	7·68		
—SR	7·65	7·4	

X	CH_3X	$R'CH_2X$	$R'R''CHX$
—SO—	7·5		
—SCN	7·39	7·02	6·52
—COPh	7·38		6·42
—Br	7·30	6·70	5·97
—NMeCHO	7·22, 7·08		
—N(CH₃)Ph	7·09		
—COBr	7·15		
—NHCOR	7·16	6·7	6·15
—Cl	7·00	6·56	5·98
⊕ —NR₃	6·67	6·60	6·50
—NCS	6·63	6·36	6·05
—OMe	6·73	6·64	6·44
—OH	6·62	6·42	6·12
—OR	6·70	6·60	6·47
—SO₃H	6·38		
—OCOR	6·29	5·80	4·9
—OPh	6·27	6·10	6·0
—OCOPh	6·00	5·68	4·8
—OCOCF₃	5·90	5·56	
—F	5·70	5·66	5·4
—NO₂	5·67	5·60	5·5

[a] Reproduced with permission from Chapman and Magnus (1966).

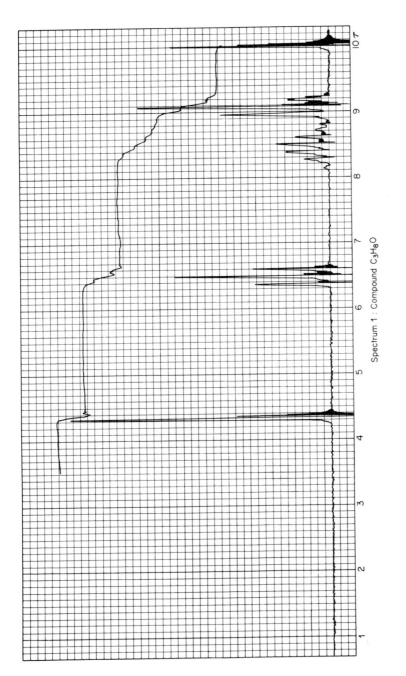

Spectrum 1 : Compound C_3H_8O

(b) Spectrum 2

The integral trace on the spectrum indicates that the multiplets centred at 7·19 and 5·82 τ represent two protons in each case. Since the resonances due to these protons are split into triplets which have equal peak separations it may be concluded that the molecule contains the structure —CH_2—CH_2—. The protons in this grouping cannot be coupled to the other protons in the molecule otherwise a more complex splitting pattern would be observed.

The resonance at highest field is a singlet of integrated intensity corresponding to three protons. It may be taken that this peak represents a CH_3— group and that the protons on this group are more than three bonds removed from the nearest neighbouring protons.

The resonance at lowest field occurs at 2·83 τ which falls in the range characteristic of protons in aromatic systems. Since the intensity of this band corresponds to five protons it can be concluded that the resonance arises from a phenyl group. The observation that the resonance is not split indicates that the five protons in the ring are in virtually equivalent environments and hence that the substituent in the ring does not markedly alter the electron density around the ring. On considering the molecular formula of the compound and the fragments so far assigned, this must mean that the fragment —CH_2—CH_2— is adjacent to the ring. This leaves two possibilities for the overall structure of the molecule:

$$C_6H_5-\underset{a}{CH_2}-\underset{b}{CH_2}-\underset{c}{O}-\overset{\overset{O}{\|}}{C}-\underset{d}{CH_3}$$

A

$$C_6H_5-\underset{a}{CH_2}-\underset{b}{CH_2}-\overset{\overset{O}{\|}}{C}-\underset{c}{O}-\underset{d}{CH_3}$$

B

The correct structure may be assigned through use of Table 4.6. It may be seen from this table that the methyl protons in structures A and B would be expected to resonate at 8·00 and 6·29 τ respectively. Since the methyl signal appears at 8·15 τ, A must be the correct structure. Table 4.7 summarizes the assignment of the spectrum.

Table 4.7

Pattern	Position τ	Splitting Hz	Protons	Assignment	Predicted position τ
Singlet	8·15	–	3	CH_3— (d)	8·00
Triplet	7·19	7	2	—CH_2— (b)	<7·38
Triplet	5·82	7	2	—CH_2— (c)	5·80
Singlet	2·83	–	5	C_6H_5— (a)	2·73 [a]

[a] Table 4.8

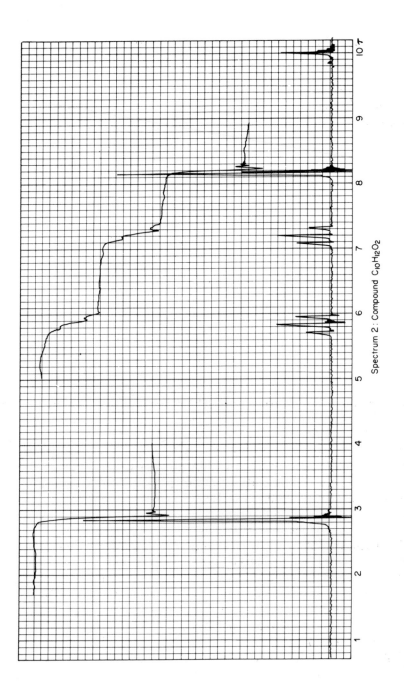

Spectrum 2 : Compound $C_{10}H_{12}O_2$

(c) Spectrum 3

A noticeable feature of this spectrum is the four sets of doublet peaks in the range 1·2–2·5 τ. The splitting, in Hertz, in these doublet peaks is shown in Fig. 4.29.

It can be deduced from the values of the coupling constants and from the fact that the peaks are doublets that (a) resonance x is split into a doublet through coupling with the single proton which gives resonance y, (b) resonance y is split into a doublet of doublets by coupling with the single protons which give resonances x and z, and (c) resonance z is split into a doublet by coupling

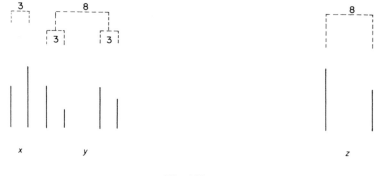

Fig. 4.29

with the proton giving resonance y. The position of these multiplets in the spectrum, viz. in the region 1·2–2·5 τ, suggests that they arise from protons in an aromatic ring. Also, the fact that the molecule contains two nitrogen atoms and five oxygen atoms and only seven carbon atoms indicates that the aromatic ring is substituted with an NO_2 group or groups. This is in accord with the large downfield shift of resonances x and y from the position of resonance of benzene protons (see Table 4.8). The aromatic ring in this compound must be tri-substituted since there are three aromatic proton signals each with a relative intensity corresponding to one proton.

The intensity of the single band at 5·86 τ corresponds to three protons and it may be assumed that this band is the resonance of a methyl group. It may be deduced from the position of the methyl resonance and from the previous information that it is bonded to an —OPh group (Table 4.6). Although the predicted resonance position for the methyl protons is at 6·27 τ, the effect of NO_2 groups in the aromatic ring would be to shift the resonance of the methyl group to lower field.

The above deductions indicate that the compound is one of the six isomers of dinitroanisole (VII–XII).

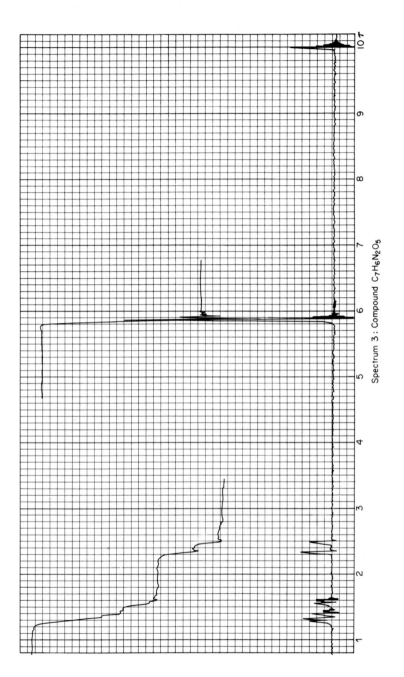

Spectrum 3 : Compound $C_7H_6N_2O_5$

Table 4.8. The effect of substitution on the proton frequency of benzene $(2 \cdot 73\ \tau)^a$

Substituent	Shifts relative to benzene (p.p.m.)		
	Ortho	*Meta*	*Para*
—NO$_2$	−0·97	−0·30	−0·42
—COOCH$_3$	−0·93	−0·20	−0·27
—COCl	−0·90	−0·23	−0·30
—CCl$_3$	−0·80	−0·17	−0·23
—CHO	−0·73	−0·23	−0·37
—COCH$_3$	−0·63	−0·27	−0·27
—COOH	−0·63	−0·10	−0·17
—CN	−0·30	−0·30	−0·30
—CONH$_2$	−0·50	−0·20	−0·20
—NH$_3^{\oplus}$	−0·40	−0·20	−0·20
—NHCOR	−0·40	−0·20	−0·30
—I	−0·30	−0·17	−0·10
—C≡C	−0·20	−0·20	−0·20
—OCOR	−0·20	+0·1	+0·2
—CHCl$_2$	−0·13	−0·13	−0·13
—SR	−0·1	+0·1	+0·2
—CH$_2$Cl	0·00	0·00	0·00
—Cl	0·00	0·00	0·00
—Br	0·00	0·00	0·00
—CH$_2$NH$_2$	+0·03	+0·03	+0·03
—CH$_2$CH$_3$	+0·07	+0·07	+0·07
—CH$_2$OH	+0·07	+0·07	+0·07
—CH$_3$	+0·15	+0·1	+0·1
—OCH$_3$	+0·23	+0·23	+0·23
—OH	+0·37	+0·37	+0·37
—N(CH$_3$)$_2$	+0·50	+0·20	+0·50
—NH$_2$	+0·77	+0·13	+0·40
—NHCH$_3$	+0·80	+0·30	+0·57

[a] Reproduced with permission from Chapman and Magnus (1966).

The compound cannot have either structure VII or VIII since in these structures two out of the three aromatic protons are in the same environment and would be expected to resonate at the same position. Structure IX can also be rejected since in this structure the aromatic protons are either *ortho* or *meta* to each other and would be coupled, whereas analysis of the splitting in the multiplets has shown that two out of the three protons are not coupled to each other. The decision as to which is the correct structure out of X, XI and XII can be made on the basis of the predicted resonance positions cal-

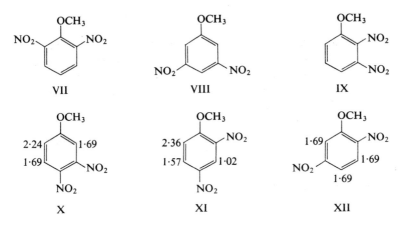

VII   VIII   IX

X   XI   XII

culated using Table 4.8. The calculated resonance positions are shown on structures **X**, **XI** and **XII** and it can be seen that structure **XI** is the only one which would be expected to give three separated multiplets. Comparison of the predicted and the actual resonance positions for the aromatic protons in 2,4-dinitroanisole shows that the values calculated using Table 4.8 can only be used as a guide when ascertaining the structure of an unknown benzenoid compound.

(d) *Spectrum 4*

The observations that the band at lowest field represents five protons, lies in the aromatic region of the spectrum and exhibits a complex spin–spin coupling pattern suggests that the molecule is a mono-substituted derivative of benzene and that the five aromatic protons are in non-equivalent environments.

Analysis of the expanded spectrum for the resonance at 7·97 τ reveals a doublet of doublet set of bands with coupling constants of 1·8 and 2·9 Hz. This shows that the three protons (methyl group) which cause this signal are coupled to the two single protons which give the resonances centred at 4·64 and 5·00 τ.

The only structures which are in accord with the above conclusions and with the molecular formula of the compound are:

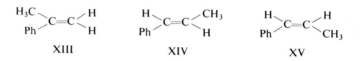

The olefinic protons in each structure are held in non-equivalent environments and would be expected to couple to the methyl group to differing extents. The similarity between the coupling constants for the interaction between the olefinic protons and the methyl protons indicates that there is likely to be the same number of bonds between the interacting nuclei in each case. This suggests that XIII is the correct structure. The olefinic system in this structure will have an associated magnetic field (see Fig. 4.10) and consideration of the geometry of the molecule shows that the aromatic protons will lie in different regions of this field and hence will be shielded or deshielded to different extents. This means that the protons in the ring will be in non-equivalent environments and will couple to each other to give a complex pattern of bands in the n.m.r. spectrum.

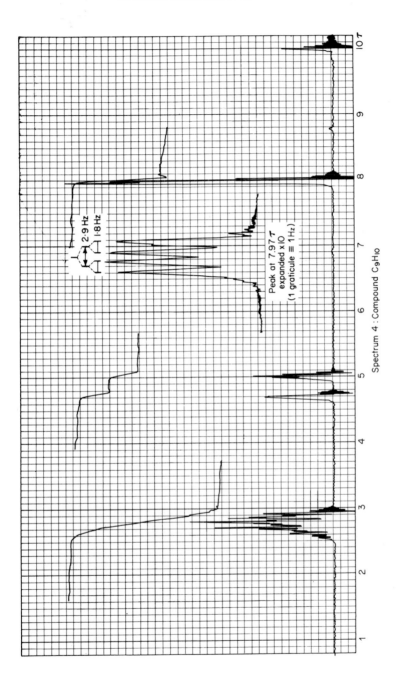

Peak at 7.97 τ
expanded ×10
(1 graticule ≡ 1Hz)

2·9 Hz 1·8Hz

Spectrum 4 : Compound C₉H₁₀

II. INSTRUMENTATION AND SAMPLE HANDLING

A. DESCRIPTION OF A SPECTROMETER

The basic requirements for an n.m.r. spectrometer are (a) a magnet producing a magnetic field of high homogeneity, (b) a sweep generator for varying the magnetic field, (c) a radio-frequency signal generator and transmitter unit, (d) a radio-frequency receiver and detector unit and (e) a recorder system. The components of an n.m.r. spectrometer are shown in the schematic diagram of Fig. 4.30.

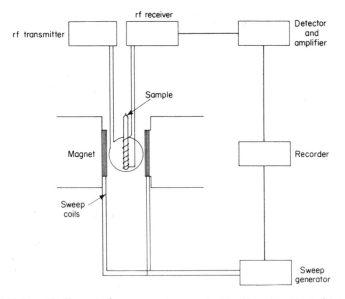

Fig. 4.30. Schematic diagram of an n.m.r. spectrometer (double-coil method of detection)

In the particular type of instrument illustrated in Fig. 4.30 the coils from the rf transmitter, the rf receiver and the sweep generator are held mutually perpendicular to each other in the pole gap of the magnet. Alternatively, the rf transmitter coil and rf receiver coil may be replaced by a single coil which is held perpendicular to the coils from the sweep generator (see later).

It may be recalled that the n.m.r. experiment is most conveniently performed by holding the field, B_1, generated by the rf transmitter constant and varying the main magnetic field B_0. The magnetic field B_0 is varied by passing the output of a linearly varying sweep generator to coils either wound around the magnet pole piece or contained in the pole gap. The output of the sweep generator is synchronized with the x axis of an oscilloscope or pen recorder. At a certain value of the magnetic field, nuclei in the sample may come into resonance and there is a net absorption of energy from the field B_1. This absorption of energy is either monitored by following the drop in the rf voltage in the tuned circuit producing the field B_1 (single-coil method) or by measuring the signal picked up in a receiver coil (double-coil method). The drop in rf voltage or the signal from the receiver coil is amplified and fed to the y axis of the recorder. The n.m.r. spectrum is thus presented as a plot of detector signal against magnetic field at constant rf transmitter frequency.

I. Magnet

Either a permanent magnet or an electromagnet may be used to provide the main magnetic field. For studies on proton resonance the field strengths which are generally used are 1·4092 and 2·3487 Tesla; these field strengths corresponding to the use of 60 and 100 MHz rf transmitters respectively. Superconducting magnets are used to provide even higher field strengths. Permanent magnets and electromagnets are comparable in that both types are capable of giving a field homogeneity of 3 parts in 10^9. In order to achieve this homogeneity the magnets have to be constructed with the pole faces parallel and almost optically flat. Homogeneity of the magnetic field in the pole gap can be improved by the use of shim coils which are located on the pole faces and through which currents may be passed in order to smooth out any gradients in the field.

2. Radio-frequency Transmitter and Receiver

In instruments with a double-coil (nuclear induction) system of detection, a receiver and a transmitter coil are located in the sample holder with the axis of the receiver coil perpendicular to the axis of the transmitter coil. With this geometrical arrangement, the background rf signal from the transmitter coil is isolated from the detector coil, and any signal picked up by the receiver coil results from the absorption of energy by the sample.

The basic circuit for the rf transmitter unit in instruments with single-coil systems is shown in Fig. 4.31. An alternating current of fixed frequency supplied from the generator passes to earth through a circuit consisting of a high resistance R and a capacitor C plus inductance coil L arranged in parallel. The coil L surrounds the sample space. The capacitor C and coil L form one arm of a bridge circuit and the background signal from the rf generator is prevented from reaching the detector by balancing the bridge. When the sample comes into resonance, power is absorbed by the sample and this results in a decrease in resistance of the generator circuit and the bridge goes out of balance. The "out-of-balance" signal from the bridge circuit is amplified and passed to the recorder.

The power level in the rf transmitter circuit can affect the appearance of the spectrum. At the resonance position the effect of the field B_1 is to cause transitions of the nuclei, both upwards and downwards, between the two energy levels corresponding to $m_I = +\frac{1}{2}$ and $m_I = -\frac{1}{2}$. The strength of the signal received at the detector is proportional to the net absorption of energy from the field B_1; that is, it is proportional to the difference between the number of nuclei in the lower and higher energy levels. At room temperature the excess relative number of nuclei in the lower level is about six in every million. If there is a net absorption of energy then this slight excess would be rapidly cancelled out unless there were some relaxation process by which the nuclei in the higher level could return to the lower level. Relaxation

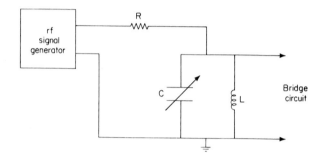

Fig. 4.31. Circuit for single-coil detection system

processes do in fact exist whereby the nuclei can revert to the lower energy state. If the rf power level is kept low the preponderance of nuclei in the lower level will be maintained by the effect of the relaxation processes. However if the power level is increased sufficiently then the number of nuclei raised to the higher level cannot be compensated for by the relaxation processes. Thus the number of nuclei in the higher level will be increased relative to the number in the lower level and the intensity of the signal at the detector will be decreased. The system is then said to be saturated. Saturation of a system can be recognized by the weakening and distortion of the peaks in the spectrum as the rf power level is raised. Saturation effects are also often observed if the rf power level is kept constant while the rate of scan of the spectrum is decreased.

3. Presentation of Spectra

An n.m.r. spectrum is a record of the signal at the detector versus the strength of the applied magnetic field B_0. Spectra are normally recorded on calibrated chart paper. For proton studies, the signal from the TMS reference is set on the zero line at the right hand side of the chart and the position, in Hertz, of any resonance relative to the reference position can be determined by counting the number of graticules on the chart between the signal and reference and multiplying by the number of Hertz each graticule represents. The x axis of calibrated charts are marked with either, or both, the chemical-shift scale, δ, or the tau scale, τ, In the former case the line for the TMS reference is marked as 0 whereas for the latter case it is marked as 10. Examples of n.m.r. spectra are shown in Spectra 1–4 on pages 251–257.

B. SAMPLE HANDLING

I. Samples

N.m.r. spectra of samples in the solid state show very broad absorption bands because of dipole–dipole interactions between molecules in the lattice. However the spectra of dilute solutions of solids in a suitable solvent exhibit sharp bands since the intermolecular interactions are averaged out as the molecules tumble around in solution. Solutions should be free from material in suspension otherwise broad bands may result from the anisotropic averaging of the field inhomogeneities when solid particles are present in the sample. Liquid samples are generally recorded as dilute solutions in a chemically "inert" solvent in order that intermolecular effects between the molecules of the sample may be reduced. The spectra of compounds which are sufficiently volatile may be obtained in the vapour state. However in order to obtain adequate signal strength, pressures greater than atmospheric have to be used.

The volumes of the sample solution and the concentration of the solution necessary to give sufficiently intense signals are dependent upon the nature of the nuclei in the sample and on the sensitivity of the instrument. For the study of protons with an instrument operating at 60 MHz, about $0.4\ cm^3$ of a 10% solution is normally adequate. The sample is contained in a thin-walled glass tube (approximately 5 to 8 mm o.d.) which can be housed in a sample holder between the pole faces of the magnet. A plastic vane may be fitted close to the top of the tube so that when the tube is in position in the holder it may be made to spin by the force of an air blast on the vane. Spinning the sample results in an averaging of the magnetic fields experienced by the nuclei, and the consequent improvement in the homogeneity of the field experienced by the sample gives an improvement in the resolution in the spectrum. Under certain conditions spinning the sample will give rise to two or more bands symmetrically placed on either side of the resonance signal. These bands are termed "spinning side bands". The displacement of these bands relative to the resonance signal is equal to the frequency of rotation of the sample and if the rate of spinning is increased the side bands become weaker and move further out from the resonance signal. Thus changing the spinning speed provides a simple means of identifying spurious signals of this type.

2. Solvents

Ideally solvents should be chemically inert and should be magnetically and electrically isotropic, i.e. spherically symmetrical in magnetic and electrical properties. For proton spectra, it is preferable that the solvent should not contain any protons or if protons are present in the solvent the signals from these protons should be well removed from signals from the sample. Solvents such as cyclohexane and carbon tetrachloride are "inert" and are magnetically and electrically isotropic but have the limitation that many compounds are not sufficiently soluble in them to give adequate spectra. In such cases a polar

Table 4.9. Approximate resonance positions for protons in some common solvents

Solvent	τ	Solvent	τ
CF_3COOH	0·17	CH_3Cl	6·95
$CHCl_3$	2·73	CH_3SOCH_3	7·45
C_6H_6	2·73	CH_3COOH	7·90 (CH_3)
CH_2Cl_2	4·70	CH_3COCH_3	7·91
H_2O	~5·2	CH_3CN	8·00
CH_3NO_2	5·67	C_6H_{12}	8·56
CH_3OH	6·53 (CH_3)	$(CH_3)_3COH$	8·73 (CH_3)

solvent may have to be used but caution should be exercised since solute–solvent interaction may result in a spectrum for the solute which is significantly different from that of the "free" solute in a non-polar solvent. For instance, the spectra of aqueous solutions may contain features due to hydrogen bonding and/or exchange processes between the solute and water molecules. Aromatic molecules such as benzene are often used as solvents and in such cases complexes may be formed between the solute and solvent in which the solute is held in a specific orientation to the aromatic ring. Owing to the diamagnetic anisotropy of the aromatic ring there may be a marked difference between the chemical shift of the protons when the sample is in an uncomplexed state and when in a solution of the aromatic hydrocarbon. In instances where the chemical shift is dependent upon the concentration of the solute, the chemical shift of the protons in the "free" state may be determined by measuring the band positions for various concentrations and extrapolating to infinite dilution (Expt. N11).

Some solvents which are commonly used in n.m.r. spectroscopy and the resonance positions of the protons in these solvents are given in Table 4.9.

3. Reference Compounds

The chemical shift of a nucleus is quoted relative to the position of resonance of a nucleus in a reference compound. The reference may be either *external* where the reference compound is in a separate container from that of the sample, or *internal* where the reference compound is mixed in solution with the sample. For both external and internal methods the reference compound should be magnetically and electrically isotropic, and for the internal method the compound should be chemically inert.

The main advantage of an external standard is that there is no contamination of the sample. However there is the disadvantage that the bulk diamagnetic properties of the sample and standard may be different and in such cases a correction has to be applied to the observed chemical shift in order to give the true chemical shift. The main advantage of an internal standard is that no correction has to be applied for differences in bulk diamagnetic properties since the sample and reference are surrounded by the same medium. The disadvantages of an internal standard are that the sample is contaminated and that there may be specific solute–solvent interactions which affect the positions of the sample and reference resonance bands to different extents.

Cyclohexane and tetramethylsilane are commonly used as reference compounds for proton-resonance studies in non-aqueous solutions, utilizing the single sharp band at 8·564 and 10·000 τ respectively as the reference. For studies in aqueous solutions, the sodium salt of 2,2-dimethyl-2-silapentane-5-sulphonic acid, $(CH_3)_3SiCH_2CH_2CH_2SO_3^-$ Na^+ (DSS), is frequently used. The resonance due to the methyl protons is taken as the reference line. At the concentrations used for reference compounds (approximately one per cent) the complex multiplet arising from the methylene protons is lost in the background noise of the spectrum. Acetonitrile (8·00 τ) and dioxan (6·32 τ) are also good internal standards for aqueous solutions.

III. EXPERIMENTS

A. INSTRUMENTATION

The homogeneity of the magnetic field in high-resolution n.m.r. spectrometers used for routine work is of the order of 1 part in 10^8. Since the homogeneity may lessen with time it is normal practice to check and adjust the homogeneity of the field each day before spectra are recorded. The procedure for checking the field homogeneity is outlined in the following experiment.

NI. Performance of an N.M.R. Spectrometer

Object

To test the homogeneity of the magnetic field in an n.m.r. spectrometer.

Theory

When a system containing proton nuclei is subjected to a magnetic field, B_0, then, according to classical theory, the nuclei will be induced to precess around the field direction and will be aligned so that the secondary fields associated with the spin of the nuclei are either parallel or anti-parallel to the direction of the field B_0. Since there will be a slight excess of nuclei in the state of lower energy, viz. the "parallel" state, there will be a resultant magnetic field from these nuclei in the direction of the field B_0. The excess nuclei in the "parallel" state, the resultant magnetic field and the precessional orbit of these nuclei are represented in Fig. N1.1. It can be envisaged that the vectors drawn in the figure to represent the individual precessing nuclei form a cone which rotates about the field axis. The axis of this cone and also the resultant magnetic field arising from the nuclei will be directed along the field axis.

In the n.m.r. experiment an oscillating magnetic field, B_1, from an rf transmitter coil is applied at right angles (x axis) to the axis of field B_0 (z axis). This field B_1 can be considered as two vectors rotating in opposite directions (page 216), one of which will rotate in the same direction as the precessing nuclei

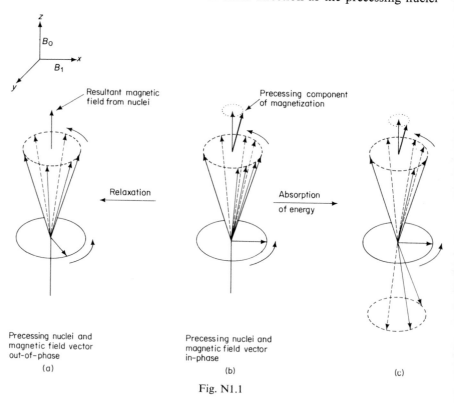

Precessing nuclei and
magnetic field vector
out-of-phase

(a)

Precessing nuclei and
magnetic field vector
in-phase

(b)

(c)

Fig. N1.1

(diagram (a) in figure). The nuclear precessional frequency may be varied by changing the value of the field B_0 and at a certain value the precessional frequency of the nuclei and the frequency of the rotating magnetic field vector will be the same. When this happens an interchange of energy occurs between the magnetic field B_1 and the nuclei, and some of the nuclei in the "parallel" state may transfer to the "anti-parallel" state (diagram (c)). When the magnetic-field vector and the precessing nuclei rotate in phase the magnetic-field vector will set up a force which acts at right angles to the resultant magnetic field arising from the nuclei. This force causes the vector representing the resultant magnetic field from the nuclei to be displaced from the field axis, whereupon the vector will precess around the field axis at the nuclear precessional frequency. This situation could also be considered as arising from the nuclei being "bunched" together as they precess around the field axis (diagram (b)). As the vector representing the resultant magnetic field from the nuclei precesses around the field axis a rotating component of magnetization is set up in the x and y directions and the alternating field set up in the y direction will induce a current in the receiver coil sited along this axis. This current may then be amplified and recorded as the n.m.r. signal.

If the field B_1 is switched off when the system is in resonance, then the component of magnetization in the x and y directions will gradually die away until the original state represented by diagram (a) is attained. The process whereby the system returns to the initial state is termed "relaxation".

When the field B_0 experienced by the sample is non-homogeneous, nuclei in different regions of the sample will have different precessional frequencies. Consequently there will be a range of values for the field B_0 over which the precessional frequency of nuclei will equal the frequency of the rotating magnetic-field vector, and the observed resonance band will be broadened because it represents a superposition of spectra for nuclei in different regions of the sample. Thus the better the homogeneity of field B_0 throughout the sample the narrower will be the resonance bands. There is however a "natural band width" beyond which the width of the bands cannot be reduced.

The "ringing" pattern (Fig. N1.2) associated with n.m.r. signals is also affected by the homogeneity of the field B_0; the better the field homogeneity the greater the extent of the "ringing" pattern. The "ringing" pattern is observed on the highfield side of the signal if the spectrum is scanned by increasing the field B_0. As the strength of the field B_0 is increased then the nuclear precession frequency increases and when this frequency equals the frequency of rotation of the magnetic-field vector a component of magnetization is set up in the x and y directions. As the field strength is increased further the system goes out of resonance and the magnetization in the x and y directions dies away as the system reverts by relaxation to the initial state. If the relaxation time is long, then as the spectrum is scanned the frequency at which the component of magnetization precesses around the field axis increases, and if this component gains 360° on the rotating-field vector the two come into phase again. The result is that the magnetization in the x and y directions is increased and the signal strength is increased. If the relaxation time of the nuclear system is sufficiently long then the precessing component of magnetization and the rotating-field vector may come into phase several times before the magnetization in the x and y directions dies away completely and an extended "ringing" pattern will be observed. If the magnetic field B_0 is non-homogeneous over the sample, then nuclei in different regions will have different precessional frequencies and since there will no longer be any phase

Homogeneous field

Non-homogeneous field

Fig. N1.2

coherence between the nuclei in the "resonant" state the rate of relaxation to the initial state will be increased. Thus the magnetization in the x and y directions will die away quickly when the system goes out of resonance and the extent of the ringing pattern will be reduced.

The homogeneity of the field experienced by the sample may be controlled by means of pairs of printed circuit coils (shim coils) which are arranged symmetrically about the sample space in the magnet gap. Passage of currents through the coils sets up separate magnetic fields which alter the different field gradients in the sample and by using the shim coils in conjunction with each other the field gradients may be smoothed out.

Apparatus

N.m.r. spectrometer (60 MHz); acetaldehyde.

Procedure

Record the spectrum of acetaldehyde with a sweep width of 500 or 600 Hz. Record the spectrum of the band at highest field using the standard instrumental operating conditions recommended in the manufacturers handbook for observing the "ringing" pattern. Alter the settings of the shim-coil controls in the order recommended in the handbook until a maximum ringing pattern is obtained. Record the spectrum for the aldehydic proton in acetaldehyde with low rf power, slow sweep conditions and a sweep width of 50 Hz.

Results and Discussion

Measure the band width at half-height for the bands representing the aldehydic proton. Compare the measured value for the band width with the manufacturer's specification and comment on the field homogeneity of the instrument. (For routine spectrometers the band width at half-height should be of the order of 0·5 Hz which is equivalent to a resolution of approx. 1 part in 10^8.)

B. SPIN–SPIN COUPLING

If a molecule containing nuclei which possess a magnetic moment is located in a magnetic field, then spin–spin coupling can often occur between the nuclei. Spin–spin coupling is not limited to nuclei of the same type and coupling can, in principle, occur between any two nuclei which have magnetic moments. The magnitude of the interaction between nuclei of different types and some of the factors which affect this interaction are examined in the experiments of this section.

N2. Factors Affecting Coupling Constants

Object

(a) To measure the coupling constants of o, m and p protons in some derivatives of benzene and to determine the dependence of the magnitude of the coupling on the number of intervening bonds.

(b) To determine the extent of coupling between proton and nitrogen nuclei in tetramethylammonium and tetraethylammonium salts.

Theory

The interaction between protons in a molecule which leads to "spin–spin" splitting of the resonance bands appears to occur via the intervening bonds between the nuclei. The mechanism of the interaction involves correlation of electron spins in the intervening bonds (page 229). The fact that the interaction occurs via the intervening bonds rather than through space is supported by the following observations.

(a) The coupling constant, J, between protons decreases as the number of intervening bonds increases.

(b) The coupling constant between vicinal protons in an olefinic system (H—C=C—H) decreases for constant bond angles as the π-bond order of the olefinic bond decreases.

(c) The coupling constant for vicinal protons in a fixed conformation changes as the dihedral angle, ϕ, changes.

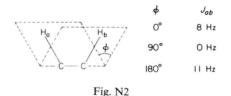

ϕ	J_{ab}
0°	8 Hz
90°	0 Hz
180°	11 Hz

Fig. N2

In alkanes the coupling constant between protons is virtually zero when there are four intervening bonds between the nuclei. However in aromatic compounds there may be appreciable coupling (0–3 Hz) between protons separated by four or five bonds. The magnitude of such long-range coupling in some derivatives of benzene is measured in this experiment.

Although the coupling between proton nuclei apparently only occurs via the intervening bonds, it has been suggested (Petrakis and Sederholm, 1961) that the coupling between fluorine nuclei involves a through-space contribution if the nuclei are in close proximity. Although this idea has been criticized (Evans, 1962) it may account for the observation that in the ^{19}F spectrum of $(CF_3)_2NCF_2CF_3$, the CF_2 band consists of a septet. This indicates that the fluorine nuclei in the CF_2 group are coupled to the six equivalent fluorine nuclei of the $—N(CF_3)_2$ group (four intervening bonds) rather than with the adjacent CF_3 group (three intervening bonds). Preferential long-range coupling is also observed for coupling between proton and nitrogen nuclei in ammonium salts. In this experiment the coupling between proton and nitrogen nuclei in tetramethylammonium and tetraethylammonium salts is examined.

Apparatus

N.m.r. spectrometer (60 MHz); 2,3,4-trihydroxyacetophenone (I), 2,4-dibromo-6-nitro-phenol (II), 2,4,5-trimethoxyacetophenone (III), tetramethylammonium iodide, bromide or chloride (IV), tetraethylammonium iodide, bromide or chloride (V), TMS.

Procedure

Prepare 10% w/v solutions of compounds I, II and III in acetone and of compounds IV and V in water.

Record the spectra of compounds I–III using TMS as internal reference and with a sweep width of 500 or 600 Hz. If the resonances arising from the aromatic protons show spin–spin splitting then re-run the spectra in the region of the aromatic resonances with a sweep width of 100 Hz.

Record the spectra of compounds IV and V with a sweep width of 600 Hz and also record the integral trace for the resonances arising from compound V. Record the spectrum of compound V with a sweep width of 300 Hz.

Results

Measure the coupling constants between the aromatic protons in compounds I, II and III. Measure the chemical shifts of the aromatic protons from the spectra and calculate the predicted chemical shifts using Table 4.8 on page 256. Tabulate the values for the coupling constants, the chemical-shift data (τ scale) for the aromatic protons and for the other protons in compounds I, II and III.

List, where applicable, the coupling constants, the chemical shifts (δ scale relative to water) and the relative integrated intensities of the resonances in the spectra of compounds IV and V. Assign the individual resonances to the protons in compounds IV and V.

Discussion

1. Comment on any correlation between the magnitudes of the coupling constants and the number of bonds between the interacting nuclei in compounds I, II and III, and on any differences between the calculated and experimental values for the chemical shifts of the aromatic protons.

2. Account for the observed splitting patterns of the multiplets in the spectra of compounds IV and V.

N3. ^{19}F–^{1}H Coupling

Object

To determine the ^{19}F–^{1}H and ^{1}H–^{1}H coupling constants in 2,2,2-trifluoro-ethanol and ethanol respectively.

Theory

The ^{19}F nucleus is the only naturally occurring isotope of fluorine. For ^{19}F the spin quantum number I equals one-half and there are thus two orientations the nucleus can take up when in a magnetic field. In this respect the ^{19}F nucleus is the same as the ^{1}H nucleus and consequently the multiplicity of the resonance of the methylene protons in compounds such as 2,2,2-trifluoro-ethanol and ethanol will be the same.

The chemical-shift values for fluorine nuclei extend over a much wider range ($\sim$500 p.p.m.) than those for proton nuclei ($\sim$15 p.p.m.). The ratio $(\nu_a - \nu_b)/J_{ab}$ for two interacting fluorine nuclei a and b is frequently large since the chemical-shift difference between a and b is usually large, and as a result ^{19}F spectra are often first order. Although the chemical shift values for fluorine and proton nuclei in corresponding environments are markedly different there is not such a noticeable difference in the corresponding coupling constants. Nevertheless ^{19}F–^{19}F coupling and ^{19}F–^1H coupling tend to be greater than the corresponding ^1H–^1H coupling.

Apparatus

N.m.r. spectrometer (60 MHz); 2,2,2-trifluoroethanol, ethanol, TMS.

Procedure

Record the ^1H spectrum of 2,2,2-trifluoroethanol and ethanol with a sweep width of 500 or 600 Hz. Assign the resonance arising from the hydroxylic proton in each spectrum and if this resonance is a multiplet add a drop of dilute mineral acid to the sample and re-record the spectrum. Determine the position of resonance of the methylene protons in each spectrum and re-record the spectrum of the methylene protons with a sweep width of 100 Hz.

Result

Tabulate the chemical shifts of all the protons in the two compounds and the values for the coupling constants between the ^{19}F nuclei and the methylene protons in 2,2,2-trifluoroethanol and between the methyl protons and the methylene protons in ethanol.

Discussion

1. Comment on the relative values of the chemical shifts of the methylene protons in the two compounds and on the values of the coupling constants recorded.

2. Comment on the multiplicity of the peak or peaks representing the resonance of the hydroxylic proton in the spectra of the two compounds and suggest a reason for the observed multiplicity.

N4. ^{13}C–^1H Coupling

Object

To interpret the spectra of *cis*-1,2-dichloroethylene, *trans*-1,2-dichloroethylene and 1,1-dichloroethylene in relation to ^{13}C–^1H and ^1H–^1H coupling.

Theory

The ^{13}C nucleus has a spin quantum number I equal to one-half and when in a magnetic field any protons adjacent to or bonded to this nucleus will experience the effect of the two spin states of the nucleus. Thus the resonances for such protons will be split into a doublet pattern because of the coupling between the ^{13}C and ^1H nuclei. This coupling is not normally observed in routine ^1H spectra as the proton-resonance bands for structures such as $R_3^{13}C$—H are very weak on account of the low natural abundance ($1\cdot1\%$) of the ^{13}C nucleus. However if the ^1H spectrum is recorded at high instrument sensitivity on liquid samples or concentrated solutions of the sample it is sometimes possible to observe the spectrum of protons coupled to the ^{13}C nucleus. For example, if the ^1H spectrum of CH_3Cl is recorded at high sensitivity then, in addition to the strong singlet band arising from the protons in the structure $^{12}CH_3Cl$, a weak doublet set of bands resulting from the protons in the structure $^{13}CH_3Cl$ would be observed. Substitution of ^{13}C for ^{12}C in methyl chloride does not markedly alter the chemical shift of the methyl protons and since the resonance of the methyl protons in $^{13}CH_3Cl$ is split into a doublet, the bands in the doublet will be almost equally spaced on either side of the band representing the protons in $^{12}CH_3Cl$. The bands on either side of the main resonance are referred to as "satellite" bands.

The spectrum for the ^{13}C nucleus in $^{13}CH_3Cl$ would appear as a quartet of bands as a result of coupling between the ^{13}C nucleus and the three equivalent protons. However ^{13}C spectra are not normally obtained in a routine manner because of the low natural abundance of the nucleus, the low natural sensitivity to n.m.r. detection and other factors.

The probability of more than one ^{13}C nucleus being in any molecule is low, especially for molecules of low molecular weight. Consequently it may be assumed in this experiment that any ^{13}C–^1H coupling observed in the ^1H spectra of the dichloroethylenes results from the structure $>^{13}C\!=\!^{12}C<$. The presence of ^{13}C nuclei in a molecule does not prevent ^1H–^1H coupling and such coupling must be taken into account when interpreting spectra.

Apparatus

N.m.r. spectrometer (60 MHz); *cis*-1,2-dichloroethylene (I), *trans*-1,2-dichloroethylene (II), 1,1-dichloroethylene (III). Note: It is often difficult for inexperienced operators to obtain satisfactory spectra showing ^{13}C satellite bands and it is advantageous to have a pre-recorded set of spectra available.

Procedure

Record the ^1H spectra of compounds I, II and III with instrumental operating conditions such that the main resonance in each spectrum is on scale. Next, record the spectra under operating conditions such that the intensity of the weakest spinning side bands of the main band is equivalent to approx. 80 per cent full-scale deflection. Re-record the spectra with the same instrumental operating conditions but with a different sample spinning speed. Compare the latter two sets of spectra and determine whether there are any bands in the spectra which have not shifted relative to the main band upon changing the sample spinning speed. If there are no bands in the spectra that can be assigned in this way as ^{13}C satellites then repeat the procedure with higher overall instrument sensitivity until the expected bands are observed.

Results and Discussion

Select the spectra of compounds I, II and III which show ^{13}C satellites and draw individual line diagrams for these spectra on the one sheet so that the spectra can be compared directly. The intensity of the peaks need not be to scale but the separation between peaks should be drawn to scale. Omit peaks in the spectra due to spinning side bands.

Given that the ^{13}C satellite bands are a result of coupling between ^{13}C and ^{1}H nuclei which are directly bonded, account for the following:

1. The spectrum of 1,1-dichloroethylene.
2. The spectrum of cis-1,2-dichloroethylene and trans-1,2-dichloroethylene taking into consideration both ^{13}C–^{1}H and ^{1}H–^{1}H coupling.

Complete Table 4.10.

Table 4.10

	$J(^{13}C-^1H)$, Hz	$J(^1H-^1H)$, Hz
1,1-dichloroethylene		
cis-1,2-dichloroethylene		
trans-1,2-dichloroethylene		

Comment on the following:

(a) The similarity between the $J(^{13}C-^1H)$ coupling constants.
(b) The difference between the $J(^1H-^1H)$ coupling constants.
(c) The difference between the $J(^1H-^1H)$ and $J(^{13}C-^1H)$ coupling constants.

N5. ^{10}B–^{1}H and ^{11}B–^{1}H Coupling

Object

To interpret the ^{1}H spectrum of the borohydride ion and to measure the coupling constants $J(^{10}B-^1H)$, and $J(^{11}B-^1H)$ in the borohydride ion.

Theory

The natural abundances of the ^{10}B and ^{11}B isotopes of boron are 19·6 and 80·4% respectively. The nuclear spin quantum number I for the ^{10}B nucleus is three while that for the ^{11}B nucleus is three-halves, and thus the ^{10}B nucleus can take up seven orientations $(2I + 1)$ in a magnetic field while the ^{11}B nucleus can take up four orientations. Any nucleus coupled to boron will experience seven different effective fields corresponding to the seven possible orientations of the ^{10}B nucleus and four different effective fields corresponding to the four possible orientations of the ^{11}B nucleus.

Apparatus

N.m.r. spectrometer (60 MHz); sodium borohydride or potassium borohydride, solution containing approximately $0.1\ mol\ dm^{-3}$ of sodium hydroxide.

Procedure

Make up a saturated solution of either sodium or potassium borohydride in the sodium hydroxide solution. Wrap a small piece of filter paper or paper tissue around the tip of the pipette and draw the solution up into the pipette so that the paper acts as a filter. Transfer the clear solution to an n.m.r. tube. Locate the resonances arising from the protons in the borohydride ion and record the spectrum with a sweep width of 300 Hz using instrumental operating conditions such that the intensities of the major peaks in the spectrum correspond to approximately 90% full-scale deflection.

Results and Discussion

Measure the coupling constants $J(^{10}B-^{1}H)$ and $J(^{11}B-^{1}H)$ from the spectrum.
  Comment on the following:
  1. The magnitudes of $J(^{10}B-^{1}H)$ and $J(^{11}B-^{1}H)$ as compared to commonly expected values for coupling constants between proton nuclei.
  2. The multiplicity of the splitting patterns observed in the spectrum, and the relative intensities of the bands in the spectrum.
  3. The expected spin–spin splitting pattern in the ^{11}B spectrum.

N6. $^{14}N-^{1}H$ Coupling and Nuclear Quadrupole Relaxation

Object

To interpret the ^{1}H spectrum of ammonium chloride, methylammonium chloride and trimethylammonium chloride.

Theory

The ^{14}N nucleus has a spin quantum number I equal to one and thus there are three possible orientations of the magnetic moment of the nucleus in an applied field. The three orientations correspond to the nucleus having the magnetic quantum numbers $m_I = +1$, 0 and -1. Any adjacent nucleus will experience three different magnetic fields arising from the different orientations of the magnetic moment of the ^{14}N nucleus. Thus the proton resonance spectrum of the system $^{14}N-^{1}H$ would be expected to consist of a triplet of bands. Sometimes however it is difficult to observe the proton resonance spectrum for an N–H system in a molecule since the bands are often considerably broadened owing to nuclear quadrupole relaxation effects in the ^{14}N nucleus (see following).

Nuclei, such as ^{14}N, which have I values greater than one-half appear to have their charge distributed over an ellipsoidal surface rather than over a spherical surface as in the case for nuclei with I equal to zero or one-half. Thus nuclei such as ^{14}N act as electric quadrupoles, i.e. if a point charge were to be brought up towards the nucleus, then for a given distance between the nucleus and point charge the force experienced by the point charge would vary for different directions of approach. The symmetry axis of the nuclear quadrupole will be collinear with the magnetic-moment vector of the nucleus. For a

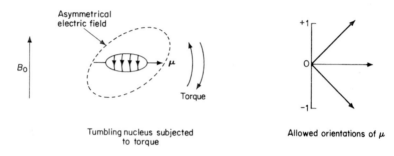

Tumbling nucleus subjected
to torque

Allowed orientations of μ

Fig. N6

compound such as methylammonium chloride the ^{14}N nucleus is surrounded by a cloud of valence electrons which is asymmetric and as the $CH_3NH_3^+$ ion tumbles around in solution a time-variable electric torque will be exerted on the nuclear quadrupole. This torque will tend to alter the orientation of the nuclear quadrupole and since the magnetic moment vector, μ, and the nuclear quadrupole are collinear, the orientation of the magnetic-moment vector will also change. As there are only three possible orientations for the magnetic moment vector of the ^{14}N nucleus in a magnetic field the result of the torque on the nuclear quadrupole is to cause the nucleus to "flip" between the three states corresponding to $m_I = +1, 0, -1$. The interchange between the states is termed "nuclear quadrupole relaxation".

As a result of the continuous flip of the nucleus between the different states, the protons attached to the ^{14}N nucleus in methylammonium chloride will experience an average of the fields set up by the ^{14}N nucleus in the different states. The effect of this is to cause broadening of the proton resonance bands.

If the ^{14}N nucleus is surrounded by an electric field which is spherically symmetrical, then there is no nuclear quadrupole and hence no nuclear quadrupole relaxation. For example, there is no quadrupole-induced relaxation in the ^{14}N nucleus in the ammonium ion and if the protons in the ammonium ion are not exchanging then the proton resonances will appear as narrow bands.

If protons are "detached" from a ^{14}N nucleus by exchange with other protons in the system, the time of attachment of a proton to a ^{14}N nucleus may be so short that the nuclear quadrupole effect and coupling between the nuclei cannot be observed. For instance, the exchange of protons between ammonia and water molecules in ordinary liquid ammonia is so rapid that separate O–H and N–H resonances cannot be observed and only a single average proton resonance is obtained.

Apparatus

N.m.r. spectrometer (60 MHz); ammonium chloride, methylammonium chloride, trimethylammonium chloride, conc. hydrochloric acid, DSS.

Procedure

Prepare near-saturated solutions of each of the compounds in approximately 5 cm³ of H_2O. Add a small amount of DSS and four drops of conc. hydrochloric acid to each of the solutions. Prepare a near-saturated solution of ammonium chloride in approximately 5 cm³ H_2O and add DSS to the solution. Record the spectrum of each of the solutions with a sweep width of 500 or 600 Hz and using instrumental operating conditions which permit the resonance arising from the protons bonded to the ^{14}N nucleus to be observed.

Results

Tabulate the following information:
1. Chemical shifts of the singlet bands and multiplet patterns relative to the DSS signal at 10 τ.
2. Splitting (in Hz) between the bands in the multiplet patterns.
3. Width (in Hz) at half peak height of the bands in the spectra.
4. Assignment of the bands in the spectra.

Discussion

Suggest reasons for the following:
1. The effect of acid on the spectrum of ammonium chloride.
2. The splitting pattern for the resonance representing the protons bonded to the ^{14}N nucleus in methylammonium chloride.
3. The multiplicity of the resonance representing the methyl protons in methylammonium chloride.
4. The difference between the widths at half peak height of the resonance lines for the protons bonded to ^{14}N nucleus in methylammonium chloride and in ammonium chloride (acidified solution).

10

5. The difference between the widths at half peak height of the resonance lines for the protons bonded to the ^{14}N nucleus in methylammonium chloride and in trimethylammonium chloride.

C. APPLICATIONS TO QUANTITATIVE MEASUREMENTS

An n.m.r. spectrum may yield quantitative information on a particular system in one of three ways: (1) the relative areas of the bands gives the relative numbers of different types of nuclei present, (2) the position of bands in systems where molecular complexes are formed yields data on the relative proportion of complexed and uncomplexed molecules present and (3) the width of the bands in systems where exchange is occurring gives information on the mean lifetimes of nuclei in particular environments.

Examples of the use of measurements of band areas, band positions and band widths in quantitative analysis are considered in the experiments in this section.

N7. Analysis of a Two-component Mixture—Keto–Enol Tautomerism
Object

To assign the resonances in the spectra of acetylacetone and derivatives of acetylacetone and to determine the relative proportions of keto and enol forms present in acetylacetone.

Theory

Derivatives of acetylacetone may undergo keto–enol tautomerism:

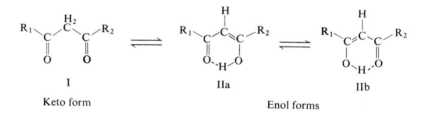

| I | IIa | IIb |

| Keto form | Enol forms |

If the rate of interchange between the tautomeric forms is slow compared with the time scale of the n.m.r. process ($\sim 10^{-3}$ s), separate signals would be observed for the protons in each of the tautomeric forms. It would be expected that the rate of interchange in the equilibrium I $\rightleftharpoons$ II will be considerably slower than in the equilibrium IIa $\rightleftharpoons$ IIb for $R_1 = R_2 = CH_3$.

The resonance position of the methylene protons in the keto form will be markedly different from the resonance positions of the methine proton and of the hydroxyl proton in the enol form, and if the corresponding bands are not overlapped by other resonances then the relative proportions of the keto and enol forms can be determined by measurement of the areas of the bands corresponding to these protons. The change in environment of protons in the groups R_1 and R_2 on interchange between the keto and enol forms will be slight compared with that for the methylene and methine protons. Consequently the difference, if any, between the resonance positions of protons in the groups R_1 and R_2 will not be so large as the difference between the resonance positions of the methylene and methine protons.

Apparatus

N.m.r. spectrometer (60 MHz); acetylacetone, diethylmalonate, ethylaceto-acetate, carbon tetrachloride, deuterochloroform, TMS.

Procedure

Prepare solutions of acetylacetone, diethylmalonate and ethylacetoacetate in carbon tetrachloride (50% v/v). Record the spectrum of the solution of acetylacetone with TMS as internal reference and using a sweep width of 500 or 600 Hz. Record the spectrum of the lowfield region on the same chart with a sweep offset of 400 Hz and increased instrument sensitivity. Record the integral traces of all the resonances in the spectrum.

Record the spectra of the solutions of diethylmalonate and ethylacetoacetate as for acetylacetone but do not record the integral traces.

Results

Assign the bands in the spectrum of acetylacetone knowing that the ratio of the integrated areas for the methyl and methylene protons in the keto form should be 3:1 and that for the methyl and methine protons in the enol form should be 6:1. Estimate the percentage keto and enol forms present.

Assign the bands in the spectra of ethylacetoacetate and diethylmalonate on the basis of the information gained from the spectrum of acetylacetone.

Complete, as far as possible, Table N7 for each of the compounds.

Table N7

Resonance position τ	Band multiplicity	Relative integrated area	Assignment

Discussion

1. Comment on the rate of interchange of tautomeric forms in the equilibria I ⇌ II and IIa ⇌ IIb (for $R_1=R_2=CH_3$). What differences might you expect if the former rate was faster and the latter rate was slower?

2. What changes would you expect in the relative amounts of keto and enol forms in acetylacetone if the spectrum was recorded in a more polar solvent such as chloroform?

3. Assign resonances of other derivatives if spectra are available.

N8. Factors Affecting Keto–Enol Tautomerism

Object

To determine the position of equilibrium between the keto–enol forms in acetylacetone and in solutions of acetylacetone in cyclohexane and between acetylacetone and triethylamine in solution.

Theory

An equilibrium exists between keto and enol forms in acetylacetone and in solutions of acetylacetone in many solvents.

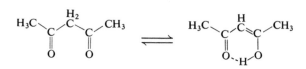

The position of equilibrium between the keto and enol forms in acetylacetone will differ from that in solution. This change in the equilibrium in going from the liquid state to solution is the result of factors such as (a) specific interaction between acetylacetone and the solvent, (b) dilution effects of the solvent on solute–solute interactions and (c) change in dielectric constant of the medium.

The occurrence of specific interaction between solute and solvent will not only affect the position of the keto–enol equilibrium but also the chemical shift of any protons in the solute or solvent directly involved in the interaction. The interaction may take the form, among other possibilities, of a hydrogen bond between the solute and solvent or charge transfer between solute and solvent (see Expt. N10).

The dilution effect of added solvent can alter the position of equilibrium by causing dissociation of intermolecularly bonded acetylacetone molecules. For instance, if a dimer of acetylacetone is formed by interactions (through hydrogen bonds or otherwise) of the types enol–enol, enol–keto or keto–keto, then dilution of the system would increase the concentration of free keto or free enol monomer present and hence alter the position of equilibrium between the free keto and enol forms.

The addition of solvent to acetylacetone will cause a change in the dielectric constant of the medium and this will contribute to the displacement of the keto–enol equilibrium. The keto form being more polar than the enol form will be relatively more favoured in polar solvents.

In this experiment the position of equilibrium between the keto and enol forms in acetylacetone is determined. Also, the effect of added triethylamine and cyclohexane on the position of equilibrium and on the chemical shift of the protons present in the two forms is examined in order to determine the extent and type of interaction, if any, between acetylacetone and these solvents.

Apparatus

N.m.r. spectrometer (60 MHz); acetylacetone, cyclohexane, triethylamine, TMS.

Procedure

Prepare the following solutions:

    (a) 0.9 cm^3 acetylacetone $+ 0.1$ cm^3 cyclohexane
    (b) 0.7 cm^3 acetylacetone $+ 0.3$ cm^3 cyclohexane
    (c) 0.5 cm^3 acetylacetone $+ 0.5$ cm^3 cyclohexane
    (d) 0.3 cm^3 acetylacetone $+ 0.7$ cm^3 cyclohexane
    (e) 0.1 cm^3 acetylacetone $+ 0.9$ cm^3 cyclohexane
    (f) 0.4 cm^3 acetylacetone $+ 0.6$ cm^3 triethylamine.

Record the spectrum of acetylacetone with TMS as internal reference and using a sweep width of 500 or 600 Hz. Re-record the spectrum of the low-field region on the same chart with a sweep offset of 400 Hz. Record the integral traces of the two spectra.

Record the spectra of solutions (a)–(e) in the same manner as the spectrum of acetylacetone. Record the integrated areas of the peaks arising from the methine and methylene protons in acetylacetone in each spectrum.

Record the spectrum of triethylamine under the same conditions.

Results

Calculate the percentage keto form in acetylacetone.

Tabulate the mole fraction of acetylacetone in solutions (a)–(e) along with the chemical shifts (in Hz) of the peaks in the spectra of these solutions. By examination of the chemical-shift values estimate the chemical shifts of the proton resonances when the mole fraction of acetylacetone is zero.

Calculate the percentage keto form of acetylacetone in each of the acetyl-acetone–cyclohexane mixtures and plot the percentage keto form against mole fraction of acetylacetone in the mixture. Extrapolate the plot to infinite dilution of acetylacetone in cyclohexane to determine the percentage keto form present under these conditions.

Assign the peaks in the spectrum of the mixture of acetylacetone and tri-ethylamine by comparison with the spectra of acetylacetone and triethylamine. Tabulate the chemical shifts of the acetylacetone peaks in the spectrum of the mixture.

Discussion

1. Comment on whether the results for the chemical shifts of the protons in acetylacetone–cyclohexane mixtures indicate the presence or absence of specific interaction between the solute and solvent.

2. If there is no interaction between acetylacetone and cyclohexane in solution, suggest an explanation for the difference between the percentage keto form present in cyclohexane at infinite dilution and the percentage present in acetylacetone.

3. Indicate by comparison of the chemical shifts of the protons in the acetylacetone–triethylamine mixture and in acetylacetone whether there is any interaction between acetylacetone and triethylamine in solution and, if so, indicate the possible site for the interaction in acetylacetone. Account for any marked differences between the spectrum of acetylacetone and that of the acetylacetone–triethylamine mixture.

N9. Analysis of Three- and Four-component Mixtures

Object

To determine the molar ratio of the components in a mixture of naphthalene, tetralin and decalin, and in a mixture of naphthalene, tetralin, decalin and cyclohexene.

Theory

If the n.m.r. spectrum of a mixture of compounds exhibits a resolved band due to one of the components and if the number of protons in the molecule giving rise to that band is known, then the integrated area of the band gives a direct measure of the relative concentration of that component in the mixture. For example, the spectrum of a mixture of benzene and cyclohexane exhibits bands at $2 \cdot 73$ and $8 \cdot 56 \tau$ from benzene and cyclohexane respectively, and the integrated area of the former band will be proportional to six protons while that of the latter band will be proportional to twelve protons. Obviously if the integrated areas of the two bands in the spectrum of the mixture are equal then the molar ratio of benzene to cyclohexane in the mixture must be 2:1.

The analysis of a mixture for which the bands in the spectra are overlapped is more complicated than the case where all the bands are resolved. However, if the spectrum of a mixture of n components exhibits a resolved band for each of $(n-1)$ components, then the complete analysis can be performed readily. The contribution, A', to the total integrated area, A'', of an overlapped band from components which exhibit resolved bands is given by the expression

$$A' = A_1\frac{x_1}{y_1} + A_2\frac{x_2}{y_2} + \ldots A_{n-1}\frac{x_{n-1}}{y_{n-1}} \qquad (N9.1)$$

where $A_1, A_2,\ldots A_{n-1}$ represent the integrated areas of resolved bands due to components $1, 2,\ldots n-1$; $y_1, y_2,\ldots y_{n-1}$ represent the number of protons in components $1, 2,\ldots n-1$ giving rise to the resolved bands, and $x_1, x_2,\ldots x_{n-1}$ represent the number of protons in components $1, 2,\ldots n-1$ contributing to the area of the overlapped band.

The contribution of the nth component to the integrated area of the overlapped band is given by $A'' - A'$, and if the number of protons giving rise to this integrated area is known then the relative concentration of the nth component can be calculated.

Apparatus

N.m.r. spectrometer (60 MHz); naphthalene, tetralin, decalin, cyclohexene, TMS.

Procedure

Prepare accurately the following mixtures with approximate molar ratios as indicated:

(a) Naphthalene: tetralin: decalin = $1:2\cdot5:1\cdot5$
(b) Naphthalene: tetralin: decalin: cyclohexene = $1:2\cdot5:1\cdot5:2\cdot5$

Record the spectrum of mixture (a) with TMS as internal reference and using a sweep width of 500 or 600 Hz. Record the integral trace of all the bands in the spectrum. Assign the observed bands to the different proton types in the components. If there is any doubt in the assignment record the spectra of two of the individual components. Repeat the procedure for mixture (b).

Results and Discussion

Tabulate the chemical shifts, integrated areas and assignments of the bands in the spectra of the mixtures. Calculate the molar ratio of the components in the mixtures and comment on the comparison between the experimentally derived values and the known values. Discuss any advantages and disadvantages of the n.m.r. method for the analysis of mixtures.

N10. Molecular Complexes
(a) Charge-transfer Complexes
(b) Hydrogen-bonded Complexes

Object

To evaluate the equilibrium constant for complex formation between either (a) mesitylene and s-trinitrobenzene or (b) acetone and chloroform.

Theory

A dynamic equilibrium will be set up in a system where a molecular complex is formed between an electron-donor component D and an electron-acceptor component A

$$D + A \rightleftharpoons DA$$

If the time for transfer from the uncomplexed to complexed state is short compared with the time required to observe resonance ($\sim 10^{-3}$ s) then the chemical shift for any proton directly involved in complex formation will occur at a value corresponding to the weighted average chemical shift of the two states. When the equilibrium is displaced in the direction of either state, the contribution of that state to the average chemical shift of the protons will increase and the position of resonance will shift toward the resonance position of the completely uncomplexed or complexed states.

The interaction between an aromatic donor, e.g. mesitylene and an aromatic acceptor, e.g. s-trinitrobenzene, results in the formation of a charge-transfer complex and in this system the proton resonances occur at higher field than for the uncomplexed component molecules. It has been suggested (Hanna and Ashbaugh, 1964) that in the case of aromatic donor and acceptor molecules the difference between the chemical shift of the protons in the complexed state and in the uncomplexed state arises because of the effect of the ring currents in the donor and acceptor molecules and because of the alteration of the paramagnetic contribution to the donor and acceptor proton shifts by the presence of the other component.

If a hydrogen bond DX...HA is formed between two molecules DX and HA then the magnetic field experienced by the proton H will be modified by the influence of the donor atom X. In the case of the acetone–chloroform system, the oxygen atom acts as donor and the hydrogen atom in chloroform as acceptor. The formation of a hydrogen bond results in a downfield shift of the proton resonance, except when the donor is an aromatic compound when an upfield shift occurs (see Expt. N11). The downfield shift of the proton resonance on formation of a hydrogen bond has been explained as being mainly due to the electrostatic nature of the bond. Atom X provides a strong electric field (negative charge) in the region of HA and as this field tends to draw the nucleus (positive charge) away from the electrons in the H—A bond this brings about a net reduction of the electron density around the proton. Another effect of the electric field is to inhibit the diamagnetic circulation (page 225) of the electrons in the hydrogen atom. This also reduces the shielding around the proton and contributes to the downfield shift.

The observed chemical shift for protons involved in complex formation is related to the chemical shift in the complexed and uncomplexed molecules by the following expression given by Becker *et al.* (1958) and by Berkeley and Hanna (1963)

$$\delta^A_{obs} - \delta^A_0 = \frac{[D]K_a}{1 + [D]K_a} \cdot (\delta^A_{DA} - \delta^A_0) \qquad (N10.1)$$

where δ^A_{obs} is the observed chemical shift of the acceptor protons in the complexing medium, δ^A_0 is the chemical shift of the acceptor protons in the uncomplexed acceptor, δ^A_{DA} is the chemical shift of the acceptor protons in the pure complex, [D] is the concentration of donor at equilibrium and K_a is the equilibrium constant for complex formation. For dilute solutions, K_a is given by

$$K_a = \frac{[DA]}{[D][A]} \qquad (N10.2)$$

where square brackets represent concentrations at equilibrium. If the initial concentration of the donor is very much greater than the initial concentration of the acceptor, then the concentration of donor at equilibrium may be taken as being equal to its initial concentration.

An alternative form of equation N10.1 is

$$\frac{\Delta}{[D]} = \Delta_0 K_a - \Delta K_a \qquad (N10.3)$$

where $\Delta = \delta^A_{obs} - \delta^A_0$ and $\Delta_0 = \delta^A_{DA} - \delta^A_0$.

It can be seen from equation N10.3 that a plot of $\Delta/[D]$ versus Δ should be a straight line and that the value of K_a is given by the slope of the line.

Apparatus

N.m.r. spectrometer (60 MHz); s-trinitrobenzene, mesitylene (1,3,5-trimethylbenzene), acetone, chloroform, carbon tetrachloride, TMS.

Procedure

MESITYLENE–S-TRINITROBENZENE SYSTEM. Prepare approximately 25 cm³ of a saturated solution of s-trinitrobenzene in carbon tetrachloride. Add a few drops of TMS to this stock solution. Pipette 4 cm³ of the stock solution into each of five 5 cm³ graduated flasks and then add in turn 0·2, 0·4, 0·6, 0·8 and 1·0 cm³ of mesitylene to the flasks by means of a micro-burette and make up to 5 cm³ with carbon tetrachloride.

Calculate the expected resonance position for the aromatic protons in s-trinitrobenzene using Table 4.8 on page 256. Record the spectrum of the stock solution with a sweep width of 600 Hz and locate the resonance of the aromatic protons. Note: The instrumental sensitivity may have to be high in order to observe the resonance. Re-record the spectrum over the region of the aromatic resonance with a sweep width of 300 Hz; alter the sweep offset (in 100 Hz units) so that the TMS resonance will be on the same chart at this sweep width and then record the spectrum of the TMS resonance. Repeat the procedure of scanning over the aromatic resonance, changing the sweep offset and then scanning over the TMS resonance for each of the solutions containing mesitylene.

ACETONE–CHLOROFORM SYSTEM. Add a few drops of TMS to 5 cm³ of chloroform and pipette 0·2 cm³ of this solution into a 10 cm³ graduated flask and make up to 10 cm³ with carbon tetrachloride. Pipette 0·2 cm³ of the TMS-chloroform solution into each of five 10 cm³ graduated flasks and then pipette in turn 1, 2, 3, 4 and 5 cm³ of acetone into the flasks and make up to 10 cm³ with carbon tetrachloride.

Record the spectrum using a sweep width of 500 or 600 Hz of the solution which does not contain acetone and locate the resonance of the chloroform proton. Re-record the spectrum over the region of the chloroform resonance with a sweep width of 300 Hz, alter the sweep offset (in 100 Hz units) so that the TMS resonance will be on the same chart at this sweep width and then record the spectrum of the TMS resonance. Repeat the procedure of scanning over the chloroform resonance, changing the sweep offset and then scanning over the TMS resonance for each of the solutions containing acetone.

Results

Calculate the concentration of mesitylene or acetone in each of the solutions in units of mol (kg solvent)$^{-1}$. (Densities: mesitylene = 0·86 g cm^{-3}, acetone = 0·79 g cm^{-3}, carbon tetrachloride = 1·59 g cm^{-3}). Measure the separation, in Hz, between the resonance position of TMS and that of s-trinitrobenzene or chloroform in each spectrum. Calculate a value for Δ for each solution and plot $\Delta/[\text{D}]$ versus Δ. Determine a value for the equilibrium constant K_a in units of mol^{-1} kg solvent. Convert this value into units of mol^{-1} dm³. Determine the value of Δ_0.

Discussion

Suggest a reason why *either* the value of Δ_0 is greater in the hexamethylbenzene–trinitrobenzene system than in the mesitylene–trinitrobenzene system *or* the value of Δ_0 is greater in the N,N-dimethylformamide–chloroform system than in the acetone–chloroform system.

N11. Diamagnetic Anisotropy Effects in Aromatic Compounds

Object

To calculate the distance between the chloroform proton and the plane of the aromatic ring in the molecular complex formed between chloroform and benzene.

Theory

The observation that the chloroform-proton signal is shifted upfield when chloroform is diluted with an aromatic hydrocarbon has been attributed to complex formation between chloroform and the aromatic hydrocarbon. It has been postulated that in the complex with benzene the hydrogen atom of the chloroform is sited above the plane of the aromatic ring in the vicinity of the six-fold symmetry axis of the ring (Fig. N11.1).

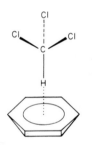

Fig. N11.1

When a magnetic field is applied to the system a diamagnetic current is induced in the plane of the ring and the secondary magnetic field created by this "ring current" opposes the applied field at points above and below the plane of the ring (see page 226). The chloroform proton is located in one of these positions and thus the resonance position is displaced to higher field.

The nature of the forces involved in the complex is not clear but it has been suggested by Reeves and Schneider (1957) that a weak hydrogen bond is formed between chloroform and the π-donor molecules. As indicated above there is an upfield shift of the proton resonance in such cases. By contrast, when hydrogen bonds are formed which involve electrons in non-bonding orbitals there is a downfield shift of the proton resonance (see Expt. N10).

The position of the proton relative to the plane of the aromatic ring in the complex may be represented diagrammatically as shown in Fig. N11.2.

If the aromatic ring is considered as a circular loop of radius r (equal to 0·14 nm) then the distance x may be determined from the following expression given by Reeves and Schneider (1957).

$$\delta = \frac{e^2 r^2}{mc^2 x^3} \qquad \text{(N11.1)}$$

where δ represents the chemical shift in p.p.m. of the chloroform-proton signal in the complex relative to that in the unassociated molecule, c the velocity of light and e and m the electronic charge and mass respectively. Thus if δ is measured experimentally, the value of x and hence the distance BH may be calculated.

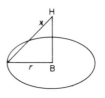

Fig. N11.2

Chloroform is weakly self-associating and allowance must be made for this effect in order to determine the true position of resonance of the chloroform proton in the unassociated state. Dilution of chloroform with an inert solvent, such as cyclohexane, causes dissociation of associated chloroform molecules and results in a shift of the proton signal to higher field. Extrapolation of the shift to infinite dilution gives the shift relative to pure chloroform and this represents the true chemical shift of an unassociated molecule. The interaction shift in chloroform–benzene systems is then the total shift at infinite dilution in this system less the shift at infinite dilution in an inert solvent. The chemical shift δ in equation N11.1 is given by the interaction shift expressed in p.p.m.

Apparatus

N.m.r. spectrometer (60 MHz); benzene, chloroform, cyclohexane, TMS.

Procedure

Pipette into 10 cm³ graduated flasks 0·5, 1·0, 1·5, 2·0, 3·0, 5·0 and 7·0 cm³ of chloroform and make up to 10 cm³ with benzene. Record the spectrum of each of the solutions and of chloroform with a sweep width of 500 or 600 Hz and using TMS as internal reference. Repeat the procedure using cyclohexane in place of benzene.

Results

Calculate the mole fraction of chloroform in each of the solutions used. (Densities: benzene 0·88 g cm^{-3}; cyclohexane 0·78 g cm^{-3}; chloroform 1·50 g cm^{-3}.)

Measure the chemical shift (in Hz) of the chloroform-proton signal in each of the solutions relative to that of the proton signal in pure chloroform.

Plot on the one graph the chloroform chemical shift (in Hz) for the solutions in benzene and in cyclohexane against the mole fraction of chloroform in each solution. Extrapolate the plots to infinite dilution and measure the chemical shift of the chloroform proton at infinite dilution for each system. Tabulate the values of the chemical shift (in Hz) at infinite dilution and also the value of the interaction shift at infinite dilution in the chloroform–benzene system. Calculate the distance between the chloroform proton and the plane of the benzene ring.

Discussion

Although the induced "ring-current" effect is operative in nitrobenzene the interaction shift for the chloroform proton in a solution of nitrobenzene is very nearly zero. Suggest an explanation for this.

N12. The Structure of Poly(Methyl Methacrylate)

Object

To determine the relative proportions of isotactic, syndiotactic and heterotactic units in a commercial sample of poly(methyl methacrylate).

Theory

The polymerization of methyl methacrylate can yield a polymer of the following basic structures:

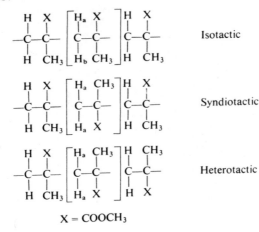

In the isotactic chain the monomer unit (in square brackets) has as nearest neighbours two other monomer units of the same configuration. Examination of the structure shows that the environments of the methyl groups is the same for each unit and hence the methyl resonance would be expected to be a single band. The environments for the methylene protons H_a, H_b differ from each other; one has X groups in the equivalent position on the adjacent carbon atom while the other has CH_3 groups in the neighbouring equivalent positions. Since the difference in environment will not be marked the resonances for the methylene protons would be expected to be of the AB type rather than of the AX type. The spectrum for purely isotactic polymer will consist of six bands: two for the methyl protons (one being for group X) and four for the methylene protons.

In the syndiotactic structure the monomer units alternate in configuration but, as can be seen from the structure, all the methyl groups are in the same environment. The spectrum for purely syndiotactic polymer would therefore consist of three bands: two for the methyl protons (one being for group X) and one for the methylene protons. The integrated areas of the bands would be in the ratio $3:3:2$.

For the heterotactic form, the central monomer unit has one neighbouring unit of the same configuration and the other of opposite configuration. Here again the methyl protons are all in the same environment as are the methylene protons. The spectrum for these two sets of protons will comprise two bands with relative areas of $3:2$.

The relative proportion of isotactic, syndiotactic and heterotactic units in poly(methyl methacrylate) depends upon the method of preparation, and for a sample comprising all three structures the n.m.r. spectrum will be a composite of the spectra of the individual forms. The position of resonance for the protons in group X would be expected to be the same for each structure but for the other corresponding protons the resonance positions would be expected to be similar but not identical because of the differences between the environment of the protons in the different structures. The integrated areas of non-overlapped bands due to the various types of protons will be a direct measure of the relative proportion of each type of structural unit present in the polymer.

Apparatus

N.m.r. spectrometer (60 MHz); sample of commercial poly(methyl methacrylate), stock samples of polymer which are mainly isotactic and syndiotactic in structure (these samples may be prepared by the methods of Goode *et al.*, 1960 and Szwarc and Rembaum, 1956), TMS.

Procedure

Prepare 10% w/v solutions of each of the samples in chloroform and record the spectra of the solutions with TMS as internal reference using a sweep width of 300 Hz. Assign the bands in the spectrum of the commercial sample by comparison with the spectra of the isotactic and syndiotactic polymer. Re-record the spectrum of the commercial sample with instrumental operating conditions such that the most intense band from the methyl protons has a peak height of approximately 95% full-scale deflection. Record the integral trace of the peaks in this spectrum.

Results and Discussion

Tabulate the chemical shifts, integrated areas (where appropriate) and the assignments of the resonances in the spectra of the three samples. Estimate relative proportions of isotactic, syndiotactic and heterotactic forms in the the commercial sample.

Suggest reasons why the bands in the spectra are relatively broad and how the experimental conditions might be altered so as to decrease the band width.

N13. Reaction Kinetics

Object

To determine (a) the catalytic-rate constant for the hydration of pyruvic acid and for the dehydration of 2,2-dihydroxypropanoic acid, and (b) the equilibrium constant for the pyruvic acid–2,2-dihydroxypropanoic acid system.

Theory

The spectrum of pure pyruvic acid consists of two resonances at $0·70 \tau$ and $8·40 \tau$ which are due to the carboxyl proton and the methyl protons respectively. The separation between the resonances is dependent upon the purity of the sample; the greater the purity the greater the separation.

In an aqueous solution of pyruvic acid the following equilibrium is established

$$CH_3COCOOH + H_2O \rightleftharpoons CH_3C(OH)_2COOH$$

The spectrum of an aqueous solution of pyruvic acid consists of three bands. The bands at $7·40$ and $8·25 \tau$ represent the resonances of the methyl protons of pyruvic acid and 2,2-dihydroxypropanoic acid respectively, while the third band which represents the resonances of the carboxyl, hydroxyl and water protons appears at variable positions in the spectrum dependent upon the composition of the mixture. This latter band is a singlet since the proton exchange rate between the three different environments is fast compared with the time of transition between different magnetic states.

The forward and back reactions in the equilibrium are acid catalysed and as the concentration of the acid present is increased so the rate of exchange between the pyruvic acid form and the 2,2-dihydroxypropanoic acid form is increased. As this happens the bands due to the methyl protons in these forms will broaden and eventually merge into a single sharp band (see Fig. 4.27). The width of the exchange broadened bands at half the band height is related to the mean lifetime, τ, of the proton in each environment by the expression

$$\frac{1}{T_2'} = \frac{1}{T_2} + \frac{1}{\tau} \qquad (N13.1)$$

where $2/T_2'$ is the width of the exchange broadened band at half-height and $1/T_2$ is the width at half-height in the absence of exchange.

The reciprocal of the mean lifetime in an environment is equal to the specific rate of the reaction. The specific rate in turn is equal to $k_0 + k_{H^+}[H^+]$ where k_0 is the spontaneous rate constant, k_{H^+} is the catalytic rate constant and $[H^+]$ is the hydrogen-ion concentration. Equation N13.1 can therefore be written in the form

$$\frac{1}{T_2'} = \text{constant} + k_{H^+}[H^+] \qquad (N13.2)$$

Thus a plot of $1/T_2'$ versus $[H^+]$ should yield a straight line with a slope equal to the value for the catalytic-rate constant. Thus by measuring the band width at half-height of the methyl resonances in the spectra of pyruvic acid and 2,2-dihydroxypropanoic acid at different acid concentrations the catalytic-rate constants for the forward and back reactions may be determined. Also by measuring the integrated areas of the methyl resonances the equilibrium constant for the system can be evaluated.

Apparatus

N.m.r. spectrometer (60 MHz); distilled pyruvic acid, concentrated hydrochloric acid ($\sim$12 mol dm^{-3}, exact concentration to be known), TMS.

Procedure

Prepare the following solutions:

(a) 0·50 cm^3 pyruvic acid + 0·50 cm^3 water
(b) 0·50 cm^3 pyruvic acid + 0·55 cm^3 water + 0·05 cm^3 HCl
(c) 0·50 cm^3 pyruvic acid + 0·50 cm^3 water + 0·10 cm^3 HCl
(d) 0·50 cm^3 pyruvic acid + 0·45 cm^3 water + 0·15 cm^3 HCl
(e) 0·50 cm^3 pyruvic acid + 0·40 cm^3 water + 0·20 cm^3 HCl
(f) 0·50 cm^3 pyruvic acid + 0·35 cm^3 water + 0·25 cm^3 HCl

Record the spectrum of pyruvic acid using TMS as internal reference with a sweep width of 500 or 600 Hz. Record the spectrum of solution (a) with instrumental operating conditions such that the more intense of the two peaks representing the methyl resonances of pyruvic acid and 2,2-dihydroxypro-panoic acid has a peak height of approximately 95% full-scale deflection. Note the instrumental operating conditions. Record the integral trace for the methyl resonances in this spectrum. Record the spectra of solutions (b)–(f) over the region of the methyl resonances using the same instrumental operating conditions as for the spectrum of solution (a).

Results and Discussion

List the chemical shifts of the signals observed in the spectrum of pyruvic acid and hence comment on the purity of the pyruvic acid.

Tabulate the chemical shifts and assignments of signals observed in the

Table N13

Solution	Concentration of HCl mol dm^{-3}	Band widths				$\dfrac{1}{T_2'}$	
		Pyruvic acid		Hydrate		Pyruvic acid	Hydrate
		Hz	Radians sec^{-1}	Hz	Radians sec^{-1}	Radians sec^{-1}	Radians sec^{-1}

spectrum of solution (a). Calculate the equilibrium constant for the pyruvic acid–2,2-dihydroxypropanoic acid system in solution (a).

Measure the band widths at half-height of the methyl resonances in the spectra of solutions (a)–(f). Complete Table N13.

Plot the values for $1/T_2'$ for pyruvic acid and 2,2-dihydroxypropanoic acid in solutions (a)–(f) against the concentration of hydrochloric acid and hence determine the catalytic rate constant, $k_{H^+}^1$, for the hydration of pyruvic acid and the catalytic rate constant, $k_{H^+}^{-1}$, for the dehydration of 2,2-dihydroxy-propanoic acid. Calculate the ratio k^1/k^{-1} and comment on the comparison between this value of the equilibrium constant and the value determined from the integral traces.

D. ANALYSIS OF SPECTRA

In many instances the value of chemical shifts and coupling constants cannot be obtained directly from the spectrum and the spectrum has to be "analysed" in order to obtain this information. If the type of system giving rise to the spectrum is known then the values for chemical shifts and coupling constants can often be determined using standard rules for the analysis. Examples of the analysis of the spectra of AB and ABX systems are given in this section.

N14. Analysis of an AB spectrum

Object

To identify the AB bands in the spectrum of ethyl cinnamate and to determine the chemical shifts and coupling constants for the AB system.

Theory

The appearance of the spectrum of an AB system depends upon the ratio $(\nu_a - \nu_b)/J_{ab}$ (page 236). As this ratio increases the spectrum approaches that of a first-order AX spectrum. If the value of $(\nu_a - \nu_b)/J_{ab}$ in an AB system is relatively large then the agreement between the values of ν_a, ν_b and J_{ab} calculated using the expressions for exact analysis of an AB system (page 240) and those derived from the spectrum on the assumption that the spectrum is first-order are reasonably good and for many purposes it is unnecessary to carry out an exact analysis.

Apparatus

N.m.r. spectrometer (60 MHz); ethyl cinnamate, TMS.

Procedure

Record the spectrum of ethyl cinnamate with TMS as internal reference and using a sweep width of 500 or 600 Hz. Assign the resonances in the spectrum. Re-record the spectrum using different operating conditions if the AB system of bands cannot be observed.

Results and Discussion

Calculate the positions (in Hz) of the AB lines in the spectrum relative to the TMS line. Calculate two sets of values for ν_a, ν_b and J_{ab} by firstly using the expressions given on page 240, and secondly, by assuming the spectrum is first order and deriving the values directly from the spectrum. Tabulate the results and comment on any differences between the two sets of values. Calculate the chemical shift values (τ) for protons A and B using the values of ν_a and ν_b determined by exact analysis. Calculate the band positions (in Hz) of the AB system relative to the TMS resonance if the spectrum were recorded using a radio frequency of 100 MHz instead of 60 MHz.

Check the assignment of the lines in the AB system by calculating the theoretical positions of the lines and their relative intensities using the formulae given in Table 4.4 on page 240. Tabulate the results to show the comparison between the calculated and observed values for the line positions and relative intensities.

N15. Analysis of an ABX Spectrum

Object

To determine the values of the chemical shift of the protons and the coupling constants between the protons in the ABX system of vinyl acetate.

Theory

As has been mentioned previously (page 243) the spectrum of an ABX system in a molecule may consist of twelve or fourteen bands, or possibly less if some of the bands overlap. The frequencies and relative intensities of the bands in an ABX spectrum have been derived in terms of the quantities v_a, v_b, v_x, J_{ax}, J_{bx} and J_{ab} (see Pople *et al.* 1959) and the results are summarized in Table N15.1. It is to be noted that band 13 has zero intensity and so the maximum number of bands that may appear in the spectrum is fourteen.

The symbols used in Table N15.1 are defined by the equations:

$$v_{ab} = \tfrac{1}{2}(v_a + v_b) \qquad \text{(N15.1)}$$

$$D_+ \cos 2\phi_+ = \tfrac{1}{2}(v_a - v_b) + \tfrac{1}{4}(J_{ax} - J_{bx}) \qquad \text{(N15.2)}$$

$$D_+ \sin 2\phi_+ = \tfrac{1}{2}J_{ab} \qquad \text{(N15.3)}$$

$$D_- \cos 2\phi_- = \tfrac{1}{2}(v_a - v_b) - \tfrac{1}{4}(J_{ax} - J_{bx}) \qquad \text{(N15.4)}$$

$$D_- \sin 2\phi_- = \tfrac{1}{2}J_{ab} \qquad \text{(N15.5)}$$

The following rules for an ABX spectrum can be drawn up from consideration of the results summarized in the table:

Rule 1. The value of v_{ab} is given by the mean of the frequencies of the eight bands in the AB region.

Rule 2. The difference in frequency between bands 1 and 5, and bands 3 and 7 equals $2D_-$ while that between bands 2 and 6, and bands 4 and 8 equals $2D_+$.

Rule 3. The AB system consists of two quartets, the frequency of the bands in one quartet being dependent upon D_-, i.e. bands 1, 3, 5 and 7, and the frequency of the bands in the other being dependent upon D_+, i.e. bands 2, 4, 6 and 8. The separation between the mid-points of the two quartets is given by $\tfrac{1}{2}|J_{ax} + J_{bx}|$.

Rule 4. The difference in frequency between bands 1 and 3, 2 and 4, 5 and 7, 6 and 8 is equal to $|J_{ab}|$.

Table N15.1

Band no.	Assignment	Frequency	Relative intensity
1	B	$\nu_{ab} - \frac{1}{4}(+2J_{ab} + J_{ax} + J_{bx}) - D_-$	$1 - \sin 2\phi_-$
2	B	$\nu_{ab} - \frac{1}{4}(+2J_{ab} - J_{ax} - J_{bx}) - D_+$	$1 - \sin 2\phi_+$
3	B	$\nu_{ab} - \frac{1}{4}(-2J_{ab} + J_{ax} + J_{bx}) - D_-$	$1 + \sin 2\phi_-$
4	B	$\nu_{ab} - \frac{1}{4}(-2J_{ab} - J_{ax} - J_{bx}) - D_+$	$1 + \sin 2\phi_+$
5	A	$\nu_{ab} - \frac{1}{4}(+2J_{ab} + J_{ax} + J_{bx}) + D_-$	$1 + \sin 2\phi_-$
6	A	$\nu_{ab} - \frac{1}{4}(+2J_{ab} - J_{ax} - J_{bx}) + D_+$	$1 + \sin 2\phi_+$
7	A	$\nu_{ab} - \frac{1}{4}(-2J_{ab} + J_{ax} + J_{bx}) + D_-$	$1 - \sin 2\phi_-$
8	A	$\nu_{ab} - \frac{1}{4}(-2J_{ab} - J_{ax} - J_{bx}) + D_+$	$1 - \sin 2\phi_+$
9	X	$\nu_x - \frac{1}{2}(J_{ax} + J_{bx})$	1
10	X	$\nu_x + D_+ - D_-$	$\cos^2(\phi_+ - \phi_-)$
11	X	$\nu_x - D_+ + D_-$	$\cos^2(\phi_+ - \phi_-)$
12	X	$\nu_x + \frac{1}{2}(J_{ax} + J_{bx})$	1
13	Combination	$2\nu_{ab} - \nu_x$	0
14	Combination (X)	$\nu_x - (D_+ + D_-)$	$\sin^2(\phi_+ - \phi_-)$
15	Combination (X)	$\nu_x + (D_+ + D_-)$	$\sin^2(\phi_+ - \phi_-)$

Rule 5. The X region is symmetrical about ν_x and the value of ν_x is given by the mid-point of the X region.

Rule 6. If the X region consists of six bands, then
(a) bands 9 and 12 are the most intense and have the same relative intensity
(b) (intensity of 10) + (intensity of 14) = (intensity of 11) + (intensity of 15) = 1
(c) separation $|9 - 12| = |J_{ax} + J_{bx}|$; $|10 - 11| = 2(D_+ - D_-)$; $|14 - 15| = 2(D_+ + D_-)$.

Rule 7. If the X region consists of four bands, then
(a) the four bands are of equal intensity (ideal case for pure ABX spectrum)
(b) (intensity of 14) = (intensity of 15) = 0
(c) separation $|9 - 12| = |J_{ax} + J_{bx}|$; $|10 - 11| = 2(D_+ - D_-)$.

The values of ν_{ab} and ν_x can be obtained directly from the spectrum while the values for D_+, D_-, $|J_{ab}|$, $|J_{ax} + J_{bx}|$ can be measured after the bands in the spectrum have been assigned. Once these values have been determined the values for ν_a, ν_b, J_{ax}, J_{ab} and J_{bx} can be calculated using standard equations (see results section). Under certain conditions the relative signs of the coupling constants J_{ax}, J_{bx} can be found.

Apparatus

N.m.r. spectrometer (60 MHz); vinyl acetate, TMS.

Procedure

Record the spectrum of vinyl acetate with a sweep width of 600 Hz. Identify the lines representing the ABX system in vinyl acetate and re-record the spectrum of the ABX system with a sweep width of 300 Hz. Note the sweep offset from the TMS reference in the latter spectrum.

Results

Plot the ABX region in the spectrum of vinyl acetate as a line spectrum on graph paper. Assign the frequency of the line at highest field as 0 Hz and label the lines in Hz relative to this reference. Note: In order to determine the chemical-shift values for the individual resonance lines (see later), the frequencies must be converted so that they are relative to the position of the TMS line in the spectrum.

ASSIGNMENT OF BANDS IN SPECTRUM. Assume that the spectrum is first order and determine the values for the chemical shifts of the protons (mid-points of quartets) and the coupling constants between the protons (line separations within the quartets; each quartet will yield two coupling constants).

Given that the magnitude of the coupling between protons in a vinyl system is generally in the order: *trans* > *cis* > geminal, assign the values of J_{13}, J_{12} and J_{23} in the structure

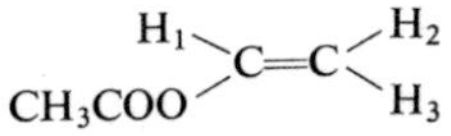

$$\underset{CH_3COO}{\overset{H_1}{\diagdown}} C = C \underset{H_3}{\overset{H_2}{\diagup}}$$

Knowing the values for the two coupling constants in each quartet assign the quartets to protons H_1, H_2 or H_3.

Complete Table N15.2.

Table N15.2

Hz	Hz	
J_{13}	ν_1	$\dfrac{\nu_1 - \nu_2}{J_{12}}$
J_{12}	ν_2	$\dfrac{\nu_1 - \nu_3}{J_{13}}$
J_{23}	ν_3	$\dfrac{\nu_2 - \nu_3}{J_{23}}$

By examination of the ratios of the difference in chemical shifts to the coupling constant in Table N15.2 assign the resonance arising from the X system. Also assign the A and B quartets in the spectrum. Note: The values of $(v_a - v_b)/J_{ax}$ and $(v_b - v_x)/J_{bx}$ will be larger than the value of $(v_a - v_b)/J_{ab}$ and the resonance common to the two largest ratios in the table must be that of the X system.

Make an assignment of the bands in the spectrum on the basis of the following:

(a) The subtraction rules

$$3 - 1 = 4 - 2 = 7 - 5 = 8 - 6$$
$$2 - 1 = 4 - 3 = 11 - 9 = 12 - 10$$
$$10 - 9 = 12 - 11 = 6 - 5 = 8 - 7$$
$$(10 - 11) = (8 - 4) - (5 - 1)$$

(b) The separation of the mid-points of the two quartets (1, 3, 5, 7 and 2, 4, 6, 8) in the AB region is equal to half the separation of lines 9 and 12 (Rules 3 and 7(c)).

(c) The value of $2(D_+ - D_-)$ calculated using Rule 2 is equal to the value calculated using Rule 7(c).

If the assignment is not in agreement with the above conditions (a)–(c) then make a new assignment and check the assignment again. Repeat the procedure until a satisfactory assignment is made.

CALCULATION OF SPECTRAL PARAMETERS. Determine the values of v_{ab}, v_x, $|J_{ab}|$, $|J_{ax} + J_{bx}|$, D_+ and D_- using the appropriate rules. Hence calculate the values of $|J_{ax} - J_{bx}|$ and $(v_a - v_b)$ using equations N15.7 and N15.8 which are derived from equations N15.2–5.

$$(v_a - v_b) + \tfrac{1}{2}|J_{ax} - J_{bx}| = (4D_+^2 - |J_{ab}|^2)^{1/2} \qquad (N15.7)$$

$$(v_a - v_b) - \tfrac{1}{2}|J_{ax} - J_{bx}| = (4D_-^2 - |J_{ab}|^2)^{1/2} \qquad (N15.8)$$

Calculate $|J_{ax}|$ and $|J_{bx}|$ from the values of $|J_{ax} + J_{bx}|$ and $|J_{ax} - J_{bx}|$ and also the values of v_a and v_b from the values of $(v_a - v_b)$ and v_{ab}. Calculate the chemical-shift values $(\delta_a, \delta_b, \delta_x)$ and also the tau values (τ_a, τ_b, τ_x) corresponding to the values of v_a, v_b and v_x.

Tabulate the values of v_a, v_b, v_x, J_{ax}, J_{bx}, J_{ab}.

DETERMINATION OF RELATIVE SIGNS OF J_{ax} AND J_{bx}. Calculate the value of $(v_a - v_b)/J_{ab}$ and if it is equal to or greater than two then determine the relative signs of J_{ax} and J_{bx} by means of procedures (a)–(c) given below. Note: If the ratio $(v_a - v_b)/J_{ab}$ is not equal to or greater than two, then the two sets of intensities calculated below will not differ from the observed spectrum by more than the experimental error.

(a) Solve equations N15.9 and N15.10 given below for ϕ_+ and ϕ_- and then calculate the relative intensities of the bands in the spectrum using the relationship given in Table N15.1.

$$2D_+ \cos 2\phi_+ = (\nu_a - \nu_b) + \tfrac{1}{2}(|J_{ax}| - |J_{bx}|) \qquad \text{(N15.9)}$$

$$2D_- \cos 2\phi_- = (\nu_a - \nu_b) - \tfrac{1}{2}(|J_{ax}| - |J_{bx}|) \qquad \text{(N15.10)}$$

(b) Solve equations N15.11 and N15.12 given below for ϕ_+ and ϕ_- and then calculate the relative intensities of the lines in the spectrum.

$$2D_+ \cos 2\phi_+ = (\nu_a - \nu_b) + \tfrac{1}{2}(|J_{ax}| + |J_{bx}|) \qquad \text{(N15.11)}$$

$$2D_- \cos 2\phi_- = (\nu_a - \nu_b) - \tfrac{1}{2}(|J_{ax}| + |J_{bx}|) \qquad \text{(N15.12)}$$

(c) Compare the relative intensities of the bands calculated by procedures (a) and (b) with the experimental relative intensities (normalized to intensity of line 9) and hence determine which set of calculated intensities is the better fit to the experimental data. Note: If the intensities calculated by procedure (a) are the better fit then J_{ax} and J_{bx} have the same sign, whereas if those from procedure (b) are the better fit then J_{ax} and J_{bx} are of opposite sign.

Discussion

1. What would be the appearance of the X region of the spectrum if $D_+ \approx D_-$?

2. Comment on the difference between the values of the chemical shifts and the coupling constants determined directly from the spectrum on a first-order basis and those calculated by analysis of the spectrum.

3. The spectrum of the ABX system of methyl vinyl ketone in benzene is shown schematically in Fig. N15. Given that $J_{ab} = 17\cdot85$, $J_{ax} = 10\cdot66$, $J_{bx} = 1\cdot15$ Hz pick out the quartets in the spectrum and assign as either A, B or X.

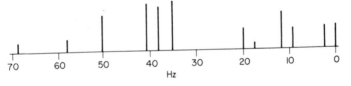

Fig. N15

REFERENCES

Becker, E. D., Liddel, U., and Shoolery, J. N. (1958). *J. molec. Spectrosc.* 2, 1.
Berkeley, P. J., and Hanna, M. W. (1963). *J. Phys. Chem.* 67, 846.
Chapman, D., and Magnus, P. D. (1966). *In* "Introduction to Practical High Resolution Nuclear Magnetic Resonance Spectroscopy". Academic, London and New York.

Evans, D. F. (1962). *Discuss. Faraday Soc.* **34**, 139.

Emsley, J. W., Feeney, J., and Sutcliffe, L. H. (1965). *In* "High Resolution Nuclear Magnetic Resonance Spectroscopy". Pergamon, Oxford.

Goode, W. E., Owens, F. H., Fellmann, R. P., Snyder, W. H., and Moore, J. E. (1960). *J. Polym. Sci.* **46**, 317.

Hanna, M. W., and Ashbaugh, A. L. (1964). *J. Phys. Chem.* **68**, 811.

Matheson, D. W. (1967). *In* "Nuclear Magnetic Resonance for Organic Chemists". Academic, London and New York.

Petrakis, L., and Sederholm, C. H. (1961). *J. Chem. Phys.* **35**, 1243.

Pople, J. A., Schneider, W. G., and Bernstein, H. J. (1959). *In* "High Resolution Nuclear Magnetic Resonance". McGraw-Hill, London.

Reeves, L. W., and Schneider, W. G. (1957). *Can. J. Chem.* **35**, 251.

Szwarc, M., and Rembaum, A. (1956). *J. Polym. Sci.* **22**, 189.

Chapter 5

Mass Spectrometry

I. PRINCIPLES OF MASS SPECTROMETRY

In mass spectrometry ions are produced from an element or compound by bombarding the molecules of the element or compound with a mono-energetic beam of electrons. The mass spectrometer then separates the ions according to their mass to charge ratios (m/e) and measures the relative abundances of each species with a given m/e value. It is also possible to obtain values for the energies involved in the production of the ions. This information on mass and abundance ratios of ions and on the energy of formation of an ion is used in the interpretation of molecular structure.

A. TYPES OF IONS PRODUCED IN A MASS SPECTROMETER

When a molecule is bombarded with electrons which have just sufficient energy to ionize the molecule, then only one type of ion is obtained in the spectrum, namely the *parent* or *molecular* ion

$$M + e \rightarrow M^{+} + 2e$$

If the electron-beam energy is increased the molecular ion can be formed with excess electronic and vibrational energy and will eventually possess sufficient energy to dissociate into fragment ions. Molecular ions are usually formed when the electron beam energy is 8–15 eV; when the energy is of the order 20–25 eV fragmentation is normally quite extensive. It is found that the relative intensities of fragment ions vary very little for beam energies between 50 and 100 eV so that a median energy of 70 eV is used for most analytical work.

I. Molecular Ions

It is important to identify the peak arising from the molecular ion since the mass number at which the molecular ion occurs will correspond to the molecular weight of the sample. Molecular ions are the ions of greatest mass which can be produced in a mass spectrometer by a simple unimolecular ionization process. Since they are not always stable they may occur with a very low intensity. Occasionally, when a beam energy of 70 eV is used to bombard the molecule, no molecular ion is detected since it is formed with so much excitational energy that it immediately fragments to produce peaks at lower m/e values. For this reason molecular-weight determinations are often performed using low electron-beam energies so that the molecular ion will not have a great deal of excess energy and can exist for the time interval necessary for detection (about 10^{-5} second). In addition the intensity of the molecular ion relative to the intensities of the other ions present is increased when the electron-beam energy is reduced.

Molecular ions of compounds will obey the "nitrogen rule" (Beynon, 1960). This rule states that if the molecular weight, calculated on the basis of the most abundant isotope of each element present, is even, then the number of nitrogen atoms in the molecule is even (or zero); if the molecular weight is odd there is an odd number of nitrogen atoms present. Since molecular ions are odd-electron ions, having lost an electron in the ionization process, they are of even mass unless there is an odd number of nitrogen atoms present. The rule is a consequence of the fact that, apart from nitrogen, the most abundant stable isotopes of all common elements of odd valency have an odd mass number and those with an even valency have an even mass number.

In a larger molecule there is normally more fragmentation than in a small molecule, so that the abundance (or relative intensity) of the molecular ion is reduced. Pahl (1954) defined the probability of decomposition of the molecular ion, W_z, as

$$W_z = \Sigma I_f / (I_p + \Sigma I_f) \qquad (5.1)$$

where I_p equals the intensity of the molecular ion and ΣI_f is the intensity of all other singly charged ions.

The stability of the molecular ion, W_p, is then

$$W_p = 1 - W_z \qquad (5.2)$$

W_p decreases as the size of the molecules in a homologous series increases. The presence of multiple bonds or a branched carbon-atom site in a molecule enhances fragmentation and results in molecular-ion peaks of low intensity.

2. Fragment Ions

Fragment ions are produced from the molecular ion by bond cleavage. When the energy of the electron beam is greater than the ionization potential of the molecule then the molecular ion may be formed in an excited state. The excitational energy is redistributed among the various bonds in the molecule and causes cleavage of a number of the bonds. Attempts have been made to predict the mode of fragmentation of the molecular ion using the quasi-equilibrium theory which is briefly discussed on page 306. The calculated results from this theory are often in good agreement with the experimentally observed fragmentation.

A possible cleavage to produce a fragment ion is

$$[CH_3-CH_2-CH_3]^+_\cdot \longrightarrow CH_3-CH_2^+ + CH_3^\cdot$$

In this and subsequent fragmentations the symbol ⌒ is used to indicate a one-electron movement and ⌒ to indicate a two-electron movement (Budzi-kiewicz *et al.* 1967). Many structural correlations have been made on the basis of the observed fragment ions in the mass spectra of molecules containing differing functional groups. Examples of major fragmentation pathways in various classes of organic compound are given later (page 313).

3. Rearrangement Ions

Rearrangement ions are formed from the molecular ion by redistribution of atoms or groups of atoms at the moment of decomposition of the molecular ion. Hydrogen-transfer rearrangements are very common, though migration of an alkyl group can also occur. Rearrangement processes often involve the elimination from the ion of an even electron molecule such as CO, HCN, or C_2H_2.

A very general hydrogen-rearrangement process is known as the "Mc-Lafferty rearrangement". This involves cleavage of the bond in the β position to a polar functional group and simultaneous transfer of a hydrogen atom from the γ carbon atom to the polar group

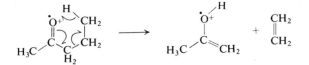

4. Multiply Charged Ions

Quite frequently peaks corresponding to doubly or even triply charged ions are found in the mass spectrum, particularly if there exists a very intense peak from a singly charged ion. The ions will be recorded at a half or a third of the m/e value of the singly charged species.

5. Odd and Even Electron Ions

All ions come into one of these categories but it is of value to discuss their significance. An ion containing an unpaired electron such as the molecular ion is referred to as an odd-electron ion. An even-electron ion is one in which the electrons are fully paired. Both types of ions may be formed in the fragmentation of molecular ions. Because all electrons are paired in an even-electron ion whereas there is an unpaired electron in an odd-electron ion, it follows that even-electron ions are more stable than odd-electron ions.

Odd-electron fragment ions are formed from the molecular ion or any other odd-electron ion by loss of a neutral molecule, e.g. H_2, CH_4, H_2O, C_2H_4 or NO. Even-electron ions are produced by loss of a neutral radical from an odd-electron ion or loss of a neutral molecule from an even-electron ion. When considering a possible fragmentation pattern for a molecule it is necessary to decide whether the species produced is an odd- or even-electron species in order to keep a "balance" of electron movements in the fragmentation process. The possible modes of breakdown of odd- and even-electron ions are listed below.

Odd-electron ion

(1) (Odd-electron ion)$\cdot^{+}$ → (even-electron ion)$^{+}$ + (odd-electron radical)$\cdot$
(2) (Odd-electron ion)$\cdot^{+}$ → (odd-electron ion)$\cdot^{+}$ + (even-electron molecule)

Even-electron ion

(3) (Even-electron ion)$^{+}$ → (even-electron ion)$^{+}$ + (even-electron molecule)
(4) (Even-electron ion)$^{+}$ → (odd-electron ion)$\cdot^{+}$ + (odd-electron radical)$\cdot$

6. Metastable Ions

The lifetime of an ion may be so short that it decomposes during its passage between the source and the collector units in the spectrometer. Ions which arise from decomposition between the source region and the magnetic analyser are known as "metastable ions"; they give rise to low-intensity broad peaks at non-integral mass numbers. They are detected at a non-integral mass since the ion m_2^{+} produced by the decomposition of m_1^{+} does not experience the full effects of the accelerating voltage which is needed to focus an ion of mass m_2^{+}. It has been shown (Hipple et al. 1946) that for the process

$$m_1^{+} \rightarrow m_2^{+} + \quad (m_1 - m_2)$$
$$\text{neutral fragment}$$

the metastable ion is observed at a mass m^* which is related to m_1 and m_2 by the equation

$$m^* = \frac{m_2^2}{m_1} \tag{5.3}$$

The intensity of the metastable peaks, which are designated by m^*, is normally of the order 0·01 to 1 per cent of the major peak in the spectrum. If the width of the exit slit of the mass spectrometer is increased there is an increase in the relative intensity of the metastable peak. A similar effect occurs when the ion-repeller voltage (page 326) is increased. Metastable ions are vital in the interpretation of mass spectra since they give information upon the loss of a neutral fragment which could be obtained in no other way. This information on neutral fragments enables peaks in a spectrum to be linked together in a fragmentation scheme for the molecule (Expt. M8).

7. Negative Ions

In addition to positive ions, negative ions may be formed from electron bombardment of the sample. They are focused by reversing the fields in the mass spectrometer. The negative-ion spectrum of a compound has fewer fragment-ion peaks than the positive-ion spectrum. In general negative-ion abundances are lower than positive-ion abundances.

Negative ions can be produced in three ways:

(i) $AB + e \rightarrow A + B^-$ (dissociative resonance capture)
(ii) $AB + e \rightarrow AB^-$ (resonance capture)
(iii) $AB + e \rightarrow A^+ + B^- + e$ (ion-pair production)

Negative fragment ions are largely of such species as O^-, OH^- and C_2H^- and so are not very useful in structural determinations.

8. Ions Formed in Ion–Molecule Reactions

Most of the ions obtained at the normal operating pressures within a mass spectrometer are formed in unimolecular processes. As a consequence the abundance of the ions is directly proportional to the pressure in the ionization chamber. Often peaks occur in the mass spectrum whose intensities vary more rapidly with pressure than intensities of other peaks in the spectrum. These pressure-dependent peaks arise as a result of collisions between two or more molecules or ions, i.e. as a result of an ion–molecule reaction. The resultant ions frequently have an m/e value greater than that of the molecular ion. They are easily distinguished from ions due to impurities since their relative abundance is pressure dependent.

Ion–molecule reactions are often studied in a mass spectrometer by operating the instrument at source pressures in the region $1-10^{-2}$ N m^{-2}.

B. QUASI-EQUILIBRIUM THEORY

The quasi-equilibrium theory is based on the work of Rosenstock and Kraus (1963) and attempts to predict the fragmentation pattern of a molecule under electron-impact conditions. It is considered that the molecular ion has excitational energy which can be distributed throughout its electronic and vibrational degrees of freedom in a random fashion. The molecular ion will then fragment when there is sufficient energy in the appropriate degree of freedom. If the fragment ion retains sufficient excitational energy further fragmentations occur. Consequently a series of ions are obtained which can be considered to be in a state of quasi-equilibrium. Dissociation within this state occurs by a series of consecutive and parallel unimolecular decompositions of the fragment ions as in the generalized example below:

$$M^+ \longrightarrow \begin{array}{ccccc} F_{11}^+ & \longrightarrow & F_{12}^+ & \longrightarrow & F_{13}^+ \\ F_{21}^+ & \longrightarrow & F_{22}^+ & \longrightarrow & F_{23}^+ \\ F_{31}^+ & \longrightarrow & F_{32}^+ & \longrightarrow & F_{33}^+ \end{array}$$

The quasi-equilibrium theory enables the rate constants for the postulated consecutive steps to be determined and hence the number of ions of each m/e can be evaluated at a given time. Thus a theoretical mass spectrum may be calculated.

It has been shown in certain cases that the theoretical mass spectrum is in semi-quantitative agreement with experiment. Qualitatively the theory enables predicted mechanisms for the formation of the ions to be tested.

C. TOTAL IONIZATION

Total ionization may be defined as the sum of all the peak heights in the mass spectrum multiplied by the sensitivity of the base peak which is the most intense peak in the spectrum. The sensitivity of the base peak is measured in units of positive-ion current per unit pressure. The value of the total ionization for a given compound is a characteristic of the particular instrument used. If the value of the total ionization for n-butane is taken as a standard then the *relative* total ionizations of other compounds become independent of the instrument used. Relative total ionization has the same value for many isomeric series and increases with the number of carbon atoms in any homologous series. The values are used to standardize mass spectra for analytical purposes.

An alternative definition of total ionization is the sum of the relative intensities of the peaks in the spectrum from a given mass, e.g. Σ_{15} represents the sum of all of the peak intensities from m/e 15 up to the molecular-ion peak. Σ_m evaluated this way is fairly constant for isomeric series whatever the structure (Hood, 1958; Otvos and Stevenson, 1956). Peak intensities are often expressed as a percentage of Σ_m.

D. IONIZATION AND APPEARANCE POTENTIALS

The ionization potential of a molecule or atom is defined as the energy required to remove an electron from the molecule or atom. Ionization occurs in accordance with the Franck–Condon principle (page 11) so that the nuclear separations in the ionized species are the same as in the neutral molecule or atom.

The appearance potential of a fragment ion is defined as the energy required to produce that ion in the mass spectrometer. Since the production of any ion other than the molecular ion involves cleavage of bonds, the appearance potential of a fragment ion will give information on bond-dissociation energies (Expt. M14).

Appearance and ionization potentials are related to the energies of the electron beams which are required to produce the ions in the mass spectrometer and can be determined experimentally by interpreting *ionization-efficiency curves*.

I. Ionization-efficiency Curves

An ionization-efficiency curve is a plot of the intensity of the ion beam (either as ion current or peak height) against the electron-beam energy. If it is assumed that the energy gained by a molecule or ion upon electron impact is equal to the energy of the electron beam, then the ionization potential or appearance potential is given by the value of the electron-beam energy at which the relevant ions are first detected (Fig. 5.1). However, since there is generally a spread of energies of the electrons in the electron beam it is not possible to determine directly the absolute value of the electron-beam energy at which ions are first produced. Consequently ionization and appearance potentials are determined by comparing the ionization-efficiency curves of the test species with that of a standard substance, usually an inert gas, whose ionization potential is accurately known.

There can be a 1 eV difference across the source between the filament and the ionization chamber called the *contact potential* which varies with time. Because the magnitude of this potential is not known it is necessary that the unknown and calibrating gases are present in the source simultaneously. In addition it is preferable that the ionization potentials of the two gases are of the same order of magnitude.

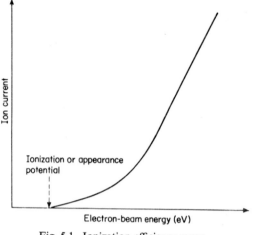

Fig. 5.1. Ionization-efficiency curve

The greatest source of error in measuring ionization and appearance potentials is in the determination of the origin of the ionization-efficiency curves since the curves approach the abscissa asymptotically. There are several approximate methods for determining the origin of the curves. These are listed below.

(a) Initial-break Method (Smyth, 1922)

The initial-break method is useful with low-sensitivity instruments and involves extrapolation of the ionization-efficiency curve back to its intersection with the electron-beam energy axis. In Fig. 5.2, X is the extrapolated value for the

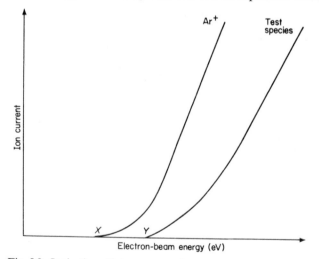

Fig. 5.2. Ionization-efficiency curves for argon and a test species

calibrant gas, e.g. argon, and Y is the value for the species under examination. The value given by X may differ from the true value by an amount $\pm Z$ eV which is then applied as a correction factor to the estimated value, Y, for the unknown gas.

(b) Linear-extrapolation Method (Vought, 1947)

In the linear-extrapolation method the linear portion of the ionization-efficiency curve is extrapolated back to the energy axis for both the unknown and reference compounds. It is then assumed that the error in the extrapolated value, X, of the ionization potential of the calibrant gas is equal to the error in Y, the extrapolated value for the species under consideration.

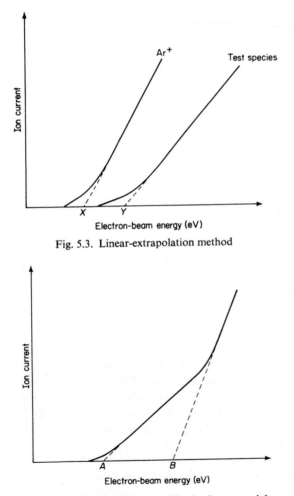

Fig. 5.3. Linear-extrapolation method

Fig. 5.4. Curve showing second ionization potential

11

This method often gives a rather high value for the ionization or appearance potential but has the advantages that it is the simplest method to use and is the only method available for determining second or higher ionization potentials. Second or higher ionization potentials are shown by changes in the slope of the ionization-efficiency curve. An example is shown in Fig. 5.4 where A is the uncorrected value for the first ionization potential and B the uncorrected value for the second.

(c) Extrapolated Voltage-difference Method (Warren, 1950)

Ionization-efficiency curves for the test species and the calibrant gas are plotted together after arbitrary scale manipulation of the ion current to make the linear portions of the curves parallel for a few electron volts above the initial break (Fig. 5.5).

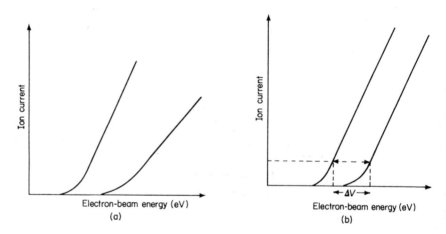

Fig. 5.5. Ionization-efficiency curves (a) with no scale manipulation (b) with scale manipulation

The differences, ΔV, between the electron-beam energies shown on the two curves in Fig. 5.5(b) corresponding to various values of the ion current are determined, particularly for those values of the ion current near to the foot of the curve. The values of the ion current, I, at which the ΔV values are determined are then plotted against the corresponding ΔV values (Fig. 5.6). Extrapolation of this plot to $I = 0$ enables the voltage difference, ΔV_0, between the ionization potential of the unknown and standard to be obtained. Thus if the value of the ionization potential for the standard gas is known, the value for the unknown gas can be obtained from ΔV_0.

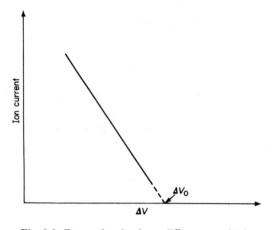

Fig. 5.6. Extrapolated voltage-difference method

E. APPLICATIONS OF MASS SPECTROMETRY

I. Bond-dissociation Energies

Dissociation as well as ionization of a molecule can occur if the bombarding electron beam has sufficient energy. The mode of ionization and decomposition of the molecule is dependent upon the value of the electron-beam energy and upon the structure of the molecule. At low electron-beam energies (6–7 eV), an electron resonance capture process may occur to produce a negative molecular ion

$$e + AB \rightarrow AB^-$$

This ion can then dissociate

$$AB^- \rightarrow A^{\cdot} + B^-$$

or

$$AB^- \rightarrow A^- + B^{\cdot}$$

At beam energies of 7–10 eV both positive and negative ions can be formed

$$e + AB \rightarrow A^+ + B^- + e$$

The relative electronegativities of A and B govern which species is positive or negative, and when ionization occurs by this mechanism the appearance potentials of A^+ and B^- are identical.

At energies of 9–20 eV or higher, the commonest ionization process is that producing a positive molecular ion

$$e + AB \rightarrow AB^+ + 2e$$

The molecular ion AB^+ can dissociate further

$$AB^+ \rightarrow A^+ + B^{\boldsymbol{\cdot}}$$

or

$$AB^+ \rightarrow A^{\boldsymbol{\cdot}} + B^+$$

The appearance potential of A^+, for the process $AB^+ \rightarrow A^+ + B^{\boldsymbol{\cdot}}$ can be measured from the ionization-efficiency curve of ion A^+. Since A^+ can only be formed after the ionization of AB has occurred, the appearance potential, $A(A^+)$, must be the sum of the energies required to dissociate AB and to ionize the radical $A^{\boldsymbol{\cdot}}$. If $I(A^{\boldsymbol{\cdot}})$ represents the ionization potential of the radical $A^{\boldsymbol{\cdot}}$ and $D(A\text{—}B)$ the bond-dissociation energy of AB, then the following relationship can be written

$$A(A^+) = I(A^{\boldsymbol{\cdot}}) + D(A\text{—}B) \qquad (5.4)$$

If the ionization potential of $A^{\boldsymbol{\cdot}}$ can be obtained from other techniques, e.g. photoionization methods, then it is possible to determine the bond-dissociation energy (Kiser, 1965) of A—B. Alternatively, if bond-dissociation energies are known, ionization potentials may be obtained from a knowledge of appearance potentials.

The thermochemical relationship given in equation 5.4 assumes that the ion and the neutral fragments are produced in their lowest energy states with no excess electronic, vibrational or rotational energy. If excess energy is present the correct expression for the appearance potential is

$$A(A^+) = I(A^{\boldsymbol{\cdot}}) + D(A\text{—}B) + \Sigma E \qquad (5.5)$$

where ΣE is the sum of all the excess energy terms. The assumption is often made that the ion and the radical are both produced with no excess energy so that ΣE is equal to 0. Reasonable values for ionization potentials have been obtained on the basis of this assumption.

2. Molecular Formulae and Molecular Weights

Mass spectrometry is very frequently used to determine the molecular weight and molecular formula of an unknown compound. If a high-resolution instrument is used then the mass of the molecular ion can be determined with such a degree of accuracy that there is only one possible combination of elements which can give that mass and hence only one molecular formula that could be possible. For example an ion of m/e 28 can arise from CO^+ of mass 27·994 914, CH_2N^+ of mass 28·018 723 or $C_2H_4^+$ of mass 28·031 299 (Masses are quoted on the basis of $^{12}C = 12\cdot000\,000$ as standard.)

Many instruments of the single focusing type (page 328) can only resolve peaks to the nearest integer value. In order to determine a molecular formula with such instruments use must be made of the intensities of the peaks at one and two mass units higher than the molecular-ion peak. These peaks are molecular ions containing the heavy isotopes of C, H, N, O, etc. and are referred to as the $M + 1$ and $M + 2$ peaks, etc. Any particular combination of elements will give a definite value for the $(M + 1)/M$ and $(M + 2)/M$ ratios. The calculation of these ratios is fairly simple but is time consuming; standard tables are available (Beynon and Williams, 1963) listing the values of $(M + 1)/M$ and $(M + 2)/M$ for combinations of C, H, O and N up to m/e 500. On consulting these tables several possible molecular formulae with $(M + 1)/M$ and $(M + 2)/M$ ratios close to the experimental values can often be obtained for an unknown substance. The correct molecular formula is then selected from these possible alternatives by an examination of the fragmentation pattern, by the use of other spectroscopic techniques and by utilization of information that is already known such as the presence or absence of N.

3. Structural Correlations and Qualitative Analysis

Qualitative analysis can be carried out by an examination of the fragmentation pattern (also called the cracking pattern) of a compound. Various catalogues (A.S.T.M.S.T.P. 356) of mass spectral data exist and in principle the fragmentation pattern of a compound is unique and therefore characteristic of that compound. Unfortunately, the relative intensities of peaks in a mass spectrum can vary by as much as 5 per cent or more for repeat scans on any instrument and the differences in relative intensities are often considerable if a comparison is made between spectra run on different instruments. The major peaks in the spectrum of any particular compound are the same on any instrument so that information on modes of fragmentation in various types of compound is of utility in the qualitative analysis of an unknown compound.

Many correlations between known structures and major fragmentation routes have been made. The reader is referred to the comprehensive books by Budziekiewicz et al. (1967) and Beynon et al. (1968) for a detailed coverage of these correlations. Using the information obtained from these correlations it is possible to predict the major peaks in the mass spectrum of a molecule of any postulated structure (Expts M5–M8). The major fragmentations encountered in molecules containing various organic groupings are outlined below.

(a) Hydrocarbons

ALIPHATIC HYDROCARBONS. Aliphatic hydrocarbons show no preferred site for the localization of the charge in the molecular ion. As a consequence cleavage occurs at each bond along the length of the chain, so that groups of peaks 14 mass units apart are observed. The most abundant ion in each group of peaks in saturated hydrocarbons is that of formula C_nH_{2n+1}. The ions of formulae C_nH_{2n+1} occur with decreasing abundance as n increases. Fragmentation is however enhanced at any branched carbon atom in the molecule producing a relatively larger peak. In general, ions of odd mass tend to be more abundant than ions of even mass. Loss of CH_3^+ does not produce a peak of high intensity unless the compound contains a methyl side chain.

The mass spectra of olefins show groups of peaks separated by C_nH_{2n-1} mass units; these peaks become less intense as n increases. A site for the stabilization of the positive charge is provided by the removal of one of the π electrons with the result that molecular ions are more abundant than in saturated hydrocarbons. Fragments of mass C_nH_{2n} are produced by the elimination of an olefin by a McLafferty rearrangement (page 303). Acetylenes show a fairly pronounced $M - 1$ peak and groups of peaks of general formula C_nH_{2n-3}.

AROMATIC HYDROCARBONS. The molecular-ion peaks from aromatic hydrocarbons are fairly intense since the loss of an electron from the π-electron system has only a small destabilizing effect on the molecule. The aromatic nucleus also acts as a centre for the promotion of bond rupture and rearrangement processes. The common fragment ions from benzene derivatives are C_3H_3, C_4H_2, C_4H_3, C_4H_5 and C_6H_5.

Monosubstituted aromatic systems of the type shown below cleave β to the ring to produce an ion of m/e 91. Appearance-potential measurements and deuterium-labelling experiments on the ion of m/e 91 have shown that all the hydrogen atoms are equivalent suggesting that the ion has a tropylium structure:

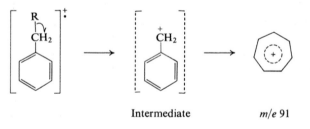

Intermediate m/e 91

The tropylium ion readily loses C_2H_2 to produce the species m/e 65.

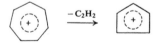

Cleavage of the structure $PhCH_2R$ at the bond α to the ring with consequent production of the ion of m/e 77 $(C_6H_5)^+$ only occurs to a small extent.

Polycyclic aromatic hydrocarbons undergo very little fragmentation due to the high stability of the conjugated ring systems but loss of acetylene does occur, resulting in ring contraction. Methyl or other alkyl substituents on the conjugated rings are cleaved from the system.

(b) Alcohols

ALIPHATIC ALCOHOLS. The molecular ion of an aliphatic alcohol is produced by removal of one of the lone pair electrons on the oxygen atom. Generally

$$R—\overset{..}{\underset{..}{O}}—H + e \longrightarrow R—\overset{+\bullet}{\underset{..}{O}}—H + 2e$$

the molecular-ion peaks are not very intense.

The most intense peak in the spectrum of an aliphatic alcohol is frequently that from the oxonium ion formed by cleavage of the bond α to the hydroxyl function:

$$\begin{array}{c} H \\ | \\ CH_2{\overset{\bullet+}{\frown}}O—H \end{array} \longrightarrow CH_2{=}\overset{+}{O}H + H^\bullet$$
$$m/e\ 31$$

In secondary and tertiary alcohols there is more than one α bond which can be broken, and in such cases the most probable fragmentation pathway is the one corresponding to the loss of the fragment of largest mass.

Cleavage of the β, γ, δ bonds in the alkyl chain also occurs but not to the same extent as for the α bond and the probability of cleavage decreases as the distance from the hydroxyl function increases.

AROMATIC ALCOHOLS. The molecular ion of an aromatic alcohol may be formed by loss of an electron from either the phenyl ring or from the lone-pair electrons on the oxygen atom. The molecular-ion peaks are normally quite intense. Phenol shows, in addition to an intense molecular-ion peak, a strong peak at $M - 28$ due to the loss of carbon monoxide

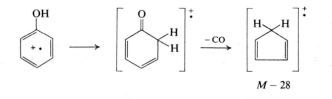

$$M - 28$$

Benzyl alcohols produce strong molecular-ion and $M - 1$ peaks. Deuterium labelling has shown that the loss of a hydrogen atom to produce the $M - 1$ species occurs from any of the carbon atoms. Two possible mechanisms for the loss of the hydrogen atom are indicated below. Carbon monoxide is lost from the $M-1$ species in the case of derivatives of benzyl alcohol followed by the loss of two hydrogen atoms to produce the ion of m/e 77

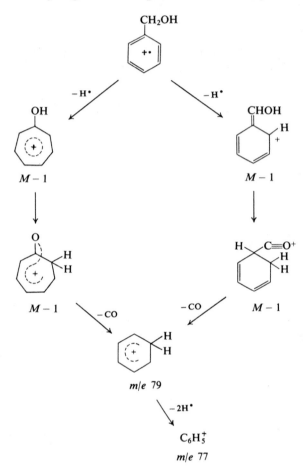

(c) Carbonyl Compounds

ALIPHATIC CARBONYL COMPOUNDS. The molecular ion of aliphatic carbonyl compounds is formed by loss of one of the lone-pair electrons from the oxygen atom

$$\begin{array}{c} R_2 \\ \diagdown \\ R_1 \end{array} C{=}O \quad \longrightarrow \quad \begin{array}{c} R_2 \\ \diagdown \\ R_1 \end{array} C{=}O^{+\cdot}$$

The most extensive cleavage is by α fission to produce the species $R_1{-}C{\equiv}O^+$ or $R_2{-}C{\equiv}O^+$. Small peaks are obtained due to cleavage of the bond β to the carbonyl function but a more important process is cleavage of the β bond and simultaneous transfer of the γ hydrogen atom to the oxygen atom, i.e. a McLafferty rearrangement process

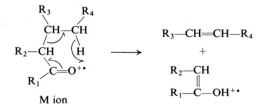

AROMATIC ALDEHYDES AND KETONES. Aromatic aldehydes and ketones show intense molecular-ion peaks. α cleavage produces a strong $M-1$ peak in benzaldehyde and an $M-R$ peak in alkyl aryl ketones. Carbon monoxide is subsequently lost from the ions of $M-1$ and $M-R$ and the ion of m/e 77 ($C_6H_5^+$) is formed. This fragments further to the $C_4H_3^+$ species as shown.

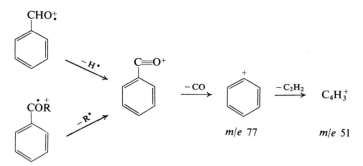

Carbon monoxide is also lost in the course of the fragmentation of the molecular ions of aromatic ketones. This process is illustrated for derivatives of benzophenone below.

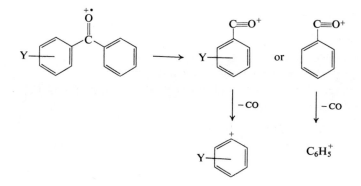

(d) Ethers

ALIPHATIC ETHERS. The molecular ion of an aliphatic ether is produced by removal of one of the non-bonded electrons on the oxygen. The major fragmentation is by cleavage of the bond α to the ether linkage. In unsymmetrical ethers α cleavage will produce different ions depending on the site of the α-bond fission; the more highly substituted fragment is lost preferentially.

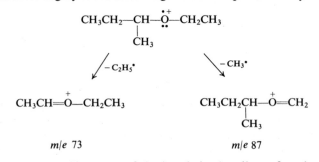

$$\underset{m/e\ 73}{CH_3CH{=}\overset{+}{O}{-}CH_2CH_3} \qquad \underset{m/e\ 87}{CH_3CH_2\underset{CH_3}{CH}{-}\overset{+}{O}{=}CH_2}$$

AROMATIC ETHERS. Cleavages of the bonds in the alkoxy function occur in aryl alkyl ethers followed by ring contraction.

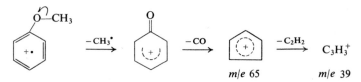

Loss of formaldehyde can also occur.

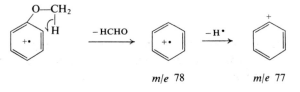

Direct loss of the alkoxy group produces the ion at m/e 77.

(e) Amines

ALIPHATIC AMINES. The molecular ion of aliphatic amines is formed by loss of one of the lone-pair electrons on the nitrogen atom.

α cleavage of the C—C bond adjacent to the amino group is the major fragmentation mechanism.

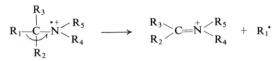

The largest alkyl group is lost preferentially in this process.

AROMATIC AMINES. Aniline has an intense molecular-ion peak. Loss of H
from the amino group produces the $M - 1$ species. Loss of HCN is a common
process in aromatic amines followed by loss of a hydrogen atom.

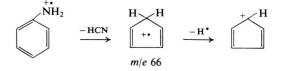

m/e 66

Cleavage of the β bond in the side chain occurs readily in alkyl-anilines with
consequent rearrangement to a tropylium-type structure.

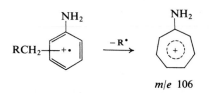

m/e 106

(f) Aliphatic Amides

The molecular ion of aliphatic amides is produced by loss of a lone-pair
electron from either the oxygen or the nitrogen atom. In primary amides, α
cleavage produces the resonance stabilized ion of *m/e* 44.

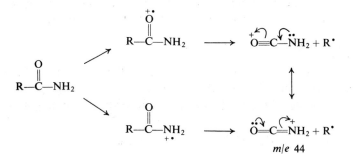

m/e 44

When three or more carbon atoms are in the aliphatic chain a McLafferty
rearrangement (β cleavage and transfer of a γ hydrogen atom) can occur with
the production of an ion of *m/e* 59.

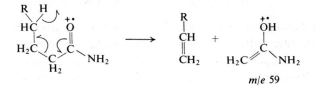

m/e 59

γ fission produces a peak at m/e 72:

$$R \overset{\gamma}{\underset{}{\Big\{}} CH_2 \overset{\beta}{-} CH_2 \overset{\alpha}{-} \overset{\overset{\displaystyle O}{\|}}{C} - NH_2$$
$$\longleftarrow m/e\ 72 \longrightarrow$$

(g) Aliphatic Nitriles

Aliphatic nitriles are best studied using high resolution mass spectrometry since difficulties in interpretation arise from the fact that some common species produced from these compounds have the same integer mass, i.e. CH_2 and N; C_2H_2 and CN.

Molecular ion peaks are weak and a low abundance $M - 1$ peak is formed by loss of an atom of hydrogen.

$$R-CH_2C\equiv \overset{+\bullet}{N} \quad \overset{-H^\bullet}{\longrightarrow} \quad R\overset{\bullet}{C}HC\equiv \overset{+\bullet}{N}$$

An $M + 1$ peak is normally present, the relative intensity of which is dependent upon the pressure of the nitrile in the mass spectrometer. This pressure dependence indicates that the ion is formed by an intermolecular process involving H abstraction.

$$R-CH_2C\equiv N^{+\bullet} \quad \overset{+H^\bullet}{\longrightarrow} \quad R-CH_2C\equiv \overset{+}{N}H$$

Cleavage of the alkyl chain occurs to produce peaks from hydrocarbon fragments although cleavage α to the nitrile group is not favoured.

(h) Halides

ALKYL HALIDES. The molecular-ion peaks of alkyl halides are fairly intense. Since halogen atoms have a high electron affinity, α cleavage occurs readily; the positive charge is normally retained on the hydrocarbon fragment.

$$R-CH_2 \overset{\curvearrowright}{\overset{\bullet+}{X}} \quad \longrightarrow \quad R-CH_2^+ + X^\bullet$$

Charge retention by the halogen-containing fragment does occur to produce peaks of low intensity.

$$R\overset{\frown}{-CH_2}\overset{\bullet+}{X} \quad \longrightarrow \quad R^\bullet + CH_2 = \overset{+}{X}$$

Fragments containing chlorine or bromine atoms are easily detected due to the characteristic pattern of isotopic peaks from these elements.

Loss of hydrogen halide also occurs by the elimination of the requisite elements from adjacent carbon atoms (1,2 elimination) or by 1,3 elimination.

AROMATIC HALIDES. Benzyl halides lose the halogen atom followed by ring expansion to give the tropylium ion structure.

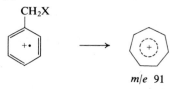

m/e 91

In ring-halogenated compounds, loss of halogen atom occurs by α fission except in fluoro compounds where the C—F bond is extremely stable. As with the alkyl benzenes, β fission of the substituent chain occurs when alkyl substituents are present.

(i) Nitro Compounds

Nitro compounds have an intense molecular-ion peak. α cleavage occurs resulting in the loss of the nitro group followed by acetylene loss.

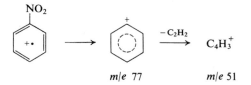

m/e 77   *m/e* 51

The other major characteristic cleavage is loss of NO. This is then followed by elimination of carbon monoxide:

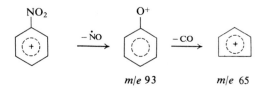

m/e 93   *m/e* 65

4. Quantitative Analysis

Quantitative analysis of mixtures can be performed by many techniques each of which has advantages for a particular mixture. In general, mixtures of liquids are most easily analysed by means of gas-phase chromatography. However, when the components of the mixture cannot be separated by chromatography, the analysis can often be achieved by means of mass spectrometry.

Mass spectrometric analysis involves the determination of the relative partial pressures of the components present in a vaporized sample of the mixture. The molar composition of the mixture can then be calculated from a knowledge of the sensitivity of the instrument to the various components.

The mass spectrometric method depends upon three assumptions:
1. The spectra of the components are linearly additive at any mass number. This condition is normally obeyed in a mass spectrometer.
2. The peak heights of the various ions are proportional to the pressure of the sample in the source. Under the normal operating conditions of molecular flow (page 334) this is the case, and if the flow through the leak is also molecular, the peak heights are directly proportional to the pressure of the sample in the reservoir (page 334). In fact the reservoir pressure is most frequently used in the calculations since it is larger and easier to measure than the pressure of the sample in the source.
3. The cracking patterns and sensitivities are invariant. In fact cracking patterns and sensitivities do vary with operating conditions, the state of the filament, and from instrument to instrument. Because of this variation in sensitivity, it is normal to calibrate the instrument by periodically measuring the sensitivity of a standard gas such as n-butane. Sensitivities of components are expressed as the height of the base peak per unit pressure or, to give an absolute value, as ion-beam current per unit pressure in ampere/N m^{-2}. Since absolute sensitivities change with time, relative sensitivities are frequently used, i.e. the ratio of the relative intensity of the base peak in the compound to the intensity of the m/e 43 peak in n-butane.

The quantitative analysis of a multicomponent mixture depends upon the selection and solution of an appropriate series of linear simultaneous equations, one for each m/e value selected. Thus the mass spectral data from a mixture spectrum can be represented by the following set of equations:

$$H_1 = h_{11}p_1 + h_{12}p_2 + h_{13}p_3 + \ldots h_{1n}p_n \qquad (5.6)$$
$$H_2 = h_{21}p_1 + h_{22}p_2 + h_{23}p_3 + \ldots h_{2n}p_n$$
$$H_3 = h_{31}p_1 + h_{32}p_2 + h_{33}p_3 + \ldots h_{3n}p_n$$
$$\vdots$$
$$H_m = h_{m1}p_1 + h_{m2}p_2 + h_{m3}p_3 + \ldots h_{mn}p_n$$

where h_{11} represents the peak height at m/e 1 due to component 1, h_{21} the peak height at m/e 2 due to component 1,..., normalized to unit pressure of each component in the reservoir and p_1, p_2,..., represent the partial pressure of each component in the mixture, and H_1, H_2,... represent the recorded peak heights at m/e 1, 2,... in the mixture spectrum.

There are thus m equations with n unknowns p_1, p_2, . . . p_n. The values of h_{mn} are found by recording the mass spectrum of each component in the mixture at a known pressure in the reservoir, and normalizing the spectrum to unit pressure of the component in the reservoir. H_1, H_2,..., are known from the mixture spectrum so that the equations are soluble for the values of p. The simultaneous equations can readily be solved for mixtures containing several components. The application of mass spectrometry to the analysis of a three-component mixture is considered in Expt. M9, where a somewhat simpler technique, dependent upon the selection of peaks which are due to one component only, is used.

5. Isotopic Abundance Measurements

Since certain elements have more than one highly abundant isotope (see Table 5.1), fragments which contain one or more atoms of any of these elements will produce a set of peaks which are characteristic of the number of atoms of that element in the fragment. The relative abundances of some common elements are listed below.

Table 5.1

	Per cent		Per cent		Per cent
^{1}H	99·985	^{2}H	0·015		
^{12}C	99·892	^{13}C	1·108		
^{14}N	99·635	^{15}N	0·365		
^{16}O	99·759	^{17}O	0·037	^{18}O	0·204
^{32}S	95·018	^{33}S	0·750	^{34}S	4·215
^{35}Cl	75·529	^{37}Cl	24·471		
^{79}Br	50·52	^{81}Br	49·48		
^{10}B	19·6	^{11}B	80·4		

As mentioned earlier $(M + 1)/M$ and $(M + 2)/M$ ratios are listed in the literature but to avoid excessive computation these are restricted to compounds containing only C, H, N and O. These tables can easily be extended by the user provided he knows what additional elements are present. When Cl or Br are present in the molecule, intense $M + 2$ and $M + 4$ peaks are obtained since the isotopes are two mass units apart. The numbers of these atoms present can be obtained by collecting together terms in a binomial expansion of the form $(x + y)^{a}$ and calculating which expansion fits the observed peak ratios. In the expansion, x is the percentage abundance of the lighter isotope, y of the heavier isotope and a is the number of atoms of the element present in the molecule.

Thus if one chlorine atom is present, then the ratio $(M:M + 2)$ of the intensities of the molecular-ion peaks will be $3:1$ as given by

$$(x + y)^{a} = (75·529 + 24·471)^{1} = (3 + 1)^{1} \equiv 3:1 \qquad (5.7)$$

When two chlorine atoms are present, the intensities of the molecular-ion peaks are in the ratio $M:M+2:M+4=9:6:1$ since $(x+y)^a=(3+1)^2$ which on expansion and collection of terms gives $9:6:1$.

Similarly, if three chlorine atoms are present the molecular-ion peaks are in the ratio $M:M+2:M+4:M+6=27:27:9:1$ from the expansion of $(x+y)^a=(3+1)^3$.

Calculations are similar for Br, B or other elements. If two different elements with large isotopic abundances are present, e.g. Cl and Br, then the molecular-peak ratios are derived from a product of the binomials for each element. The expression is of the form $(x+y)^a(l+m)^b$ where x, y, l and m are the isotopic abundances of the two elements and a and b are the number of atoms of the two elements present in the molecule.

6. Kinetic and Mechanistic Studies

Mass spectrometry can be used in kinetic and mechanistic studies (Expt. M15). Since unstable intermediates may be detected in a mass spectrometer, the technique is of value in the study of radical reactions. The usefulness of the technique lies in the fact that the concentration of many of the unstable species present in the sample from the reaction vessel can be determined. The time intervals at which the reactive species are examined from the initiation of the reaction can be varied by altering the distance between the reaction vessel and the source, or by using a variable rate of gas flow. Consequently the kinetics for the disappearance and for the formation of radicals may be obtained.

Mass spectrometry can also be used to study heterogeneous reactions. If the reactive surface is used as a filament inside the ion source, radicals or reaction intermediates originating from the surface of the catalytic filament will enter the ion beam and will be ionized and detected in the normal fashion.

II. INSTRUMENTATION AND SAMPLE HANDLING PROCEDURES

A. DESCRIPTION OF A MASS SPECTROMETER

Two types of instruments are used in mass spectrometry, viz. mass spectrographs and mass spectrometers. Both types of instrument produce positive and negative ions from an atom or molecule and analyse them according to their mass to charge ratio (m/e). The mass spectrograph brings the separated ions to a focus on a photographic plate whereas the mass spectrometer focuses the ions onto a collector which detects the signal electrically. Instruments can be operated so that either positive or negative ions are focused at any given time.

All instruments consist of three basic components: the ion source, which produces the ions in a gaseous state from the sample; the analyser, which resolves the total ion beam from the source into ion beams corresponding to the component m/e ratios; and a detector system, which records the intensity of each of the resolved ion beams. These components are shown in schematic form in Fig. 5.7.

There are several differing types of sources, analysers and detectors which are used in mass spectrometers or spectrographs; the major categories of each will be discussed.

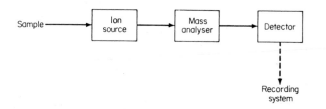

Fig. 5.7. Schematic diagram of a mass spectrometer

I. Ion Sources

Sources, in general, need to meet the following requirements. Firstly, the spread of energies in the ion beam from the source should be as small as possible in order to aid resolution in the instrument, and to make appearance-potential and ionization-potential measurements meaningful. Secondly, there must be a minimum of background ions produced from material adsorbed onto the surfaces of the source. Thirdly, the efficiency of ionization of the sample in the source chamber should be high, in order that positive-ion currents generated at the detector are as large as possible. Also there should be no discrimination against particular substances by specific variations in ionization efficiency. Fourthly, since sources require cleaning and filament replacement fairly frequently, they should be constructed so that removal and re-attachment is easy.

(a) Electron Bombardment Source

The electron bombardment source, developed by Dempster and Nier, is most widely used since it is of the most general applicability.

The sample in the vapour state, in the pressure range 10^{-2}–10^{-5} N m^{-2}, is admitted to the source at right angles to the electron beam from a tungsten or rhenium filament (Fig. 5.8). Solid samples are inserted by means of a direct insertion probe which enters into the electron beam region where the sample is partially vaporized. The molecules are then ionized by the electron beam.

A collimated beam of electrons is obtained by drawing the electrons across to a positively charged "trap" plate through a slit system. Various grid plates can be inserted to control the energy of the electron beam. A weak magnetic field in the direction *B* helps to collimate the electron beam. B is an ion-repeller plate which is positively charged in order to repel the ions out of the source region towards the analyser. The source temperature can be varied by means of insulated electric heaters.

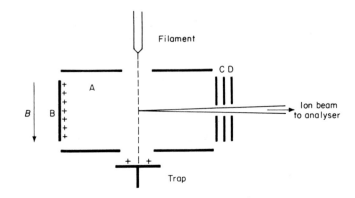

Fig. 5.8. Electron bombardment source

Grids C and D have a potential difference of up to 8000 V which serves to accelerate the ions in the ion beam. The resultant ion beam is divergent. The beam produces a positive ion current of 10^{-10}–10^{-16} ampere. After acceleration, the ions in the beam have almost identical kinetic energy for each m/e value.

Tungsten is widely used as a filament material but it has the disadvantage that it reacts with samples and also with background oxygen and nitrogen to form oxides and nitrides. Since rhenium does not form oxides and nitrides readily, it is often used as a filament material although it suffers from the disadvantage that it is more volatile than tungsten.

(b) Vacuum Spark Ion Source

The vacuum spark ion source is operated by discharging a condenser between two electrodes constructed from the material which is to be analysed. Thus the source is useful when metals, alloys or semi-conductors have to be ionized. A disadvantage is that the ions which are produced have a large energy spread.

(c) *Photoionization Source*

In the photoionization source the sample is ionized by electromagnetic radiation, normally in the ultraviolet region. An advantage of this technique is that, by using monochromators, radiation of a particular frequency can be selected and so it is possible to obtain an almost mono-energetic ion beam. Consequently appearance potentials can be determined fairly accurately and the fine structure in ionization-efficiency curves can be investigated. Modern developments of this source involve using a laser beam to achieve radiation of precisely known frequency and so a more mono-energetic ion beam is obtained.

(d) *Thermal Emission Ion Source*

The thermal emission ion source is used to ionize solids which can be deposited on a platinum or tungsten filament. When the filament is electrically heated, a certain amount of the solid is vaporized in the form of positively charged ions.

The source is useful for the detection of elemental impurities such as Al, Ca, Ga, Hf, Li, Na, Ti and U.

(e) *Field Ionization Source*

The field ionization source operates on the principle that substances which are adsorbed on to a surface may be desorbed as positive ions if the surface is placed in a strong electric field (10^8 V cm^{-1}). Fields of this order can be obtained by applying a high potential to a fine point or sharp razor edge. Positive ions from the material under investigation can be produced by coating the material on the point or edge or by allowing the vapour of the sample to pass close to the edge.

The mass spectra obtained, when this type of source is used, are much simpler than those from an electron bombardment source since the spectra consist mainly of molecular ions.

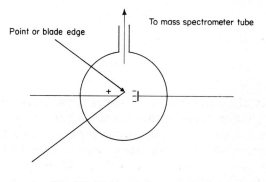

Fig. 5.9. Field emission source

2. Mass Analysers

All sources produce positive ions which are then separated according to their m/e ratios by means of various types of mass analyser. Mass analysers operate in one of three ways: (a) deflection of the ions in a magnetic field, (b) utilization of a radiofrequency voltage and (c) measurement of the time of flight of the ions. Methods (a) and (c) are most commonly used.

In a single focusing instrument, ion beams which have either a velocity spread or contain ions travelling in differing directions can be focused. If ions all with the same initial direction but varying in velocity are focused the instrument is said to be velocity focusing. Most single focusing instruments are direction focusing, i.e. a beam of ions of virtually the same initial kinetic energy for any given mass but slightly differing directions is focused by the mass analyser. Double focusing instruments utilize two analysers to achieve both velocity and direction focusing with a consequent increase in resolving power.

(a) Analysers using Deflection in a Magnetic Field

Analysers using deflection in a magnetic field give direction focusing of the ion beam. The ion beam from the source is accelerated through a potential V (of the order 2 kV or more) so that ions of each mass will have approximately equal kinetic energies (KE $= \frac{1}{2}mv^2$). The kinetic energies of the ions are equal to the electrical work done on the positive ions by the accelerating voltage, V, i.e.

$$\tfrac{1}{2}mv^2 = eV \tag{5.8}$$

where e is the electrostatic unit of charge.

When the ions enter a magnetic field, B, acting at right angles to their direction of motion, they experience a force which causes them to take up a semi-circular path of radius r. The magnetic force, acting centripetally on the particle, may be equated to the centrifugal force opposing it

$$Bev = mv^2/r \tag{5.9}$$

Combining equations 5.8 and 5.9 gives

$$m/e = \frac{B^2 r^2}{2V} \tag{5.10}$$

The principles of operation of mass analysers are based on equation 5.10.

The first direction focusing instruments utilized a 180° deflection in a magnetic field and this is currently used in some low-resolution instruments. There is, however, no reason why sector shaped magnetic fields should not be used, and some instruments use a 60 or 90° magnetic deflection, one advantage of which is that the source and collector are not in the magnetic field. In this arrangement a smaller pole gap is possible in the magnet. Schematic representations of 180 and 60° single focusing mass analysers are given in Fig. 5.10.

In all magnetic analysers only those ions which have a flight path of radius of curvature corresponding to r in equation 5.10 can pass through the slit systems to reach the collector. It follows from equation 5.10 that the m/e value corresponding to the radius r is determined by the magnetic-field strength B and the accelerating voltage, V. In an instrument which operates by magnetic scanning, the strength of the electric field, V, is kept constant, while the m/e value for the ion incident upon the collector is altered by varying the magnetic field B. In electrical scanning instruments, the spectrum is scanned by keeping the magnetic field, B, constant and altering the electrostatic accelerating voltage, V. It is more usual to scan the spectrum magnetically. If the spectrum is scanned rapidly then lags in the decay time of the magnetic field occur (hysteresis effects) and so if the spectrum must be scanned rapidly, electrostatic scanning at constant magnetic-field strength is used.

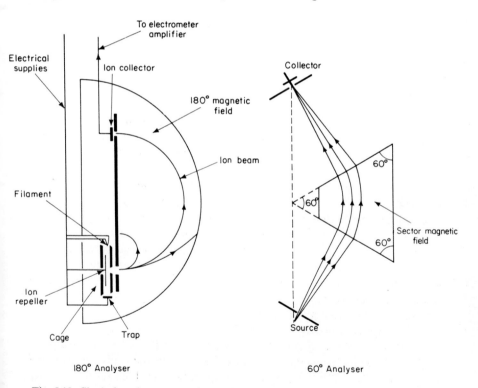

Fig. 5.10. Single focusing mass analysers. Adapted with permission from AEI Scientific Apparatus Ltd. Manchester

Since ion beams are not velocity focused in a magnetic analyser it is desirable that the source should provide a mono-energetic ion beam. In this way only ions of a single m/e value will reach the collector at any given instant.

In a double focusing instrument an electrostatic analyser is used in addition to the magnetic analyser. The effect of the electrostatic analyser is to focus ions which vary in energy at any given mass. The electrostatic analyser is a velocity analyser and precedes the mass analyser. In this way it is possible to select ions of a given energy which are then mass analysed by deflection in a magnetic field. In the electrostatic analyser, a radial electric field is applied across the

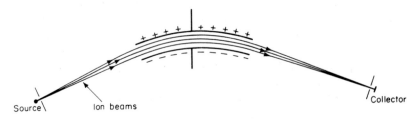

Fig. 5.11. Electrostatic analyser

analyser. The positive ions will follow an arc on entering the electric field and the radius of this arc will be given by

$$r = 2V/X \qquad (5.11)$$

where V is the accelerating voltage applied to ions leaving the source and X is the strength of the radial electric field. Ions of any given mass which differ slightly in energy at C (Fig. 5.12) can emerge from slit D as a virtually mono-energetic ion beam and will then enter the magnetic mass analyser.

Depending upon the arrangement of the electrostatic and magnetic analysers, the geometry of the ion focusing surface differs. Two basic arrangements which are used are the Mattauch–Herzog and the Nier–Johnson arrangements. The Nier–Johnson arrangement is shown diagrammatically in Fig. 5.12. In this arrangement, the focusing surface is conic and paraboloidal in two directions.

Because of the geometry of the focusing surface of the Nier–Johnson arrangement, all ion beams are measured at one point. Ion beams differing in m/e value are successively focused onto this point by varying either the accelerating voltage V or the strength of the magnetic field B.

Resolution is improved in either single or double focusing instruments by using narrow collecting slits. Very narrow slits will reduce the positive-ion currents to values below those detectable by conventional valve electrometers (approximately 10^{-14} ampere). Consequently, for high-resolution work electron-multiplier tubes are used which will detect currents as low as 10^{-18} ampere.

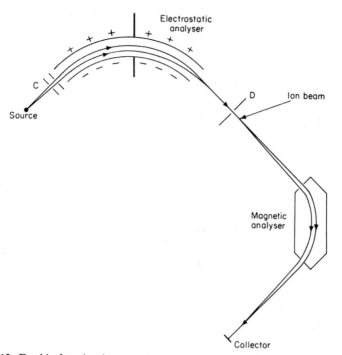

Fig. 5.12. Double focusing instrument. Adapted with permission from AEI Scientific Apparatus Ltd. Manchester

(b) Time of Flight Analysers

In time of flight analysers, it is arranged that all the ions leaving the source have the same kinetic energy. This means that ions of differing mass will travel with differing velocities. The time taken for an ion to travel a fixed distance is related to the mass of the ion; a light ion will travel faster than a heavy ion and so will reach the detector first. The most widely used time of flight instruments are those utilizing a linear flight path for the ions (see Fig. 5.13). In this type of instrument the time of flight, t_f, is related to the accelerating voltage, V, and the distance of travel of the ion, d, by the expression

$$t_f = d(1/2V)^{1/2}(m/e)^{1/2} \tag{5.12}$$

The ions are produced in pulses in the source chamber and after traversing their flight path are detected and amplified by an electron multiplier device. The output signal from the electron multiplier is fed to a CRO and recorded on a rotating drum camera. Scanning of the spectrum is very rapid, e.g. the time interval between ion pulses is of the order 0·1 second and the interval between the arrival of adjacent ions from any one pulse at the collector is approximately 0·05 microsecond.

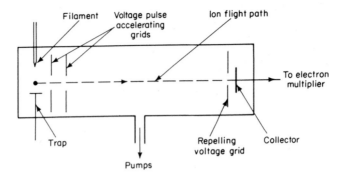

Fig. 5.13. Linear time of flight instrument

Time of flight instruments are ideally suited to the study of very rapid reactions. A disadvantage is that this type of instrument is of limited resolution compared with instruments using a magnetic analyser.

3. Detectors and Recorders

Detection of the positive-ion beam from the mass analyser can be either photographic or electrical. The photographic method is satisfactory for accurate measurement of the m/e value of an ion, but the accuracy of relative abundance measurements is low since there are variations in the sensitivity of the emulsion for differing m/e values of the ions and times of exposure. The lowest detectable ion current using a photographic plate is about 10^{-15} ampere. (Ion currents are usually in the range of 10^{-9}–10^{-18} ampere.)

Electrical detection enables a high degree of precision in relative abundance measurements to be obtained. The detector is often a long, narrow cylinder closed at one end (Faraday cup) which will collect positive ions without loss of the secondary electrons which are emitted from a metallic surface when it is bombarded by a positive-ion beam. Electrometer valves or electron multipliers are used to amplify the signal. The output currents from the amplification system are recorded as a galvonometer deflection on photographic or ultra-violet paper, a pen recorder trace or on an oscilloscope. Peak digitizers can also be used to feed the output directly to a computer.

4. Resolution

Peaks representing the ions of m/e m_1 and m/e m_2 are said to be resolved when $\Delta H/H$ (as defined in Fig. 5.14) is equal to or less than 0·1. The resolving power of the instrument is then given by the expression $m_1/\Delta m$ where $\Delta m = m_1 - m_2$.

Single focusing instruments have resolving powers in the range 100–7000, whereas a typical double focusing instrument has a resolving power of the order 10 000–30 000.

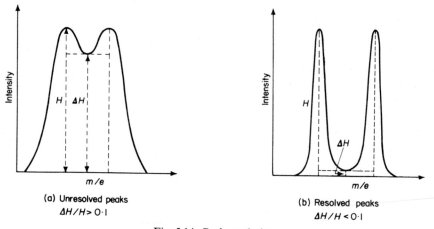

(a) Unresolved peaks
$\Delta H/H > 0.1$

(b) Resolved peaks
$\Delta H/H < 0.1$

Fig. 5.14. Peak resolution

5. Presentation of Spectra

Instruments scan a mass spectrum, by variation in magnetic-field strength or accelerating voltage, from high to low mass or vice versa. The assignment of m/e values to peaks is normally carried out by counting from a peak of known m/e value in the background spectrum of the instrument. The background spectrum exists because of minute traces of air and adsorbed organic material in the instrument, and gives peaks from air at m/e 18 (H_2O), 28 (N_2), 32 (O_2),

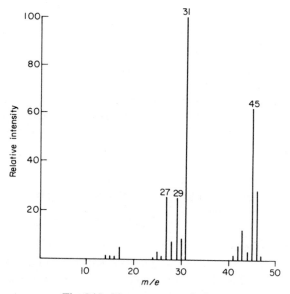

Fig. 5.15. Mass spectrum of ethanol

40 (Ar) and 44 (CO_2) and other peaks from the adsorbed organic material. A standard compound such as heptacosaperfluoro-tri-n-butylamine is introduced if there is any doubt over integer mass allocation. This compound gives a fragmentation pattern which has been well characterized.

The peak heights of each peak in the spectrum are obtained from the pen or galvonometer trace. The peak heights must then be corrected by subtracting any contribution made by the instrument background. Peak heights are generally expressed as percentages of the most intense peak in the spectrum so that there is no dependence on the amount of sample injected. The most intense peak in the spectrum is termed the "base" peak.

The information from a mass spectrum is presented in either tabular form or as a line diagram. A line diagram representing the mass spectrum of ethanol is shown in Fig. 5.15.

An alternative presentation of the intensities of the peaks is as $\%\Sigma_m$ which has been described earlier (page 307).

B. SAMPLE HANDLING

Ions are obtained from the sample in the vapour state. The flow of this vapour within an instrument can be either molecular or viscous in character. Molecular flow is the condition normally used in mass spectrometry since this flow condition is favoured at the low pressures which exist in the instrument. A molecular flow condition implies that collisions between the molecules and the walls are more frequent than collisions between the molecules themselves. Under molecular flow conditions, flow velocity parallel to the tube walls is the same over the whole cross-section of the tube, the quantity of gas flowing is proportional to the pressure differential in the tube and is inversely proportional to the square root of the mass of the gas (i.e. proportional to $1/m^{1/2}$).

Normally the sample is injected into the mass spectrometer through a low porosity silicon carbide leak. Since molecules of low molecular weight will pass through the leak more rapidly then molecules of higher molecular weight under molecular flow conditions, it is necessary to attach a large "reservoir" of vaporized sample to the leak so that the change in reservoir composition, over the time required for scanning, is small. The construction of the inlet system used depends upon the volatility of the sample to be analysed; various types are considered below.

I. Cold-inlet System

The cold-inlet system is used for gases or volatile liquids. A generalized diagram is given in Fig. 5.16. The vapour of the substance to be analysed is expanded into the manometer section and then into the reservoir. The vapour

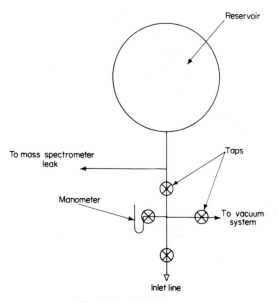

Fig. 5.16. Cold-inlet system

from the reservoir then enters the mass spectrometer tube through a low porosity leak.

2. Heated-inlet System

The heated-inlet system is similar to the cold-inlet except that the temperature is thermostatically controllable up to 300–400°C. This inlet is used for rather less volatile liquids.

3. Gallium-inlet System

An alternative method for the introduction of involatile liquids into the source is to use a gallium-inlet sinter. The sample is drawn into a capillary tube (or hypodermic syringe) and placed into molten gallium which covers a sintered disc as shown in Fig. 5.17. The sample is drawn out of the capillary tube due to the vacuum in the reservoir. The temperature can be varied from 40°C (gallium melts at 30°C) up to about 300°C. The sinter provides a method of rapid introduction of the sample into the reservoir.

4. Continuous-inlet System

The continuous-inlet system enables samples to be directly introduced to the mass spectrometer from a gas stream which is at any pressure up to one atmosphere. The system consists of a finely drawn section of stainless steel capillary tubing which is placed into the high pressure flow stream of the

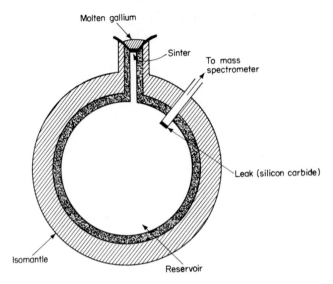

Fig. 5.17. Gallium-inlet system

sample. The system is evacuated by means of a rotary pump and the pressure at the low-pressure end of the capillary is sufficiently low to allow the normal passage of the sample through the low porosity leak into the mass spectrometer tube.

5. Direct Insertion Probe

Solids can be introduced into the mass spectrometer by means of a direct insertion probe. The solid sample is placed into a small capillary at the end of the probe and the probe is inserted into the spectrometer, via a vacuum lock, to bring the sample close to the electron beam. An alternative method of loading the sample on the probe is to drop a solution of the sample onto a porous ceramic probe tip and to evaporate off the solvent.

The direct insertion probe obviates the problem of thermal decomposition which can occur in the reservoir of a heated-inlet system. It is the only effective means of introducing a solid into a mass spectrometer.

6. G.L.C. Sampling

The sample inlet system of a mass spectrometer can be constructed so that samples eluted from a gas phase chromatographic column can be injected directly into the mass spectrometer. Spectrometers which are capable of scanning the spectrum rapidly are required for this technique since the concentration of the compound in the effluent stream changes rapidly with time.

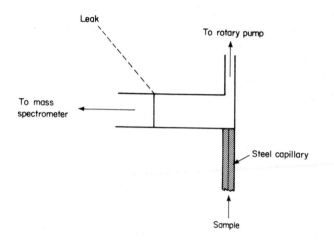

Fig. 5.18. Continuous-inlet system

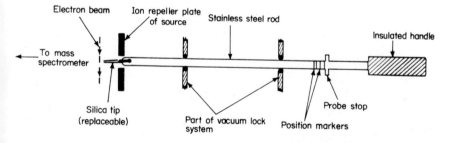

Fig. 5.19. Direct insertion probe (with permission from J. M. B. Bakker)

7. General Requirement

The high vacuum required in a mass spectrometer tube is usually achieved by means of oil diffusion pumps backed by two-stage rotary pumps. Normally a cold trap is fitted between the diffusion pump and the mass spectrometer tube so that back-streaming of the diffusion pump oil is reduced. Seals are either glass to metal, metal to gold, or metal to rubber O rings. The mass spectrometer tube system is kept clean by periodic "baking" up to temperatures of 400°C, whilst still pumping, in order to remove adsorbed materials from the walls.

III. EXPERIMENTS

A. PERFORMANCE OF INSTRUMENTS AND EXPERIMENTAL TECHNIQUES

The experiments in this section are designed to give familiarity with the operation of typical low-resolution mass spectrometers. All of the experiments which follow are designed for single focusing instruments since instruments of the double focusing type are not normally available for student use.

MI. Performance of a Low Resolution, Small Mass-Range Mass Spectrometer

Object

To check the performance of a low-resolution 180° deflection mass spectrometer.

Theory

Many simple instruments have a fixed magnetic field and in such instruments ions are focused onto the collector by altering the accelerating voltage. Ions of mass m/e are collected for a given set of conditions in accordance with the equation

$$m/e = \frac{r^2 B^2}{2V}$$

where V is the accelerating voltage, and r is the radius of the ion path in the magnetic field of strength B. Fig. M1 shows the focusing effect of the 180° magnetic field on the ions. It should be noted that all ion paths are semicircular.

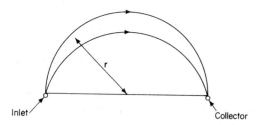

Fig. M1. Focusing effect of a 180° magnetic field

Apparatus

Low-resolution 180° mass spectrometer; vacuum line; coal gas.

Procedure

1. Start up procedure. Evacuate the spectrometer system and activate the control unit as suggested in the manufacturer's handbook.

2. Blank spectrum. Check that the tube pressure is less than 10^{-4} N m^{-2}. With the instrument at a high sensitivity, scan the spectrum either manually or automatically over the mass range to cover m/e values from 12–100. Examine the spectrum for the presence of the following background peaks:

m/e	2	12	16	17	18	28	32	44
Species	H_2^+	C^+	O^+	OH^+	H_2O^+	CO^+	O^+	CO_2^+
						N_2^+		

Note: If the instrument has been recently baked out the whole of the background spectrum may well be negligible even on the most sensitive amplifier ranges.

3. Inlet system. The inlet system for the analysis of the coal gas sample may be a cold or hot "box" or a continuous capillary inlet. The various types of inlet systems are described in the instrumentation section.

4. Mass spectrum of the sample. Fill a sample pipette (normally a glass bulb with two taps, one at either end) with coal gas by blowing out the air with a stream of coal gas. Inject the coal gas sample into the reservoir of the cold- or heated-inlet system and at the same time ensure that the pressure registered on the tube ionization gauge does not rise above 10^{-2} N m^{-2}. Alternatively use the capillary-inlet system inserted directly into a stream of coal gas. Examine the spectrum for peaks at the following m/e values, m/e 2, 12, 15, 16, 28, 44 and note that these peaks are due to H_2^+, C^+, CH_3^+, CH_4^+, N_2^+ and CO_2^+ respectively.

Results

1. Mass calibration. On a small mass range instrument, the accelerating voltage control is often calibrated directly in mass units. Allocate the m/e values by marking each output peak on the chart with the mass value recorded on the accelerating voltage control.

2. Relative intensities. Choose a suitable amplifier sensitivity range and convert all the intensities of the peaks in the spectrum of the coal gas sample to the intensities which would be obtained if all the peaks were recorded at the chosen sensitivity. Note: The different sensitivity ranges are necessary so that ion currents in the range 10^{-10}–10^{-15} ampere can be conveniently measured.

3. Resolution. Note: Peaks are adequately resolved when the valley height between adjacent peaks of equal height is less than 10%, i.e. $\Delta H/H < 0.1$ (see Fig. 5.13). A convenient peak for checking resolution in the coal gas spectrum is that of m/e 44.

Determine the peak height at m/e 44 together with the heights of the tail values at m/e 39.5 and 40.5. Calculate the height of the valley, ΔH, as twice the average value of the tail heights. Calculate $\Delta H/H$.

Discussion

Comment on (a) the resolution of the instrument and on (b) its immediate past operating history in the light of the blank spectrum.

M2. Performance of a Single Focusing Large Mass Range Instrument

Object

To check the performance and to establish operating procedures for a low-resolution 90° deflection mass spectrometer.

Theory

The ions formed in an electron bombardment source are accelerated by a voltage of 500–8000 V towards a magnetic field (Fig. M2). The sector magnetic field causes the ions to travel along different circular arcs dependent upon their m/e ratios. The equation governing the motion of the ions is the same as given in Expt. M1

$$m/e = \frac{r^2 B^2}{2V}$$

where the symbols represent the same quantities as in Expt. M1. Thus the different ions can be focused onto the collector plate by adjustment of either the magnetic field or the ion accelerating voltage. The normal mode of opera-

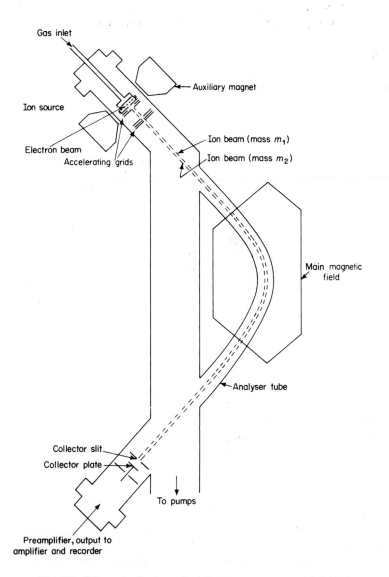

Gas inlet

Auxiliary magnet

Ion source

Electron beam

Accelerating grids

Ion beam (mass m_1)

Ion beam (mass m_2)

Main magnetic field

Analyser tube

Collector slit

Collector plate

To pumps

Preamplifier, output to amplifier and recorder

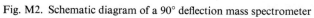

Fig. M2. Schematic diagram of a 90° deflection mass spectrometer

tion is to use full accelerating voltage and to scan the spectrum by variation of the magnetic field. By progressively reducing the ion accelerating voltage the mass range scanned can be extended but this results in a loss of resolution.

Apparatus

Low-resolution mass spectrometer; heptacosaperfluorotri-n-butylamine (reference compound), cyclohexene.

Procedure

1. Setting up for operation. This type of instrument is more complicated than the simple 180° deflection instrument described in Expt. M1. There are many control variables and it is not possible to generalize upon the operating instructions. The operator is referred to the instruction manual applicable to the particular instrument.

2. Inlet systems. All of the inlet systems described in the instrumentation section can be used on this type of instrument. The choice of a particular system depends upon the volatility of the sample.

3. Run a blank spectrum, a spectrum of cyclohexene and a spectrum of the reference compound. Note: Various hydrocarbon oils are also used as reference compounds.

Results

1. Allocation of mass. Allocate m/e values to the peaks in the spectrum of cyclohexene by comparing the cracking pattern of cyclohexene with the spectrum of the reference compound. Alternatively, count the peaks in the cyclohexene spectrum from a known low mass peak which is often present such as m/e 18 (H_2O^+) m/e 28 (N_2^+) or m/e 32 (O_2^+).

2. Relative intensities. The output from the detector is normally monitored by means of a multichannel galvonometer recorder on to photographic or ultraviolet paper. Calculate range factors by comparing peak heights for a given m/e value on the galvonometer traces of varying sensitivity from the spectrum of cyclohexene. Hence calculate the relative intensities of the peaks in the spectrum of cyclohexene.

3. Resolution. Measure resolution as in Expt. M1. The resolving power is $M/\Delta M$ where M is equal to the mass of a given ion. Calculate the resolving power of the instrument using peaks at the high mass end of the spectrum of the reference compound.

Discussion

Comment on the feasibility of mass allocation by the method of "counting peaks".

B. QUALITATIVE ANALYSIS

The mass spectrum of a compound often yields valuable information on the structure and molecular weight of the compound. It is sometimes possible to analyse a mixture of compounds qualitatively but the technique has most value when applied to a single pure substance. The following experiments deal with single substances only. Similar methods are used for mixtures but interpretation of the results is more difficult due to superimposition of the peaks from fragment ions.

M3. Fragmentation patterns

Object

To determine the fragmentation patterns of dichloromethane and chloroform.

Theory

All compounds have a characteristic mass spectrum. Mass spectra are characterized by:

(a) *The molecular (or parent) ion peak* which occurs at the m/e value corresponding to the molecular weight of the compounds. Since the molecular ion is not necessarily stable this peak is not always observed.

(b) *The base peak* which is the most intense peak in the spectrum and is used for sensitivity determinations (page 354). In some compounds the base peak in the spectrum is in fact the molecular-ion peak.

(c) *The pattern of smaller fragmentation peaks* which is characteristic of any particular compound. The intensities of these peaks relative to the base peak are reasonably constant for a given set of operating conditions.

(d) *Isotopic peaks* which are due to the existence of more than one isotope of atoms. The m/e values of peaks which are quoted are normally those for the most abundant isotope. Isotopic contributions from ^{13}C, ^{37}Cl and ^{81}Br in particular give rise to diagnostic peaks in the spectrum.

All of the categories of peaks listed above constitute the fragmentation (or cracking) pattern of the sample.

Apparatus

Low-resolution mass spectrometer (mass range 2–200), cold- or gallium-inlet system; dichloromethane, chloroform.

Procedure

Adjust the controls of the mass spectrometer to obtain the following source parameters: electron-beam voltage 70 eV, ion-repeller voltage $\simeq +1$ V, trap current $\simeq 50$ μA.

Scan a blank spectrum over the mass range 12–120. Inject sufficient of the organic compound through a cold-inlet or gallium-inlet system to give a pressure in the mass spectrometer tube not greater than 2×10^{-4} N m^{-2}. Scan over the mass range of the compound. Pump away the first sample, continue pumping for several minutes and then run another blank spectrum before injecting a sample of the second compound.

Results

Convert all peak intensities to values corresponding to the same amplifier or galvonometer sensitivity range. Express the intensities of the peaks either as output meter current or as peak heights. Tabulate the results as in Table M3.

Table M3

m/e	Blank intensity	Sample intensity	Difference
12	—	8	8
13	—	19	19
14	84	198	114
15	—	583	583

Evaluate the fragmentation pattern by arbitrarily assigning the major peak an intensity of 100 and expressing the other peak heights as percentages of the major peak intensity. Report the results in tabular form and in line-diagram form as shown in Fig. M3.

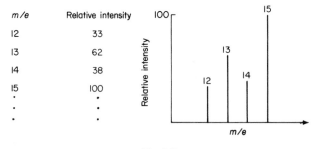

m/e	Relative intensity
12	33
13	62
14	38
15	100
.	.
.	.
.	.

Fig. M3

Discussion

1. Indicate the fragmentation process which leads to the ion corresponding to the major peak in the spectrum.

2. Identify the species giving rise to peaks in the spectrum with relative intensities greater than 10%.

3. Is there any advantage in expressing the results as a percentage of the total ion current, i.e. $100I/\Sigma I$.

M4. Determination of Molecular Formulae

Object

To demonstrate the use of isotopic molecular-ion peaks and of mass spectral indices in the determination of the formula of a compound.

Theory

Since most elements have more than one isotope any molecule having a molecular ion of m/e M will also have ions of m/e $M + 1$ and $M + 2$ due to the heavier isotopes which are present. The ratios of the intensities of the M, $M + 1$ and $M + 2$ peaks can be calculated from probability theory. There are several texts (Beynon and Williams, 1963; American Society for Testing and Materials, 1963) containing the results of such calculations; the results are listed as tables of ratios of $(M + 1)/M$ and $(M + 2)/M$ for different molecular formulae containing C, H, N and O in all combinations up to m/e 500. The experimental values of $(M + 1)/M$ and $(M + 2)/M$ can be determined and hence a molecular formula for a compound established by reference to the calculated ratios in the standard texts.

Apparatus

Low resolution mass spectrometer (mass range up to 100); acetone, ethanol.

Procedure

Inject a sample of ethanol into the mass spectrometer by means of a cold-inlet or gallium-inlet system. Run the mass spectrum up to about 10 mass numbers greater than the molecular weight. If necessary increase the sensitivity when scanning the $M + 1$ and $M + 2$ peaks. Repeat the procedure using a sample of acetone.

Results

Tabulate the heights of the M, $M + 1$ and $M + 2$ peaks for the two compounds. Evaluate the ratios $(M + 1)/M$ and $(M + 2)/M$ and compare with the set of values given below.

m/e	$100(M + 1)/M$	$1000(M + 2)/M$	Ratio
C_2H_6O	2·296	2·152	10·67

The figures for m/e 58 and atomic composition C_3H_6O are

m/e	$100(M+1)/M$	$1000(M+2)/M$	Ratio
C_3H_6O	3·377	2·400	14·07

Discussion

1. What other structure besides that of acetone is possible for a compound having a molecular formula C_3H_6O? Look up the spectra of acetone and other compounds of molecular formula C_3H_6O in the mass spectral indices and confirm that a positive identification of acetone is possible.

2. Attempt to allocate the major peaks in the experimental spectra ($>10\%$ relative intensity) on the basis of the structures of acetone and ethanol. Indicate the fragmentation routes which lead to the major ions.

C. STRUCTURAL CORRELATIONS

The fragmentation pathways of many different classes of compounds have been extensively investigated and the general mechanisms for fragmentation of molecules in any particular class of compounds have been well documented. It is of value to know the manner in which a particular type of molecule would be expected to fragment under electron impact since this information can be utilized in predicting the spectrum of any structure which is postulated for an unknown compound. The experiments in this section illustrate the manner in which aldehydes, alcohols, halogenated compounds and esters are fragmented upon electron impact. General texts on the subject of structural correlations in organic chemistry are those of Budzikiewicz *et al.* (1967), McLafferty (1966) and Beynon *et al.* (1968).

M5. Fragmentation Patterns in Aliphatic Aldehydes

Object

To correlate the mass spectra of aliphatic aldehydes with their structure and to identify an unknown aldehyde.

Theory

The molecular ion of an aldehyde (or indeed any carbonyl compound) is formed by the removal of one of the lone-pair electrons on the oxygen atom.

$$e + \begin{array}{c} H \\ \diagdown \\ R \diagup \end{array} C{=}O \longrightarrow \begin{array}{c} H \\ \diagdown \\ R \diagup \end{array} C{=}\overset{\cdot+}{\underset{\cdot\cdot}{O}} + 2e$$

The molecular ion may fragment in the following ways:

α CLEAVAGE. Cleavage of the bond between the carbonyl group and the H atom, and between the carbonyl group and the group R_1 results in the formation of two different ions.

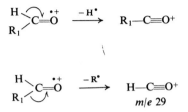

or

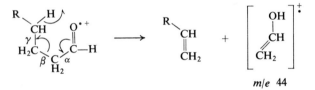

β CLEAVAGE AND MIGRATION OF A γ HYDROGEN ATOM. When the alkyl group, R_1, adjacent to the carbonyl function consists of four or more carbon atoms a rearrangement ion may be formed by β cleavage and the transfer of a γ hydrogen atom to the oxygen atom (McLafferty rearrangement). The mechanism which has been postulated for the process involves a cyclic intermediate state.

In the above process the charge can remain on either fragment with the result that an ion of m/e $M - 44$ is also produced.

β CLEAVAGE WITHOUT TRANSFER OF A γ HYDROGEN ATOM. It is possible for cleavage of the bond β to the functional group to occur without a simultaneous

$$R—CH_2—\overset{+}{CH}—O \quad \longrightarrow \quad R^+ + CH_2{=}CH—O$$

transfer of a γ hydrogen atom. This process results in the formation of an ion of m/e $M - 43$.

LOSS OF NEUTRAL FRAGMENTS. Peaks are frequently observed in the spectra of aldehydes due to loss of C_2H_2 and H_2O fragments from the molecular ion. Such peaks are observed at m/e $M - 28$ and $M - 18$.

Apparatus

Low-resolution mass spectrometer (mass range up to m/e 200); formaldehyde, acetaldehyde, propionaldehyde, n-butyraldehyde, 2-methylproponal, 2-methylbutanal, 3-methylbutanal, an "unknown" aliphatic aldehyde.

Procedure

Run a blank spectrum under normal operating conditions at a beam energy of 70 eV. Inject an aldehyde sample into the spectrometer via a cold- or heated-inlet system or gallium-sinter system. Scan the spectrum over the mass range for the compound. Pump the sample away and run another blank spectrum. Repeat this process for the other aldehydes.

Results

Correct each aldehyde spectrum for background by subtracting the appropriate blank spectrum. Evaluate the cracking pattern for each aldehyde and express it as a line diagram. Identify the molecular ion and draw up possible fragmentation routes which lead to the formation of the ions corresponding to the six major peaks in each spectrum.

Discussion

Comment on the presence or absence of the m/e 44 peak in each of the spectra. Identify the "unknown" aldehyde and give reasons for your identification.

M6. Fragmentation Patterns in Alcohols

Object

To correlate the mass spectra of aliphatic and aromatic alcohols with their structure and to identify an unknown alcohol.

Theory

ALIPHATIC ALCOHOLS. Ionization-potential measurements have shown that, under electron bombardment conditions, a saturated aliphatic alcohol loses an electron from one of the lone-pair electrons on the oxygen atom.

$$R\overset{..}{\underset{..}{O}}H + e \longrightarrow R\overset{\cdot+}{\underset{..}{O}}H + 2e$$

The molecular ion can fragment by loss of a hydrogen atom to form an oxonium ion at m/e $M - 1$.

$$R-CH_2\overset{\cdot+}{\underset{..}{O}}H \xrightarrow{\ -H^{\cdot}\ } R-CH=\overset{+}{O}H$$

Ions at m/e $M - 2$ and $M - 3$ also occur due to the further loss of hydrogen atoms. These ions are of the form $R-CH=O^{+\cdot}$ and $R-C\equiv O^{+}$. α cleavage occurs readily in the molecular ion to produce a peak corresponding to the ion m/e 31.

$$[R\!-\!\{CH_2OH]^{\cdot}_{+} \xrightarrow{\ -R^{\cdot}\ } CH_2=\overset{+}{O}H$$
$$m/e\ 31$$

A series of peaks is often observed in the spectra of primary alcohols due to cleavage in the β, γ, δ, etc. positions in the alkyl chain.

In certain cases elimination of H_2O from the molecular ion can occur to give an ion at $m/e\ M - 18$.

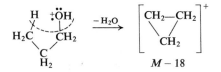

$$M - 18$$

AROMATIC ALCOHOLS. Molecular ions can be formed either by removal of an electron from the lone pair on the oxygen atom or from the aromatic ring. In general the aromatic nucleus acts as a centre for the retention of the positive charge. α cleavage to the ring will produce the phenyl cation of m/e 77. Deuterium labelling has been used to a large extent to elucidate breakdown processes, e.g. it has been shown that loss of a hydrogen atom from the ring produces the $M - 1$ peak and not loss from the side chain.

Apparatus

Low-resolution mass spectrometer (mass range up to 200): methanol, ethanol, n-propanol, n-butanol, benzyl alcohol, o-methyl benzyl alcohol and an "unknown" alcohol.

Procedure

Run mass spectra of all of the compounds at an electron beam energy of 70 eV, injecting the sample by means of a cold or heated batch inlet system or a gallium-inlet sinter. A blank spectrum should be obtained before each compound is injected so that an allowance can be made for memory effects.

Results

Correct the spectrum of each alcohol for the background by subtracting the appropriate blank spectrum. Express the spectra as relative intensities. As far as possible suggest fragmentation modes and associated electron movements within the molecule in order to obtain the six major peaks in each spectrum. Note that small molecular-ion peaks and fairly intense $M - 1$ peaks are obtained, that α cleavage is common and that in aliphatic alcohols the base peak is often at m/e 31.

Identify the unknown alcohol from a consideration of its fragmentation pattern.

Discussion

How might the spectrum of a secondary or tertiary alcohol compare with that of a primary alcohol?

M7. Fragmentation Patterns in Halogen Compounds

Object

To correlate the mass spectra of some chlorine-containing compounds with their structure.

Theory

Ionization-potential measurements have shown that the molecular ion of a halogen compound is formed by removal of an electron associated with the halogen atom.

$$R—CH_2—X \rightarrow R—CH_2—X^{+ \cdot}$$

α cleavage of the molecular ion will occur as represented below

$$R—CH_2—X^{\cdot +} \rightarrow R^\cdot + CH_2{=}X^+$$

This cleavage occurs to the greatest extent in iodine-substituted compounds and to the least extent in fluorine-substituted compounds. α cleavage may also produce an $M - 1$ species by loss of an α hydrogen atom.

Other noticeable fragmentations in halogen compounds are loss of halogen, loss of HX and loss of H_2X. Rearrangement ions are negligible. Because halogens are strongly electronegative there is a general lack of positive ions containing a halogen atom and an abundance of negative ions which do contain the halogen atom. These negative ions are not detected under the normal operating conditions.

As in other benzyl derivatives, benzyl chloride can fragment to form the tropylium ion as shown below. Subsequent loss of acetylene from the tropylium

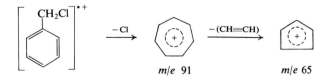

m/e 91 $\qquad$ m/e 65

ion produces the cyclopentadienyl ion. Ring-halogenated compounds produce peaks due to loss of halogen and to cleavage of the ring.

The peak ratios for the molecular-ion peaks and major fragment peaks in aliphatic and aromatic halides will indicate the number of halogen atoms substituted into the compound (Expt. M11). If the compound contains one chlorine atom then the ratio of the molecular ion peaks $M : M + 2$ will be $3 : 1$; for two chlorine atoms $M : M + 2 : M + 4 = 9 : 6 : 1$ and for three chlorine atoms the ratios will be $M : M + 2 : M + 4 : M + 6 = 27 : 27 : 9 : 1$.

Apparatus

Low-resolution mass spectrometer (mass range 12–400); benzene, benzyl-chloride, chlorobenzene, p-dichlorobenzene, ethyl chloride, n-propyl chloride, chloroform.

Procedure

Before each compound is injected run a blank spectrum. Run spectra of the compounds at an electron-beam energy of 70 eV. Note: Benzene is included so that its ring breakdown can be compared with that of the ring-substituted halogenated compounds.

Results and Discussion

Obtain the cracking pattern for each compound. Attempt to make generalizations about the mechanisms producing the major peaks in the aliphatic and aromatic halogen compounds.

Measure the M, $M + 2$, $M + 4$, etc. ratios on each compound and comment on the comparison between the experimental and calculated values.

M8. Structural Correlations in Alkyl Esters and the Use of Metastable Ion Transitions

Object

To correlate the mass spectra of alkyl esters with their structure and to allocate metastable peaks.

Theory

β cleavage with transfer of a γ hydrogen atom is a common process in long-

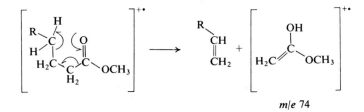

m/e 74

chain esters. The ion of m/e 74 is often the base peak in the spectrum. Cleavage of the carbon chain will produce ions of general formula $[(CH_2)_nCOOCH_3]^+$.

α cleavage occurs readily in alkyl esters with the production of four types of ions.

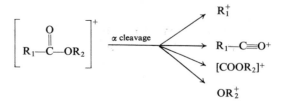

Metastable ion peaks are formed when an ion of mass m_1 decomposes between the source and the magnetic analyser with the production of an ion of mass m_2 and a neutral fragment. This decomposition of ions in the "field free region" of the mass spectrometer occurs quite frequently since the half life of many of the species is of the order of 10^{-6} second while the time taken for an ion to reach the collector is of the order 10^{-5} second. The decomposition can be represented as

$$m_1^+ \rightarrow m_2^+ + (m_1 - m_2)$$

Some of the ions of mass m_1^+ and some m_2^+ reach the collector and so a peak is recorded at both $m/e\ m_1$ and m_2. In addition there are small, diffuse metastable peaks which are often seen at non-integral mass numbers. The presence of a metastable peak shows that the ion m_2^+ has been formed from the ion m_1^+ in a one-step process. The metastable peak occurs at the m/e value given by the relationship

$$m^* = m_2^2/m_1 \tag{M8.1}$$

Metastable peaks are of use when the origins of specific ions are under investigation. If the mass number for the metastable ion is known, then the masses m_1 and m_2 required to fit equation M8.1 can be found, and it is thus proven unambiguously that the ion of mass m_2 is formed from m_1. Thus it can be seen that metastable ion transitions are of importance when investigating the fragmentation pattern of any compound.

Apparatus

Low-resolution mass spectrometer; methyl acetate, ethyl acetate, isopropyl acetate, n-butyl acetate.

Procedure

Run mass spectra of each of the compounds using an electron-beam energy of 70 eV and preceding each compound by a blank spectrum. Repeat each spectrum using wide collecting slits in the instrument in order to increase the intensities of the metastable peaks.

Results and Discussion

Obtain the cracking patterns of the esters. Record the mass numbers of the metastable ion peaks and attempt to assign the parent and daughter ions from which these have come. The assignment can be done by trial and error using the formula $m^* = m_2^2/m_1$, by using nomograms (Beynon, 1960) or by using a table of metastable transitions (Beynon *et al.* 1965). If there are a large number of metastable peaks the most satisfactory method is to use a simple computer program to allocate the peaks.

D. QUANTITATIVE ANALYSIS

In the analysis of mixtures it is useful to achieve an initial separation of the components, for example by gas-phase chromatography. The components can often then be identified and the relative concentrations determined more readily. If an initial separation is not feasible, the mass spectrum of the mixture can be used to determine the amounts of the components present since, to a first approximation, the mass spectrum of the mixture will be the sum of the spectra of each component.

With a multi-component mixture a fairly involved set of linear simultaneous equations is obtained, each equation in the set corresponding to the height of a peak in the spectrum. These equations are soluble using the methods of matrix algebra and a fairly reliable quantitative analysis is possible. The experiment below deals with a simple three-component analysis where the use of matrix algebra is unnecessary.

M9. Analysis of a Three-component Mixture

Object

To obtain the relative concentrations of the components in a mixture of benzene, dichloromethane and toluene.

Theory

The mass spectrum of the mixture will contain major peaks which are characteristic of each component. Ideally the peaks chosen for measurement should be due solely to one component. The peak heights are measured and converted into a partial pressure by using the appropriate sensitivity factor (i.e. the height of the peak per unit pressure). If this is done for a unique peak for each compound the sum of the three partial pressures should equal the total sample pressure. If it is not possible to choose a peak unique to each component, a correction must be made for the contribution from the fragmentation of the component of higher mass to the lower mass peaks (see procedure section).

Apparatus

Low-resolution mass spectrometer (mass range 12–110); cold- or heated-inlet system; benzene, toluene, dichloromethane.

Procedure

Record the mass spectrum of benzene.

Determine the sensitivity of the mass spectrometer to benzene by setting the amplifier to a range on which 10 % full-scale deflection is obtained for the peak at m/e 78. Record the ion current I_1, and the pressure, P_1, on the ion gauge (either on the ionization gauge attached to the mass spectrometer tube or to the reservoir). Admit more benzene to give almost full-scale deflection using the same amplifier range and record the ion current I_2 and the pressure P_2. The sensitivity of the instrument with respect to benzene may then be calculated from

$$S = \frac{I_2 - I_1}{P_2 - P_1} \times Z \qquad (M9.1)$$

where Z is the ion current corresponding to full-scale deflection on the amplifier range chosen. (This value is given in the instrument handbook.)

Record the mass spectrum of dichloromethane and note the intensity of the peak at m/e 78. Determine, as for benzene, the sensitivity of the mass spectrometer for dichloromethane using the peak at m/e 84.

Repeat the above procedure for toluene using the peak at m/e 92 to determine the sensitivity.

Admit the three-component mixture to the mass spectrometer using a small amount of the sample and ensuring that complete vaporization occurs. Note: This is necessary so that the composition of the vapour in the mass spectrometer is the same as the original liquid mixture. Measure the ion currents of the peaks at m/e 78 (from benzene), m/e 84 (from dichloromethane) and m/e 92 (from toluene) and the pressure of the mixture on the ion gauge.

Results

Correct for the contribution from dichloromethane to the peak at m/e 78 in the mixture spectrum. The ratio of this peak to the peak at m/e 84 can be obtained from the dichloromethane spectrum and allowance made for this contribution in the mixture spectrum by subtraction of the contribution to m/e 78 by simple proportion.

Correct for contributions to the peaks at m/e 78 and 84 from toluene in a similar fashion to the correction for dichloromethane.

Evaluate the partial pressures of benzene, dichloromethane and toluene in the mixture spectrum by dividing the measured ion currents at m/e 78, 84 and 92 by the appropriate sensitivity factors.

The sum of the partial pressures of benzene, dichloromethane and toluene should be equal to the total pressure of the mixture. Calculate $\%P/\Sigma P$ for each component where P is the partial pressure. Note: $\%P/\Sigma P$ is equivalent to expressing the results as mole %.

E. ISOTOPIC ABUNDANCE

Isotopic abundance measurements are useful for the determination of the amount of D, ^{18}O and ^{15}N in organic compounds and for the determination of the number of chlorine, bromine or boron atoms in a molecule.

MI0. Composition of an H_2O/D_2O Mixture

Object

To determine the relative amounts of deuterium oxide and protium oxide in a sample of deuterated water.

Theory

The determination of deuterium in a hydrocarbon compound is most frequently carried out by converting the compound into carbon dioxide and water and then converting the "water" into gaseous hydrogen and deuterium. The relative amounts of H_2 and D_2 are then estimated from the relative intensities of the corresponding peaks in the spectrum of the mixture. Water itself is normally considered unsuitable for direct injection into the mass spectrometer due to adsorbtion and exchange effects on the walls of the spectrometer. The direct determination of deuterium in water has, however, been carried out by modifying the mass spectrometer so as to reduce memory effects and production of H_3O^+ ions (Washburn *et al.* 1953). The experiment described here will give only approximate results since these precautions are not taken and the experiment assumes that oxygen is monoisotopic as the ^{16}O isotope.

Possible fragments in the spectrum are given below where oxygen is considered as the ^{16}O isotope only:

m/e	1	2	16	17	18	19	20	22
Species	H^+	D^+	O^+	OH^+	OD^+	HDO^+	D_2O^+	D_3O^+
					H_2O^+	H_3O^+		

The contribution of H_2O^+ to the m/e 18 peak in the spectrum of the mixture and hence the proportion of m/e 18 due to OD^+ may be obtained from a knowledge of the OH^+ to H_2O^+ ratio in normal water by scanning the peaks at m/e 17 and 18.

A difficulty is that H_2O is already present as background material in the instrument with the result that the following exchange can occur

$$D_2O + H_2O \rightarrow 2\,HDO$$

This exchange reaction would lead to a false figure for the composition of the mixture; the erroneous result may be overcome by flushing with the heavy water sample until a constant set of peak ratios are obtained.

Apparatus

Low-resolution mass spectrometer (mass range 12–22), cold-inlet system; dueterium-enriched water sample.

Procedure

Run a blank spectrum at 70 eV. Obtain the value of the ratio of the peaks at m/e 17 and 18 from the blank spectrum.

Inject a sample of heavy water through the cold inlet system. Measure the peak heights at m/e 17, 18, 19 and 20. Evacuate the mass spectrometer and inject further samples of heavy water until constant ratios of the peaks at m/e 17, 18, 19 and 20 are obtained. Note: This normally requires four or five flushes with the vapour of the heavy water sample.

Results

Tabulate the peak heights at m/e 17, 18, 19 and 20. Evaluate the ratio of the peaks at m/e 17 and 18 in all the spectra, including the blank spectrum.

Water: Let the m/e 17:18 peak ratio be $1:a$.

D_2O: Let the heights of the peaks at m/e 17, 18, 19 and 20 be w, x, y and z.

In D_2O the m/e 18 peak must be due to H_2O^+ and OD^+ where $w.a$ is due to H_2O^+ and $x - (w.a)$ is due to OD^+.

Calculate the ratio $H_2O:HDO:D_2O = w.a:y:z$.

Discussion

What assumption has been made about the relative rates of fragmentation of H_2O into OH^+ and D_2O into OD^+?

MII. Isotopic Abundance

Object

To determine the isotopic abundance ratios of nitrogen, oxygen, argon and bromine.

Theory

The isotopic abundances of easily vaporizable elements may be determined fairly readily using a mass spectrometer. When this information is coupled with the precise isotopic mass determination possible with a double focusing instrument, values of the atomic weights of the elements may be obtained.

Apparatus

Low-resolution mass spectrometer (mass range up to m/e 100); nitrogen, oxygen, argon and bromine.

Procedure

Scan a blank spectrum over the mass range appropriate to each element. Admit each sample in turn through a cold-inlet system and scan manually over the m/e values indicated below. Use the most sensitive range possible for

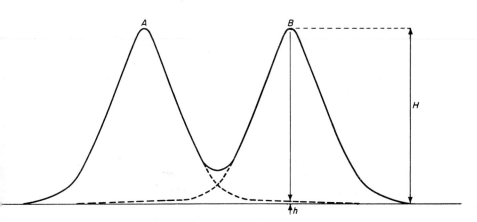

Fig. M11. Correction of peak height

each peak and repeat each reading four times. The tube pressure should be about 10^{-5} N m^{-2} for each sample.

Allow for any tailing from adjacent peaks by scanning over the peak on either side of the peak under study and subtracting the tail height as indicated in Fig. M11. The corrected height of the peak B is equal to $H - h$.

Nitrogen: Scan the peaks at m/e 28 and 29 which are from $^{14}N_2^+$ and (^{14}N-^{15}N)$^+$ respectively.

Oxygen: Scan the peaks at m/e 32, 33 and 34.

Argon: Scan the peaks at m/e 36, 38 and 40.

Bromine: Scan the peaks at m/e 79 and 81.

Results

Tabulate the values of the peak height obtained at each m/e value. Correct the peak heights by subtracting the contributions from the background. Note: The background spectrum should be less than 1 % of the isotope peaks. Correct for tailing from any adjacent peaks.

13

Evaluate the mean value of the isotopic ratios as indicated below.

Nitrogen 14:15 [The ratio 14:15 $\equiv$ (2 $\times$ m/e 28):29]

Oxygen 16:17 and 16:18 [The ratio 16:17 is equal to (2 $\times$ m/e 32):33 and 16:18 $\equiv$ (2 $\times$ m/e 32):34]

Bromine 79:81

Convert the ratios of the peak heights to percentage abundances of the isotopes.

Discussion

1. The atomic weights of the elements may be obtained by calculating the weighted means of the contributions from the isotopes. Thus if the peak ratio of m/e 35:37 is taken as 3:1 then the percentage abundances of ^{35}Cl and ^{37}Cl are 75 and 25%. The exact isotopic masses are $^{35}Cl = 34 \cdot 9689$ and $^{37}Cl = 36 \cdot 9659$, thus the atomic weight of chlorine is given by the weighted mean

$$AW = \frac{75 \times 34 \cdot 9689 + 25 \times 36 \cdot 9659}{100}$$

2. Given the following isotopic masses determine from the relative-abundance measurements the value of the atomic weights of nitrogen, oxygen, argon and bromine.

$$^{14}N = 14 \cdot 0031 \qquad ^{15}N = 15 \cdot 0001$$
$$^{16}O = 15 \cdot 9949 \qquad ^{17}O = 16 \cdot 9991 \qquad ^{18}O = 17 \cdot 9992$$
$$^{36}Ar = 35 \cdot 9675 \qquad ^{38}Ar = 37 \cdot 9627 \qquad ^{40}Ar = 39 \cdot 9624$$
$$^{79}Br = 78 \cdot 9183 \qquad ^{81}Br = 80 \cdot 9163$$

3. Comment on the comparison between the experimental values and those given in tables of atomic weights.

F. THERMODYNAMIC STUDIES

Mass spectrometry may be used for the determination of thermodynamic quantities such as latent heats, and ionization and appearance potentials. Appearance potentials are used to obtain values of bond-dissociation energies and other general thermodynamic data.

M12. Latent Heat of Vaporization of Liquids

Object

To determine the latent heat of vaporization of toluene.

Theory

The Clausius–Clapeyron equation expresses the saturated vapour pressure of a liquid as a function of the absolute temperature

$$\frac{d\ln p}{dT} = \frac{\Delta H}{RT^2} \qquad\qquad (M12.1)$$

where p is the saturated vapour pressure at the temperature T (°K), R is the gas constant and ΔH is the latent heat of vaporization. If ΔH is assumed to be a constant independent of temperature, then

$$\ln p = -\frac{\Delta H}{RT} + \text{const} \qquad\qquad (M12.2)$$

or

$$\log p = -\frac{\Delta H}{2 \cdot 303 RT} + \text{const} \qquad\qquad (M12.3)$$

In order to obtain a value for ΔH it is necessary to measure a quantity proportional to p at a series of temperatures. ΔH is then obtained from the slope of the plot of $\log p$ against $1/T$.

The height of a peak in a mass spectrum is directly proportional to p if the pressure in the ionization chamber is directly proportional to the saturation vapour pressure. This proportionality is achieved by interposing between the equilibration vessel and the ionization chamber a low porosity, molecular flow leak. This leak can be the normal silicon carbide leak in the batch inlet system (page 335).

The mass-spectrometric method of determining the latent heat of vaporization has the advantate that the sample need not be pure since measurements are carried out on one peak only, normally the molecular-ion peak. Another advantage is that peak heights can be measured over a wide range of p values and consequently the accuracy in the determination of the slope of the plot of $\log p$ against $1/T$ is increased.

Apparatus

Low-resolution mass spectrometer (mass range up to 100), thermometer (−5–100°C), thermostat, subsidiary inlet reservoir A (see Fig. M12), toluene.

Procedure

Set up the subsidiary inlet reservoir, (A), in which the liquid and its vapour can equilibrate at various temperatures.

Place toluene in A and attach to the cold-inlet system of the mass spectrometer. Freeze the liquid in A using a cold trap and evacuate. Close tap T_1.

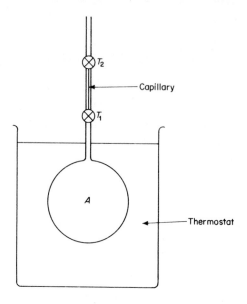

Fig. M12. Subsidiary inlet reservoir

Fill the thermostat with an ice/water slurry at about 10°C and allow time for equilibration. Record the blank spectrum, with T_1 closed, to ascertain that no background peak occurs at m/e 92 (the molecular-ion peak of toluene).

Open T_1 to expand the equilibrated vapour sample into the capillary between T_1 and T_2. Close T_1 and open T_2 to expand the vapour into the reservoir of the inlet system of the mass spectrometer. Record the peak height at m/e 92 and the temperature of the thermostat. Evacuate the inlet system up to T_1. Repeat the procedure for thermostat temperatures of 20, 25, 30, 35 and 40°C.

Results

Tabulate values of the height of the peak at m/e 92 and temperature (°K). Plot a graph of log (peak height) against $1/T$ and evaluate ΔH from the slope of the plot.

Discussion

How would you modify the procedure for a compound of low volatility?

M13. Ionization Potentials

Object

To determine the first ionization potential of nitrogen.

Theory

The ionization potential is equal to the electron-beam energy at which peaks first appear in the spectrum. The first ionization potential of a compound can be derived from the ionization-efficiency curve for the compound, i.e. from plots of the ion current (or peak height) against electron-beam energy, eV (page 326). Argon is used as an internal reference for the calibration of the electron-beam energy scale and is present at the same time as the compound to be studied to avoid contact potential effects (page 307).

Apparatus

Low-resolution mass spectrometer (mass range up to 50); nitrogen, argon.

Procedure

Scan a blank spectrum up to m/e 50 in order to determine that the peaks at m/e 28 (N_2) and 40 (Ar) are not also present from background material. Admit argon and then nitrogen through the cold-inlet system from a gas container as in Expt. M1.

Focus manually on m/e 40 (Ar^+) using a convenient amplifier sensitivity range to obtain approximately full-scale deflection. Reduce the electron-beam energy in steps of 5 eV and note the value when the output ion current (or peak height) of ion m/e 40 is equal to zero. Increase the electron-beam energy in steps of 0·25 eV up to 5 eV above the first detectable ion-current reading measuring the ion current or peak height for each step. Increase the beam energy in steps of 2 or 5 eV until full-scale deflection on the original amplifier range is again obtained.

Repeat the process for m/e 28 (N_2^+).

Results

Tabulate electron-beam energies and the corresponding ion currents.

LINEAR-EXTRAPOLATION METHOD (page 309). Plot ionization-efficiency curves for the ions of m/e 40 and 28. Extrapolate the linear portion of the argon curve and hence determine the apparent ionization potential of argon. The correct value is $I(Ar^+) = 15·76$ eV. Use the error in this value to correct the apparent ionization potential of nitrogen obtained by linear extrapolation of the nitrogen curve.

EXTRAPOLATED VOLTAGE-DIFFERENCE METHOD (page 310). Plot on the one graph the ionization-efficiency curves for m/e 40 and 28. If necessary multiply one of the ion-current scales by a convenient factor to make the first part of the linear portions of the two curves approximately parallel. Determine values of the energy difference (ΔV) between the two curves corresponding to given values of I. Obtain ΔV for about 10 values of the ion current particularly near the foot of the curves.

Plot I against ΔV and extrapolate to $I = 0$. Obtain ΔV_0, the voltage difference between $I(Ar^+)$ and $I(N_2^+)$. Calculate the value of $I(N_2^+)$.

Discussion

Comment on the departure from linearity of the ionization-efficiency curves at relatively high values of the electron-beam energy.

M14. Bond-dissociation Energies

Object

To determine the bond-dissociation energy of the C—CHO bond in benzaldehyde.

Theory

The mass spectrum of benzaldehyde shows a strong peak at m/e 77 due to cleavage of the bond α to the carbonyl function.

$$[Ph—CHO]^{\ddagger} \rightarrow Ph^+ + CHO^{\cdot}$$

The appearance potential (page 307) of this peak, $A(Ph^+)$, is the sum of the energies needed to ionize and then dissociate the benzaldehyde molecule. The process can be represented as

$$Ph—CHO \rightarrow [PhCHO]^+ \rightarrow Ph^{\ddagger} + CHO^{\cdot}$$

The overall process is energetically equivalent to

$$Ph—CHO \rightarrow Ph^{\cdot} + CHO^{\cdot}$$
$$Ph^{\cdot} \rightarrow Ph^+$$

Thus if $I(Ph^{\cdot})$ is the ionization potential of the phenyl radical and $D(Ph—C)$ is the bond-dissociation energy of the Ph—CHO bond in the neutral molecule then

$$A(Ph^+) = I(Ph^{\cdot}) + D(Ph—C) \qquad \text{(page 312)}$$

Since $I(Ph^{\cdot})$ is known, the value of $D(Ph—C)$ may be calculated if the value of $A(Ph^+)$ is measured experimentally.

Apparatus

Low-resolution mass spectrometer (mass range up to 100); argon, benzaldehyde.

Procedure

Run a blank spectrum at 70 eV and check that there are no residual peaks at m/e 40 and 77 due to background material. Fill a gas container (Expt. M1) with argon at atmospheric pressure. Attach the argon container to the cold-inlet system of the mass spectrometer and admit sufficient argon to give a pressure in the mass spectrometer tube of about 1×10^{-5} N m^{-2}. Admit benzaldehyde vapour through the cold-inlet or gallium-inlet system until the pressure is about 2×10^{-5} N m^{-2}. (Sufficient of each component must be present to produce approximately full-scale deflection for the peaks at m/e 40 and 77 on a convenient amplifier range.)

Focus the mass spectrometer on the peak at m/e 40. Decrease the electron-beam energy in steps of 0·5 eV until the ion beam cannot be detected (i.e. zero deflection on the amplifier range chosen) and record the ion current in each step. Then increase the electron-beam energy in 0·5 eV steps and again record ion currents.

Repeat the procedure for the peak at m/e 77.

Results

Tabulate values of the ion current and electron-beam energy for the ions at m/e 40 and 77. Draw ionization-efficiency curves for Ar$^+$ and Ph$^+$. Given that $I(\text{Ar}^{\bullet})$ is 15·755 eV calculate the value of $A(\text{Ph}^+)$ by the linear-extrapolation method.

Given that $I(\text{Ph}^{\bullet})$ is 9·4 eV (Vedeneyev *et al.* 1966) calculate $D(\text{Ph—C})$.

Discussion

Comment on the value obtained for $D(\text{Ph—C})$ in comparison with $D(\text{C—C})$ which is $2·5–2·9 \times 10^5$ J mole^{-1} in an aliphatic side chain to a phenyl ring.

M15. Reaction Kinetics

Object

To study the thermal decomposition of nitrous oxide in the temperature range 700–750°C and to determine the energy of activation of the reaction.

Theory

The homogeneous thermal decomposition of nitrous oxide proceeds at a measurable rate in the temperature range 700–800°C and can be represented by

$$2N_2O \rightarrow 2N_2 + O_2$$

The reaction can be followed by monitoring the partial pressure of nitrous oxide by means of a mass spectrometer.

The decomposition is first order and the appropriate integrated rate equation is

$$k = \frac{1}{t} \ln \frac{a}{a-x} \tag{M15.1}$$

where a is the initial concentration of nitrous oxide, $(a–x)$ is the concentration at time t and k is the rate constant for the decomposition.

If the sensitivity of the mass spectrometer to the ion of m/e 44 (N_2O^+) is known, the height of the peak at m/e 44 may be converted into the partial pressure of nitrous oxide present, and if the total pressure of the mixture in the spectrometer is known then the concentration of nitrous oxide may be evaluated. However the term on the right hand side of equation M15.1 contains a ratio of concentrations and it is only necessary to measure a parameter proportional to concentration in order to evaluate k. Since the height of the peak at m/e 44 is proportional to the nitrous oxide concentration, a in equation M15.1 may be taken as the initial peak height and $(a - x)$ as the peak height at time t. Thus a plot of t against log (peak height) should be a straight line of slope $-2·303/k$ from which k may be evaluated.

An approximate value for the energy of activation, E, for the decomposition is obtained from the equation

$$\ln\frac{k_2}{k_1} = -\frac{E}{R}\left[\frac{1}{T_2} - \frac{1}{T_1}\right] \tag{M15.2}$$

where k_1 and k_2 are the rate constants at temperatures T_1 and T_2 respectively.

Apparatus

Low-resolution mass spectrometer (mass range up to m/e 50), high-vacuum system, furnace and thermocouple; nitrous oxide, argon.

Procedure

The high-vacuum system is shown in Fig. M15.

Admit argon from a sample pipette (Expt. M1) containing argon at atmospheric pressure to the reservoir A via T_2 with T_3 open and T_1, T_4 closed until the pressure on the manometer is 6×10^4 N m^{-2}. Nitrous oxide is obtainable in convenient amounts in a "Sparklet bulb" as supplied by British Oxygen Co. Ltd. and may be admitted to the reservoir A directly from this bulb by means of a commercial "Sparklets Corkmaster" attached to T_2 with T_3 open and T_1, T_4 closed until the pressure on the manometer is 6×10^4 N m^{-2}. Close T_2 and T_3.

Open T_1 and evacuate the system up to T_3. Run a blank spectrum to obtain a correction factor for the peak heights at m/e 40 and 44.

KINETIC RUN. Adjust the resistance on the furnace until the temperature of the reaction vessel is 700°C. Close T_4 and expand the mixture of nitrous oxide and argon into the reaction vessel to a pressure of about 5×10^4 N m^{-2} (registered on the manometer) by control of T_3. Close T_3 and evacuate the system up to T_1. Close T_4, T_5 and T_6.

Expand a sample from the reaction vessel into the space between T_4, T_5 and T_6, noting the time. Open T_6 and expand the sample into the mass spectrometer. Record the peak heights at m/e 40 and 44. Repeat this procedure at intervals of 15 minutes to obtain at least six sets of peak height values. In between each reading evacuate the system down to T_1 keeping T_3 closed. Repeat the experiment at 730 and 760°C.

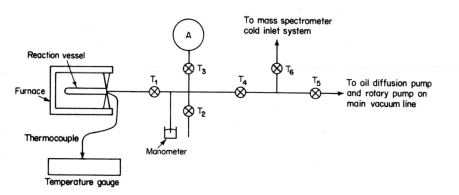

Fig. M15. High-vacuum reaction system

Results

Tabulate values of peak heights at m/e 40 and 44 for the various time intervals at each temperature, after correction for background material at the appropriate m/e values.

The argon acts as an internal calibrant to allow for variation in amount of sample. Choose one value of the argon content as a standard and correct all peak heights by simple proportion to obtain the values of the peak heights for the standard argon content.

Plot the graphs of t against log (peak height at m/e 44) for each temperature and obtain the values of the rate constant from the slopes of the graphs.

Evaluate the energy of activation, E, for the reaction.

Discussion

Normally the reaction is followed by measuring the increase in the total pressure during the reaction. What advantages or disadvantages are inherent in the mass-spectrometric method compared with other methods of monitoring chemical reactions?

REFERENCES

American Society for Testing and Materials (1963). "Index of Mass Spectral Data". A.S.T.M. S.T.P. 356.

Beynon, J. H. (1960). In "Mass Spectrometry and its Applications to Organic Chemistry". Elsevier, Amsterdam.

Beynon, J. H., and Williams, A. E. (1963). In "Mass and Abundance Tables for use in Mass Spectrometry". Elsevier, Amsterdam.

Beynon, J. H., Saunders, R. A., and Williams, A. E. (1965). "Table of Metastable Transitions for use in Mass Spectrometry". Elsevier, Amsterdam.

Beynon, J. H., Saunders, R. A., and Williams, A. E. (1968). In "The Mass Spectra of Organic Molecules". Elsevier, Amsterdam.

Budzikiewicz, H., Djerassi, C., and Williams, D. H. (1967). In "Mass Spectrometry of Organic Compounds". Holden-Day, San Francisco.

Hipple, J. A., Fox, R. E., and Condon, E. U. (1946). Phys. Rev. 69, 347.

Hood, A. (1958). Analyt. Chem. 30, 1218.

Kiser, R. W. (1965). In "Introduction to Mass Spectrometry and its Applications". Prentice Hall, Englewood Cliffs, N.J.

McLafferty, F. W. (1966). In "Interpretation of Mass Spectra". Benjamin, New York.

Otvos, J. W., and Stevenson, D. P. (1956). J. Am. Chem. Soc. 78, 546.

Pahl, M. (1954). Z. Naturforsch, 9b, 418.

Rosenstock, H. M., and Kraus, M. (1963). In "Mass Spectrometry of Organic Ions" (F. W. McLafferty, ed.). Academic, New York and London.

Smyth, A. D. (1922). Proc. R. Soc. 102, 283.

Vedeneyev, V. I., Gurvich, L. V., Kandratyev, V. N., Medvedev, V. A., and Frankevich, Ye, (1966). In "Bond Energies, Ionisation Potentials and Electron Affinities". Arnold, London.

Vought, R. H. (1947). Phys. Rev. 71, 93.

Warren, J. W. (1950). Nature, 165, 810.

Washburn, H. W., Berry, C. E., and Hall, L. G. (1953). Analyt. Chem. 25, 131.

Appendix

Tables of Units

Table 1. Values of fundamental constants in SI units

Name	Symbol	Value
Avogadro number	N_A	$6.022\ 52 \times 10^{23}\ \mathrm{mol^{-1}}$
Velocity of e.m. radiation in vacuo	c	$2.997\ 925 \times 10^8\ \mathrm{m\ s^{-1}}$
Planck constant	h	$6.625\ 6 \times 10^{-34}\ \mathrm{J\ s}$
Gas constant	R	$8.314\ 3\ \mathrm{J\ mol^{-1}\ deg^{-1}}$
Boltzmann constant	k	$1.380\ 54 \times 10^{-23}\ \mathrm{J\ deg^{-1}\ molecule^{-1}}$

Table 2. Relationship between SI units and cgs units

| Physical quantity | | SI system | | cgs units | | To convert cgs units to SI units multiply by: |
Name	Recommended SI symbol	Name	Unit	Name	Unit	
Length	l	metre	m	centimetre	cm	10^{-2}
				micron	μ	10^{-6}
				ångström	Å	10^{-10}
Mass	m	kilogramme	kg	gramme	g	10^{-3}
Reduced mass	μ	kilogramme	kg	gramme	g	10^{-3}
Volume	V	cubic metre	m^3	cubic cm (ml)	cm^3	10^{-6}
				cubic dm (litre)	dm^3	10^{-3}
Velocity	u	metre per second	$m\ s^{-1}$	centimetre per second	$cm\ s^{-1}$	10^{-2}
Angular velocity	ω	radian per second	$rad\ s^{-1}$	radian per second	$rad\ s^{-1}$	1
Frequency	ν	hertz	Hz	cycle per second	s^{-1}	1
Angular frequency	$\omega = 2\pi\nu$	hertz	Hz	cycle per second	s^{-1}	1
Momentum	p	newton second	Ns	gramme centimetre per second	$g\ cm\ s^{-1}$	10^{-5}
Angular momentum	L	joule second	Js	gramme square cm per second	$g\ cm^2\ s^{-1}$	10^{-7}

Quantity	Symbol	SI unit		Other unit		Factor
Force	F	newton	N	dyne	g cm s^{-2}	10^5
Pressure	P	newton per square metre	N m^{-2}	dyne per square centimetre	dyn cm^{-2}	10^{-1}
				atmosphere	atm	101 325
				mm Hg (Torr)	mm	133·325
Density	ρ	kilogramme per cubic metre	kg m^{-3}	gramme per cubic centimetre	g cm^{-3}	10^{+3}
Energy and Work	E, W	joule	J	erg	g cm^2 s^{-2}	10^{-7}
				litre atmosphere	1 atm	101·328
				calorie	cal	4·184
				electron volt	eV	$1·602 \times 10^{-19}$
Magnetic flux density	B	tesla	T	gauss	G	10^{-4}
Temperature	T	degree kelvin	K	degree kelvin	K	1
Extinction coefficient	ϵ	square metre per mole	m^2 mol^{-1}	litre per mole per centimetre	1 mol^{-1} cm^{-1}	10^{-1}
Dipole moment	P	coulomb metre	C m	debye	10^{-18}esu cm	$3·333 \times 10^{-30}$
Magnetic moment	μ	ampere square metre	A m^2	nuclear magneton	$5·050 \times 10^{-27}$ A m^2	$5·05 \times 10^{-27}$
Moment of inertia	I	kilogramme square metre	kg m^2	gramme square centimetre	g cm^2	10^7
Ionic strength	I	mole per kilogramme	mol kg^{-1}	mole per kilogramme	mol kg^{-1}	1

List of Experiments

Author Index

Numbers in italics are those pages on which References are listed.

Subject Index

A

Absorbance, definition of, 68
Absorption,
  cells, calibration of, 79
  n.m.r., in, 219
Absorption bands,
  areas of, 222, 244
  "*d*" complexes and, 61, 62
  intensity of, 41, 235–237
  multiplicity in, 246–247
  shifts in, 46
Absorption spectra (*see also* Electronic spectra, Vibrational spectra, N.m.r. spectra),
  aromatic compounds, of, 54–56, 91–93
  atomic, 4
  benzene, of, 49–50
  calculation of, 88–89
  carbonyl compounds, of, 45–46
  conjugated dye, of, 87–91
  diatomic molecules, of, 39–42, 115–121
  electronic spectroscopy, in, 65
  energy functions of, 40
  ethylene, of, 47–49
  ethylenic compounds, of, 50–54
  infrared spectroscopy, in, 7, 115, 121
  molecular, 4
  n.m.r., in, 220
  vibrational quantum number in, 40, 42
Acetic acid, infrared spectrum of, 196, 197
Acetone,
  carbonyl stretching in, 144
  chloroform, exchange with, 246
  hydrogen-bonded complexes with, 284–287
  n-propanol mixture, in, 205–206
Acetonitrile, as n.m.r. reference compound, 265
Acetylacetone, n.m.r. spectrum of, 278–282
Activation energy, of nitrous oxide decomposition, 363–365

Alcohols,
  charge transfer donors, as, 59
  fragmentation patterns in, 348–349
  mass spectra of, 315–316
Aldehydes,
  fragmentation patterns in, 346–348
  mass spectra of, 317
Alkali halide discs, 164
Alkanes, as charge transfer donors, 59
Alkyl esters, mass spectra of, 351–353
Alkynes, as charge transfer donors, 59
Allowed (active) electronic transitions, 31, 34, 45, 50
Allowed (active) vibrational transitions, 31, 34, 120, 123, 125, 132
Amides, mass spectra of, 319–320
Amines,
  charge transfer donors, as, 59
  mass spectra of, 318–319
Ammonium chloride, n.m.r. spectrum of, 275–278
Analysis of mixtures,
  electronic spectroscopy, by, 100
  infrared spectroscopy, by, 205, 206, 208
  mass spectrometry, by, 321, 353
  n.m.r. spectroscopy, by, 278, 280, 282
Analysis of n.m.r. spectra, 239
  AB system, 240, 294
  AB_2 system, 241
  ABX system, 243, 295
  AMX system, 242
  AX system, 239
  AX_2 system, 241
Angular,
  momentum, 214–217
  velocity, 215
  wave function, 20–21
Anharmonic oscillators, 10–11, 18, 115–121
Anharmonicity constant,
  infrared spectrum, from, 116, 120, 121
  Raman spectrum, from, 132
Anisotropy, 228